Logic, Epistemology, and the Unity of Science

Volume 56

Logic, Epistemology, and the Unity of Science aims to reconsider the question of the unity of science in light of recent developments in logic. At present, no single logical, semantical or methodological framework dominates the philosophy of science. However, the editors of this series believe that formal frameworks, for example, constructive type theory, deontic logics, dialogical logics, epistemic logics, modal logics, and proof-theoretical semantics, have the potential to cast new light on basic issues in the discussion of the unity of science.

This series provides a venue where philosophers and logicians can apply specific systematic and historic insights to fundamental philosophical problems. While the series is open to a wide variety of perspectives, including the study and analysis of argumentation and the critical discussion of the relationship between logic and philosophy of science, the aim is to provide an integrated picture of the scientific enterprise in all its diversity.

This book series is indexed in SCOPUS.

For inquiries and submissions of proposals, authors can contact Christi Lue at christi.lue@springer.com

Paolo Bussotti · Brunello Lotti

Cosmology in the Early Modern Age: A Web of Ideas

Springer

Paolo Bussotti
Dipartimento di Studi Umanistici e del Patrimonio Culturale
University of Udine
Udine, Italy

Brunello Lotti
Dipartimento di Lingue e Letterature Straniere, Formazione, Comunicazione e Società
University of Udine
Udine, Italy

ISSN 2214-9775 ISSN 2214-9783 (electronic)
Logic, Epistemology, and the Unity of Science
ISBN 978-3-031-12194-4 ISBN 978-3-031-12195-1 (eBook)
https://doi.org/10.1007/978-3-031-12195-1

This Springer imprint is published by the registered company Springer Nature Switzerland AG
The registered company address is: Gewerbestrasse 11, 6330 Cham, Switzerland

Preface

Astronomy is more ancient than physics. The cosmological research belongs to astronomy and deals with the origin of the universe, including its physical structures, geometry and dynamics. Therefore, maybe for this reason, until nowadays, the history of astronomy (and its cosmological accounts) still seems to be separated from the history of physics studies. In this sense, this remarkable Paolo Bussotti and Brunello Lotti's book deserves a special attention. In fact, it is rare that a history and epistemology of science book, including its historical methods, deals with the history of cosmology/astronomy.

In the seventeenth century, the study of the development of the cosmological conceptions is surely one of the most important subjects for the history of astronomy and science of the early modern age. There are several possible historical-inquiring approaches. For example, one might choose either to offer an intellectual history of science and ideas of all the scientists who at that time faced this subject or to focus on those most important in context. Bussotti and Lotti have chosen the latter approach. Thus, in this book, the reader will find an initial chapter on Copernicus, the fundamental reference point for the scientists who, in the following century, addressed cosmological (geometrical and physical) problems. The chapters on Kepler, Galileo, Descartes, Huygens, Newton and Leibniz follow. This convincing choice allows to:

(1) show the inner connections between cosmology and the other physical, mathematical–geometrical aspects of a scientific theory; including its development in context.
(2) stress the links between cosmology and other views which can be defined, in a general meaning, as *philosophical*, because of they were regarded as such in that period of time and because of their interest in human being, i.e. epistemological, ontological, metaphysical and theological aspects.

The authors' research brilliantly shows a development of an interdisciplinary approach to the study of cosmology of the early modern age, focusing on the selected seven main scientists living in that period. This certainly offers an utterly detailed historical account as clear as possible without dispersing in general considerations.

This choice also allowed them to share their purpose, both historiographic and theoretical, in order to present this crucial development of the cosmology within its scientific and non-scientific conceptualizations, for example, in order to better understand the geometrical structure of the universe, its kinematic and metaphysics (e.g. see Kepler) and its dynamic and gravitation (e.g. see Leibniz, Newton), etc, thus effectively an examination of the ways in which the most outstanding scientists reasoned is surely the correct choice.

Finally, the main theses and methodology of inquiring claimed in this book can be summarized like this:

(a) In the development of cosmology—as well as in the development of any section of a scientific theory—the relations between the scientific and epistemological aspects of a single scientist's theory can be of help to improve the knowledge of the previous and/or the following one. For example, for a scientist such as Kepler, it is hard to grasp his cosmological ideas without also taking into account his metaphysical views concerning the harmonies of the universe. In fact, these harmonies include geometrical, musical, astrological and theological aspects. One cannot guess his cosmology only by studying his kinematics and dynamics.
(b) The case study of Leibniz is somehow similar, though Kepler was basically a scientist, while in Leibniz, the philosophical arguments are, on many occasions, preponderant.
(c) Newton is an entirely different case: undoubtedly, he had strong theological convictions, but his cosmology is self-contained; that is, it can be explained only based on his physical and mathematical presuppositions, though this does not exclude that the theological aspect had an influence on such presuppositions.
(d) As a final point, the authors also are convinced that, in spite of the fact that philosophical views play often an important role in the development of scientific theories, the history of science in context has been overestimated by a certain historiography. In fact, coming back to Kepler, it is true that the influence of his metaphysical conceptions was fundamental in his cosmology, but even more fundamental was his will to find a theory, which fits the data, his conviction that no serious physics and cosmology are possible without mathematization and his profound interest in understanding the nature in itself. Because of all these reasons, Bussotti and Lotti, e.g. claim that Kepler can be considered a scientist with a rather modern way of thinking, or at least that this modern way of thinking is even more important than the other aspects of Kepler's complex cultural character. Of course, for other scientists as Leibniz, it is true that the contextual aspects are as important as the scientific ones. However, the authors think that a part of historiography of science has the tendency to overestimate them.

I do not want to prevent the reader from the pleasure to scroll through the book's pages as if this text were an adventure within early modern science. Thus, just few lines to remark that authors' arguments are convincing, profound and well argued. They showed both a thorough knowledge of the technical aspects of science and of the philosophical–epistemological context. Under this respect, the collaboration between a historian of science and mathematics, as Bussotti, and a historian of philosophy, as Lotti, has given an outstanding result. As historian of physics–mathematics, I was impressed by the profoundness with which the authors explain the section on Newtonian concept of inertia by connecting it to the notion of absolute space. They expounded the link between Newton's absolute space and Newton's cosmology. Bussotti's and Lotti's work represents a fundamental clarification and progress in Newtonian historiography, as well. The ideas presented in this book are absolutely consistent with those that I and Bussotti are developing as to Newtonian science (Geneva edition), on which we have already published several papers.

I recommend the reading of this book to graduate and Ph.D. students, to specialists in history of science, astronomy and philosophy and to all the readers who desire to have a vibrant clarification of the image of the cosmology in the early modern age.

Lille, France
July 2022

Raffaele Pisano

Introduction

Rational reflection on the universe is at the root of Western philosophy. As it is well known, the first Greek thinkers, who, taking leave of myth, began to elaborate philosophical inquiry, expressed their insights and ideas in the form of general statements on the essence of nature and on the nature of the universe. The enigmatic splendour of the Greek *physis* was the self-imposing object of their theories. The deforming historiographic tradition that labelled them "Pre-Socratics" is a long-standing obstacle to our understanding of their greatness, as they were the first to move towards that rational investigation of nature which links the Greek legacy to modern science. A theoretical physicist of the calibre of Carlo Rovelli writes in *Reality is not what It Seems: the Journey to Quantum Gravity* (2014), that "covariant quantum fields have become today the best description that we have of the apeiron, the primal substance of which everything is formed, hypothesized by the man who could perhaps be called the first scientist and the first philosopher, Anaximander". What matters in this statement is that a prominent contemporary physicist recognizes that the first Greek cosmological intuitions were the matrix of that mindset which guides physical and cosmological research also today. Even more significantly, we draw from his statement the evidence that cosmology is a field where, even in our age of ever-increasing specialization, science and philosophy necessarily meet. From the perspective of the history of philosophy, we are convinced that cosmology is one of the most productive regions of philosophical thinking because, in conceiving the structure of the world, the conceptual creativity of philosophy is oriented scientifically and is not wasted in fanciful and arbitrary speculation.

The motivation of this book lies mainly in the attempt to investigate how cosmology in the period of the "scientific revolution" displayed deep and multifaceted interrelations between scientific notions and philosophical ideas. This is not surprising since in the early modern centuries natural sciences constituted a branch of philosophy itself and were comprised under the denomination of "natural philosophy" (*philosophia naturalis*). To be more precise: in the early modern age, cosmology was developed by outstanding scientists, who were also philosophers or whose doctrines had profound philosophical implications. We propose an itinerary which includes Copernicus, Kepler, Galileo, Descartes, Huygens, Newton and Leibniz. We present the cosmological views of each of these thinkers and, when necessary, outline their most relevant physical theories. Furthermore, we outline the philosophical and epistemological implications of their scientific works. Our aim is to develop this topic in order to enhance the connection, in the rise of modern science, between scientific theories and related philosophical thinking. This connection is well known, but what is interesting, from the perspective both of the philosophy of science and of the history of science, is that in early modern authors connection and distinction work together. Mathematical knowledge, inchoate physical theories, traditional or innovative metaphysical views, theological and religious backgrounds, methodological procedures and assumptions, all contribute to shape cosmological theories from Copernicus to Newton and Leibniz. At the same time, these elements are not confusedly merged but interact in a complex manner. The critical awareness of their cognitive and logical differences varies according to authors, periods and even within texts by the same author.

We chose to investigate cosmology as it is a paradigmatic example of the intersections between scientific theories and philosophical issues: it was a subject dealt with by the most significant philosophers, at least until the beginning of the nineteenth century; at the same time, the foundations of classical scientific cosmology were the heliocentric system, the three laws of Kepler, and Newton's theory of gravity. By referring to current categories and distinctions, one might claim that the protagonists of the scientific revolution were both physicists and philosophers: they were, in fact, "natural philosophers" to use their self-styled definitions. Some of these authors were at the same time philosophers and scientists even in the current meaning of these terms (Descartes and Leibniz); however, also those authors whom we consider to be scientists (Copernicus, Kepler, Galileo, Huygens and Newton) or who were physicists in the full current acceptation of this term (Huygens and Newton) are significant in different ways for the philosophical implications of their thought. Moreover, they perceived themselves as "natural philosophers", as is shown by the very title of Newton's masterpiece: *Philosophiae Naturalis Principia Mathematica* (*Mathematical Principles of Natural Philosophy*).

Our research has been a typical interdisciplinary effort to reconstruct a conceptual web pervaded by various intellectual attitudes and drives. Epistemologically, the history of cosmology in the sixteenth and seventeenth centuries offers an ideal research field to grasp the subtle and complex interplay between science and philosophy that is at the core of two contemporary disciplines such as the philosophy of science and the history of science. The history of cosmology is also an example of how the history of philosophy and the history of science converge in the effort to clarify topics of common interest to both areas of research. Interdisciplinarity is a key concept and a much-abused slogan in many different intellectual fields today. In most cases, interdisciplinarity is merely desired or proclaimed, without being actually implemented. In this work, we present a case study which is paradigmatic of a fruitful interdisciplinary approach between the history of exact sciences and the history of philosophy. This book is thus conceived as an attempt to offer new insights in both areas with the aim of highlighting the connections among scientific and philosophical concepts. More generally, we aim to propose and describe a model, a paradigm for research in which interdisciplinary aspects are preponderant. In this perspective, particular care has to be taken in presenting the historical development of a discipline or of a set of concepts in the most objective possible way. We mean that the connections between two or more subject matters, though being the scope of an interdisciplinary approach, must not be either overlooked or underlined beyond what they really were both historically and conceptually. When we look at the links between science and philosophy, several features must be carefully considered:

(1) As we said, the present-day neat distinction between science and philosophy did not exist in the early modern centuries, in which science was conceived as natural philosophy. However, the distinction between natural philosophy and metaphysics (*philosophia prima*) was perfectly clear at that time.
(2) When examining natural philosophy, it is possible to distinguish notions that belong to traditional doctrines from new notions that characterize the rise of modern science, the so-called scientific revolution to use an old-fashioned definition that is fully justified if one thinks of the ideal path leading from Copernicus to Newton.
(3) The intertwining between metaphysical and scientific ideas has to be analysed both from a historical and from a theoretical point of view: (a) historically, it is necessary to distinguish concepts which were elaborated because of a decisive influence of one discipline on another from concepts which, actually, were the outcome of creativity within one field independently of the connections with other disciplines. This is not always possible because, for example, in the case of fundamental notions such as that of "force" or "energy" in physics, it is not easy to distinguish the sources and threads at the origin of these concepts. Taking into account this difficulty, each researcher should try to determine the limits within which a concept has a real interdisciplinary origin. Without a carefully detailed analysis, one risks pointing out connections where they do not exist or the reverse. The first aim of an interdisciplinary approach is, therefore, to determine the level of connection among the disciplines which are the subject of the research. Let us give an example to clarify our thought: one of the main concepts expounded by Isaac Newton in his *Principia* is that of absolute space. Newton thought that this concept played an important role within a well-founded natural philosophy. He even proclaimed that the whole of his treatise was written to determine real motions on the basis of their causes and effects and vice versa. Absolute space is, therefore, a notion which, according to Newton, has a physical relevance. However, it is also a concept connected to Newton's metaphysics. An interdisciplinary approach should establish if Newton's natural philosophy influenced his metaphysics; if the converse was true; if there was a reciprocal influence between natural philosophy and metaphysics and, besides this, what was the role of absolute space within physics and metaphysics. In connection to this: (b) theoretically, it is always possible to distinguish the scientific meaning of notions involved in physical, astronomical and cosmological theories from their metaphysical interpretations. The fact that often metaphysical interpretations of some concepts were suggested by the same thinkers who elaborated

and/or employed those concepts in their scientific theories should not mislead the interpreter into confusing these two levels. To give a clear example connected to the previous one: the fact that Newton believed that the structure of the solar system was not the effect of natural forces but the product of the will and intelligence of God does not in any way affect the scientific meaning of his celestial mechanics, that is of the equations by means of which he was able to describe the planetary orbits as the result of two vector magnitudes, the inertial velocity of the planets and their centripetal attraction towards the sun. A metaphysical reading of scientific theories is always possible; it may be more or less theoretically interesting or arbitrary, but in any case, it is not conceptually entailed by the scientific theories and it is not conceptually necessary in order to guarantee the epistemic coherence and validity of scientific theories.

The scientific revolution and, in particular, the problems connected to cosmology are, therefore, one of the cases in which an interdisciplinary approach is useful to achieve a more complete and satisfactory picture of the developments of our conception of the universe and of man's place in it. In this case, if the scientific aspects are kept separate from the philosophical ones, both can be understood only in a limited sense. The explanation of the philosophical and historical context in which scientific concepts such as those of force, motion, inertia, space and time were developed is indispensable in order to grasp their complex features. On the other hand, with the rise of modern science, concepts such as time, space and motion, which traditionally belonged to philosophy, begin to be examined in a more technical way and become part of the conceptual apparatus of physics. As a consequence of this historical process, today it would be meaningless to deal with these concepts philosophically without making reference to their scientific context.

We do not aim to discuss in this book the relations between philosophy and science from a general epistemological perspective. Rather our research has driven us to propose an intellectual path concerning these relations with reference to early modern cosmology, a case in which they are evident and significant. Thus, we hope that what is presented here might be a source of inspiration for further research. We have focused on concepts which play a fundamental role in the theories of the authors we have analysed. These concepts represent the basis on which their cosmologies were built. One of them is inertia. We have tried to offer complete interpretations of inertia in Galileo, Descartes and Newton. We have also outlined this notion in Kepler's theory. Another idea to which we have dedicated our analyses is that of force: we have explained the meaning of this concept in Kepler, Huygens, Newton and Leibniz. As regards to the notions of inertia and force, we have offered new insights. Readers who are interested in the conceptual developments of cosmology will find that in our book we clarify the difference between a purely kinematical and a dynamical astronomy: we begin analysing the complex relation between the kinematical astronomical model proposed by Copernicus and his convictions as to terrestrial physics, which gave rise to a fascinating and complex theoretical structure. Such a structure became even more complex and interconnected in Kepler, an author whose theory is determined by the convergence of kinematical, dynamical, metaphysical and theological arguments. We will focus on the interrelations of these arguments. In this regard, the reader might find a comparison between Kepler's and Leibniz's conceptions interesting because they share several similarities. Therefore, our book can be read from different perspectives and may be of interest to different readers: historians of science, philosophers, physicists and astronomers. We hope that our reconstruction of the interdisciplinary aspects of early modern cosmology offers a more general model applicable to other topics as well.

With regard to the contents of this book, after an introduction to the concept of cosmology, we examine the relevant aspects of the thought of seven authors: Copernicus, Kepler, Galileo, Descartes, Huygens, Newton and Leibniz. Nobody will doubt that these were the greatest scientists and cosmologists of their age. It is thus impossible to offer a complete treatment of their ideas. We have therefore chosen to focus on those aspects of their doctrines which were most significant for an enquiry into the interconnections between philosophical and scientific concepts in their cosmologies. We will see that the general view that these scientists had of the universe and of the connections among different levels of reality (mathematical, physical, metaphysical and theological) played a role in the development of their doctrines; but this role was different for each of them. For example, Newton's scientific ideas are fully understandable without any direct reference to metaphysics and theology (though the theological implications of his thought were important for Newton himself and for the intellectual history of his age), whereas the whole of Leibniz's dynamics depends on a complex interplay between his physical and metaphysical conceptions, so that, without knowing the latter aspect, it would be difficult to grasp the former. This is an example of the type of issues the reader will find in this text.

We have analysed the theories of seven of the most important scientist-philosophers, as we wanted to investigate the interplay of concepts between science and metaphysics. We, for example, excluded philosophers such as Cusanus, Patrizi, Giordano Bruno, Spinoza, Gassendi, Fontenelle and astronomers such as Rheticus, Tycho Brahe, Longomontanus, Horrocks, Hevelius, Hooke, Halley, Cassini, Bullialdus, Picard. This was due to the fact that our aim was not to write a history of cosmology or of astronomy in the sixteenth and seventeenth centuries. As the title of the book suggests, our purpose was to highlight the "web of ideas" behind cosmology. In this perspective, the case studies we analysed concerned the main protagonists of the rise of modern science from natural philosophy. They are the most appropriate cases to stress the complex intertwining among the different aspects (scientific, epistemological, metaphysical and theological) that characterized the emergence of a new view of the universe, following the Copernican revolution and the origins of modern physical science.

Each chapter of this book is dedicated to one of the seven aforementioned authors: after a short introductory profile of the author in question which outlines the most significant features of his thought,[1] a first section concerns his scientific discoveries and theories, while a second section regards his main philosophical issues. The first section is generally longer than the second because the discussion of scientific issues required a more detailed examination. Both in the first and second sections, particular care is taken to underline the interdisciplinary relevance of the topics under scrutiny. The book is the result of a joint effort and every chapter has been considered and discussed together in all its sections.[2]

Even if in our reconstruction we worked on well-known sources, we think that our research is original both as a synthetic overview of early modern cosmology (that is lacking in the scholarly literature) and as an analytical examination of the elements that were intertwined in early modern cosmology. This book is addressed to historians, philosophers and scientists.

References

Craig, E. (ed. in chief, 1998), *Routledge Encyclopedia of Philosophy*. London-New York: Routledge, 10 vols.

Gillispie, C.C. (ed. in chief, 1970–1980), *Dictionary of Scientific Biography*, New York: Charles Scribner's Sons, 16 vols.

[1] In the case of Copernicus alone, we have provided in the introductory section a short exposition of the intellectual context of his formation and activity, which is missing for the remaining authors. The reason for this choice is that Copernicus' astronomical revolution is the first decisive step towards that momentous change of "paradigm" which led to the rise of modern science. Thus, we thought it useful to cast a glance at the cultural milieu in which Copernicus developed his revolutionary views. As for the remaining authors readers can find the basic facts of their intellectual biographies in any good work of reference: see, e.g., Gillispie (ed. in chief), 1970–1980 and Craig (ed. in chief), 1998.

[2] This book is an entirely collaborative work of the two authors. For the mere purpose of the Italian academic evaluation system, we indicate that Paolo Bussotti is the author of the introductory profiles in Chaps. 2, 3, 4, 6, 7, of the entire Chap. 1, and of the Sects. 2.1, 3.1, 4.1, 5.1, 6.1, 7.1, 8.1; Brunello Lotti is the author of the Introduction, of the introductory profiles in Chaps. 5 and 8, of the Sects. 2.2, 3.2, 4.2, 5.2, 6.2, 7.2, 8.2, and of the Conclusion.

Contents

List of Figures

Chapter 1
The Elements of a Cosmological Model

In the early modern ages, precise quantitative studies in astronomy concerned only the solar system (or the terrestrial system, if one refers to geocentric astronomy), and those bodies, like comets, which move—at least in some parts of their trajectories—in the proximity of the solar system. This was the case until Einstein's general relativity. Although Galileo in his *Sidereus Nuncius* (1610) proved that the Milky Way was a cluster of stars, the mathematical means available to astronomers were not yet adequate to study the physical structure of a galaxy. Furthermore, knowledge of the existence of galaxies apart from the Milky Way is extremely recent: on 30 December 1924 Edwin Hubble announced that Andromeda is not a nebula within our galaxy but another galaxy, different from the Milky Way.

However, several philosophers speculated on the possible structure of the universe beyond the solar (or terrestrial) system. We may recall here Cusano and Giordano Bruno in the fifteenth and sixteenth centuries. While Cusano was primarily driven by mystical speculations, Bruno was also influenced by Copernicus. This shows that astronomical-physical cosmology and philosophical cosmology are somehow related.[1] The nature of this relation is certainly worth pursuing. In order to do so, it may be useful to illustrate some of the features of what may be called a "cosmological model". We should first of all bear in mind that in our tradition mathematics became a relatively autonomous theoretical science from Euclid onwards, while physics was an integral part of philosophy: it was regarded as such even during and after the so-called "scientific revolution" and until Newton, as it is evidenced by the title of his above-mentioned masterpiece.

First of all, let us clarify the different aspects of a cosmological model:

(a) *Kinematical aspect*. This concerns the possibility of describing and of predicting the movements of the bodies belonging to the system under examination: if one deals with the solar system, the reference is to the movements of the planets. Within a kinematical model, no research is conducted concerning the forces

[1] For the many facets of the history of cosmology, see the comprehensive North (2008).

P. Bussotti and B. Lotti, *Cosmology in the Early Modern Age: A Web of Ideas*,
Logic, Epistemology, and the Unity of Science 56,
https://doi.org/10.1007/978-3-031-12195-1_1

which determine the motions. Furthermore, from a kinematical point of view, a physical model of the form of the universe is not required. Ptolemy's astronomy in his *Almagest* is a typical example of a kinematical model. Ptolemy does not address the problem of whether his model is the true one. He is only interested in the fact that the model is able to offer geometrical explanations for planetary motions, as they are seen from the Earth, and reliable predictions of the planets' positions. Ptolemy possibly believed in the reality of his model, but this issue is not discussed in the *Almagest*, whose original Greek title is precisely *Syntaxis Mathematica*. He describes his physical model of the cosmos in another work, the *Planetary Hypotheses,* and the two models do not (and cannot) coincide.

(b) *Structural aspect*. This concerns the form of the universe. For example, Aristotle in *De Coelo* imagines that the motions of the planets depend on the fact that each of them is set on the equator of a sphere whose axis rotates because of the movement of a more external sphere, which is in turn moved by another external sphere. By means of a system of mutually nested spheres, it is possible to give a qualitative explanation of planetary motions. Thus, for each body of the solar (Earth) system, a series of spheres exists that determine its movement. The sphere of the fixed stars is the most external one and is responsible for the daily motion of the celestial bodies. Aristotle and the philosophers who adhered to this model did not claim to offer a quantitative treatment of the phenomena, but, rather, they devised a mechanism which, in their view, constituted the real structure of the world and that could explain—at least in principle—the motions of the celestial bodies. Astronomers conceived an analogous structural model. Already in the fourth century BC the great Greek mathematician Eudoxus had proposed a nested-spheres model, by means of which he tried to offer precise predictions concerning planetary motions. Ptolemy himself wrote the *Hypotheses Planetarum*, where he proposed a more in-depth and precise model of the universe as composed of nested spheres, and Peuerbach in the fifteenth century gave an even more refined version of such a theory in his *Theoricae Novae Planetarum.*[2] These astronomers believed that the celestial spheres were existing entities. Every sphere contained only one body, consisting of one of the known planets, the Sun, and the Moon. The only object beyond the system was the sphere of the fixed stars. In the early modern period, Descartes's vortex theory, which stemmed from an entirely different approach to physics, provided another structural model: he believed that vortexes were existing entities.

(c) *Dynamical aspect*. Here it is not only a question of formulating a kinematical explanation or of conceiving an idea of the form of the universe, but of determining the *physical causes* of the motions of the celestial bodies. Kepler was probably the first who fully grasped the importance of this issue, but it was Newton who gave the solution: celestial bodies move as they do because of a centripetal force that acts on them, given some initial inertial conditions for their motion. Such a force is gravity. Newton did not propose a new structural model

[2] See Peuerbach (1473), posthumous. Peuerbach's work was written in 1454, see Malpangotto (2012).

of the cosmos: from this point of view his cosmos is substantially Kepler's cosmos with the planets rotating in their elliptical orbits according to Kepler's laws, but the fundamental result obtained by Newton is to explain *why* these orbits have that form and can have no other.

Therefore, in a cosmological model all these three aspects may be present.[3] It should be noted that, while the first aspect is not necessarily connected to philosophical assumptions, the second and the third ones belong to natural philosophy (i.e. to physics) and are thus connected also to metaphysical presuppositions or interpretations. Generally speaking, a dynamical model is also kinematical, if the treatment is not limited to rational mechanics, but also examines the motions of single existing bodies, whereas the problem of the universe's real form—structural model—is something relatively independent of kinematics and dynamics. Or better, if we want to describe the structural aspect with mathematical means, we face a difficult dynamical issue concerning the more-than-two-bodies problem, but until Lagrange (with the partial exception of Newton) the treatment of this structural aspect was mostly qualitative.

A further remark: cosmology is not cosmogony, or better, cosmogony is a part of cosmology. Until the 1950s–1960s the general problem of the universe's origin could not possibly be dealt with in a concrete physical manner and, of course, it remains a tough challenge also for contemporary physics. Kant –Laplace's hypothesis on the origin of the solar system is historically significant since it represents the first precise attempt to develop a "partial cosmogony", the one relating to the solar system.

In early modern times we find types of cosmology such as that of Bruno which were a mere product of speculative philosophy. On the other hand, we find also cosmologies that were closer to our scientific idea of cosmology (see Descartes' vortex theory). Nonetheless, even 'scientific cosmologies' were regarded as part of natural philosophy, so that it would be inappropriate to separate completely a scientific from a philosophical cosmology. Moreover, the transformation of the physical world-view after Copernicus had repercussions for philosophy in general, particularly with regard to the manner in which the role of man in the cosmos was conceived. At the same time, from Galileo and Kepler onwards, as a consequence of the use of mathematics in physical investigations (contrary to the Aristotelian tradition), physics emerged as an autonomous science and developed within 'natural philosophy'. Therefore, cosmology became gradually characterized by contents and methods that do not need to be framed within a general philosophy. However, its connection to philosophy persisted mainly through the philosophical interpretation of the most fundamental concepts involved in cosmology such as space, time, matter, cause and effect.

The best way to address these complex issues is to refer to important case studies that distinguish the most significant stages in the development of cosmology in the

---

[3] For this distinction, see Bussotti (2015), Bussotti and Pisano (2017). An interesting work on the history of scientific model is Knobloch (2008).

early-modern age. We have chosen seven subjects: (1) Copernicus and Copernicanism; (2) Kepler; (3) Galileo; (4) Descartes and his mechanistic image of the world; (5) Huygens; (6) Newton; (7) Leibniz.

For each author, we propose a section concerning the scientific explanations of the subject and a philosophical section, where the connections between science and philosophy will be pointed out. The scientific exposition aims at clarifying the new cosmological theories in their conceptual structure, while the philosophical exposition pays more attention to the intellectual context in which such theories were conceived and developed.

References

Bussotti, P., & Pisano, R. (2017). Historical and philosophical details on Leibniz's planetary theory as physical-structural model. In R. Pisano, M. Fichant, P. Bussotti, & A. R. E. Oliveira (Eds.), *The dialogue between sciences, philosophy and engineering. New historical and epistemological insights. Homage to Gottfried W. Leibniz 1646–1716* (pp. 49–92). London: College Publications.

Bussotti, P. (2015). *The complex itinerary of Leibniz's planetary theory: Physical convictions, metaphysical principles and keplerian inspiration.* Basel: Springer-Birkhäuser Verlag.

Galilei, G. (1610). *Sidereus Nuncius* (pp. 55–96). EN, III, Parte Prima.

Galilei, G. (1890–1909). *Edizione Nazionale delle Opere.* Direttore A. Favaro, (20 vols). Firenze: Barbèra. Abbreviated as EN.

Knobloch, E. (2008). Modelle in der Geschichte der Wissenschaften. In U. Dirks & E. Knobloch (Eds.), *Modelle* (pp. 85–101). Frankfurt am Main: Peter Lang.

Malpangotto, M. (2012). The early manuscripts of Georg von Peuerbach's *Theoricae Novae Planetarum*: Presentation, description and elaboration of a book. *Revue D'histoire Des Sciences, 65*, 339–380.

North, J. (2008). *An illustrated history of astronomy and cosmology.* Chicago: University of Chicago Press.

Peuerbach, G. (1473). *Theoricæ novæ planetarum, id est septem errantium siderum nec non octavi seu firmamenti.* Nürnberg.

Chapter 2
Copernicus' Astronomical Revolution

From a historical perspective, it is worth focusing on the context in which Copernicus lived and worked as well as on his scientific points of reference.[1] After that, the conceptual nature of his "revolution" will be addressed.

Copernicus was born in Torun in 1473, today a city in Poland, but then belonging to Prussia. He was educated in Krakow, a city which was a centre for research in astronomy. Copernicus possibly followed several courses on this subject: here, Albert Pniewy gave lectures on Sacrobosco's *Sphera*; Albert of Brudzewo lectured on Peuerbach's *Theoricae planetarum*; Bernard of Biskupie specialised in and taught the study of the eclipses; Martin of Olkusz studied the complex problem of the calendar and gave lectures on Regiomontanus' calendar. Finally, Albert of Szamotuly taught astrology and, in particular, he commented on Ptolemy's *Tetrabiblos*. Furthermore, the then modern—though not universally accepted—"physics of *impetus*" was taught in Krakow. This city was thus a stimulating environment for a young scholar and its cultural life was comparable to that of the most advanced Italian cities; it is no coincidence that Krakow University had several contacts with Italy. At various times, between 1496 and 1503, Copernicus spent long periods in Italy, staying in Bologna, Rome, and Padua, where he followed the lectures of the celebrated astronomer Domenico Maria da Novara. When Copernicus returned to Poland, he was engaged in several important administrative, religious and diplomatic jobs, but he devoted most of his time to writing his masterpiece *De Revolutionibus Orbium Coelestium* (*On the Revolutions of the Celestial Spheres*).[2] Though cautiously, he informed his correspondents on the development of his work, which was concluded around 1537–1538. Interestingly enough, the first exposition of the Copernican system was not published by its author, but by his friend and "disciple" Georg Joachim Rheticus

[1] At the end of our Introduction we explained the reasons why we have chosen to outline briefly Copernicus' cultural milieu, which we have not done for the other authors.

[2] The English translation of *De Revolutionibus* is due to Rosen: see Copernicus (1543, 1978). Rosen's translation is available in pdf version at people.reed.edu › ~ wieting › mathematics537 › DeRevolutionibus.

P. Bussotti and B. Lotti, *Cosmology in the Early Modern Age: A Web of Ideas*, Logic, Epistemology, and the Unity of Science 56,
https://doi.org/10.1007/978-3-031-12195-1_2

in his *Narratio Prima* (1540).[3] The *De Revolutionibus* appeared in spring 1543. Copernicus received a printed copy of it on 24 May 1543, the same day on which he died of a cerebral haemorrhage.[4] Besides *De Revolutionibus,* Copernicus wrote two other works on astronomy: the so called *Commentariolus* and the *Epistula de octava sphaera contra Wernerum* (1524). We will focus on *De Revolutionibus.*

Since the environment in which Copernicus worked was particularly stimulating for the study of astronomy, it is worth highlighting the ideas on cosmology that were known at the time. The geocentric system was not questioned. Albert of Brudzewo asserted: "All the planets in their motion have connections with the Sun's motion".[5] However, this claim is so generic that, in a sense, it is compatible also with the Ptolemaic system, because, in this system, the inferior planets have the peculiar property that the centre of their epicycle rotates in their deferent in one sidereal year. This depends on the fact that, in reality, their orbit is included in that of the Earth and, hence, their elongation with respect to the Sun is always limited. Their angular position seen from the Earth is, thus, always close to that of the Sun. Furthermore, any inferior and superior planet makes a complete rotation in its epicycle in a synodic year, that is in the period necessary for the planet—seen from the Earth—to assume the same angular position with respect to the Sun-Earth line.

Copernicus was not the first to think of a heliostatic system. Some Greek scientists had already advanced this hypothesis. To understand Copernicus' ideas, it is not relevant to determine to what degree a heliostatic system had been developed in Greece, but we need to know what Copernicus' ideas were on this topic. In the introductory letter to the Pope Paolo III, he mentioned Nicetus (in fact, Iceta)—drawing this piece of information from Cicero—as the scholar who had guessed that the Earth might move. It was from a work by Plutarch that Copernicus learnt that Philolaos the Pythagorean had claimed that the Earth as well as the Sun and the Moon rotate obliquely around a common fire; Plutarch also reported that Heraclides Ponticus and Ecphantus the Pythagorean ascribed a rotational movement to the Earth, but not a translational one (in substance, diurnal, but not annual motion, see Copernicus, 1978, Preface, p. 5). Copernicus also mentioned Aristharcus of Samos, the most significant proponent of the heliocentric system in Greece. Since no work by these authors concerning a possible heliocentric system had survived, Copernicus was the first to propose a structured heliocentric (or better heliostatic) system. Thus, he is to be considered as the inventor of this system. The true reference point for Copernicus

---

[3] For a Latin version with an exhaustive introduction in English, see Rheticus (1540, [2015]).

[4] Essential bibliographical reference to Copernicus: Freely (2014), Goddu (2010), Hoyle (1973), Omodeo (2014), Repcheck (2007), Somervill (2005). Though recent publications are far more specialised and richer in details, a good general survey of Copernicus' cultural milieu is still provided by Francesco Barone in his introduction to the Italian translation of Copernicus' scientific works (Copernicus, 1979). The publication of the German edition of Copernicus' works, edited by Menso Folkerts, Stefan Kirschner and Andreas Kühne, has been completed in 2019 with the appearance of vol. 4. See: https://www.gn.geschichte.uni-muenchen.de/forschung/datenbanken/copernic/cedition/index.html.

[5] Birkenmajer (1924, p. 95); see also Rybka (1964, p. 164).

was his ideal adversary, Ptolemy, whom Copernicus mentions innumerable times and whom he considers as the greatest astronomer.

Some aspects of Copernicus's work will be analysed under the threefold perspective described in the previous section: kinematical, structural and dynamical models.[6]

2.1 *De Revolutionibus Orbium Coelestium*: A Text that Changed the Image of the Universe

A short exposition of the content of the work is necessary: the published version of *De Revolutionibus* opens with the famous preface in which Andreas Osiander argues that the new Copernican system must be considered as a mere mathematical hypothesis, and not as a description of a physical reality. Copernicus introduces his work with an epistle to Pope Paolo III (Alessandro Farnese), where he expounds the reasons in favour of a heliostatic system. The first book is the most complex one, since it deals with the following issues: the form of the universe; the form of the Earth; the composition of circular motions; the refutation of the ancients' arguments against the Earth's mobility; the possible motions that can be ascribed to the Earth. The book closes with a series of theorems concerning spherical geometry, which recur throughout the treatise. The second book deals with the Earth's daily motion and related phenomena. The determination of the right ascension and declination of celestial bodies in a heliostatic system is explained; the variability of the shadows during the day and at different latitudes, and the length of the day are discussed. The third book deals with the annual motion of the Earth around the Sun and with the solar theory. Relying upon some inaccurate observations, Copernicus accepted the erroneous idea that the precession motion was not regular. In order to explain this alleged irregularity, he devised a third motion of the Earth, which, in fact, does not exist. Copernicus believed that our planet was carried by its sphere, which rotates with it around the Sun. As a consequence, the Earth's axis would describe a cone during the Earth's annual motion. As we shall see, this would have implied a series of phenomena which are not coherent with the observations, while the observed phenomena are explicable only if the Earth's axis remains parallel to itself during the annual rotation. Copernicus thus devised the third motion of the Earth—which is, in fact, a motion of the axis—whose aim is to maintain the axis parallel to itself. This motion also plays a fundamental role in Copernicus' explanation of the equinoxes precession. He also examined the motion of apses (which does exist).

The fourth book deals with the Moon theory and its irregularities. In the fifth book, Copernicus illustrates the motion in longitude of the planets, that is their motion along their orbits assuming that the orbits are coplanar to the ecliptic. Therefore,

[6] On Copernicus as a scientist, see Bortz (2013), Dreyer (1953) [1906], Gingerich (2004), Goddu (2010), Koyré (1973) [1961], Kuhn (1957), Rosen (1984, 1995), Swerdlow and Neugebauer (1984), Westman (2011).

Copernicus explains phenomena such as the retrograde motions of the planets other than the Earth, which are connected to their motion in longitude. As a matter of fact, the rotational planes of the other planets do not coincide with the ecliptic. Copernicus thus added the sixth book, where he addressed the question of the planets' latitudes, that is to say the difference between the position of the planet in its real orbit and its position if the plane of its orbit coincided with the ecliptic.

With the exception of the first, all the other books examine the same contents as Ptolemy's *Almagest*, though considered in a heliostatic rather than in a geostatic system. It should be recalled that the Copernican system also resorted to the traditional deferent-epicycle model, which was required to explain the irregularities of the planets' motions. The treatment of books II–VI is completely mathematical, while philosophical, methodological and physical considerations are introduced only in the first book. The book on the latitudes is, in practice, identical to the 13th book of *Almagest*.

2.1.1 Copernicus' System of the World

It is certain that Copernicus's planetary system is not dynamical, it is kinematical. There is no trace of a force that might determine the motions of the planets around the Sun. However, there is no doubt that Copernicus believed in the physical reality of his system. Osiander's preface—clearly written out of theological prudence—does not catch the real spirit with which Copernicus wrote his masterpiece. The argumentations adduced in the first book indicate that Copernicus thought that he had presented the real system of the world. These arguments are physical, logical and philosophical. Some of them are very convincing, others less so. Let us see why.

In the fifth chapter of the first book, Copernicus poses the problem of whether a circular motion might affect the Earth. First of all, he clarifies the relative nature of any explanation of motion: depending on the observer's position, each apparent movement can be explained either by taking into account that the observed object moves or that the observer moves. It follows that the daily motion of the celestial sphere can also be explained supposing that the Earth rotates around its axis in the opposite sense. As for the planets, they apparently move closer and further from Earth and have a retrograde motion. These apparent motions are compatible with Copernicus' view that the Earth is not at the geometrical centre of the system, since they might be explained as the result of the combined real motions of the Earth and of the planets around the Sun. This argument is very subtle and can be interpreted in the following terms: if the planets rotated at uniform speed in perfect circles around the Earth as their centre, it would be implausible (though not impossible) that such a perfect appearance would be determined by anything other than the Earth's rest. However, since the planets' motions are not perfectly regular, there still is the possibility that the Earth is not at rest.

Copernicus also observes that the sky is the container and the Earth is the contained: hence, there is no reason to suppose that it is the container that moves,

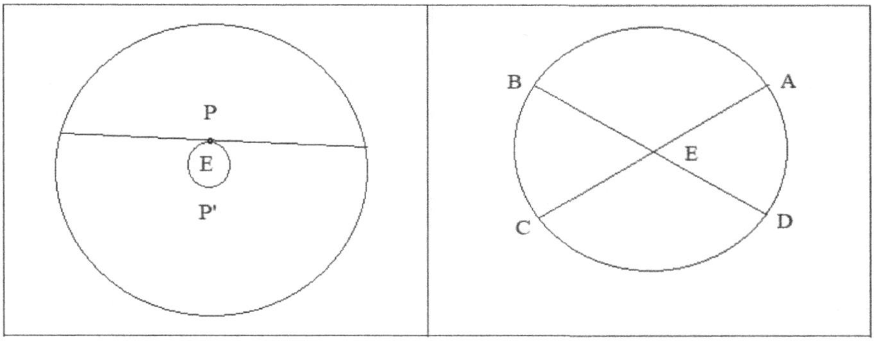

Fig. 2.1 **a** If the size of the Earth (*E*) were not negligible compared to that of the sky, the horizon would not be a great circle and the parts *P* and *P′* of the firmament would not be equal. **b** The figure used by Copernicus to prove that the part *ABC* of the firmament is equal to the part *CDA*. Thus, the horizon is a great circle of the sky and, hence, the size of the Earth (*E*) is negligible with respect to that of the sky. The figures are ours

rather than the contained. This argument makes sense only in the traditional image of the universe as a finite sphere that Copernicus maintained despite his innovations. However, in Copernicus's cosmology the distinction between container and contained is not merely evidence of a traditional view of the universe, but takes a new meaning when related to a geometrical proof concerning the proportion between the size of the Earth and the dimension of the celestial sphere. In the sixth chapter, Copernicus proves the immensity of the so-called Sky of the Fixed Stars with respect to the Earth through the following, acute reasoning based on the assumption that the ecliptic is a great circle of the sky: this proof aims to demonstrate that the horizon divides the celestial sphere into two equal parts. As a consequence, the size of the Earth is negligible if compared to that of the Sphere of the Fixed Stars: if it were not so, then the horizon would not divide the sky into two equal parts (see Fig. 2.1a), whereas the proportion between the Earth and the Sky is the one referred to in Fig. 2.1b.

He argues that: be *ABCD* the horizon and *E* the Earth; if, through a diopter or a horoscopic instrument or water level placed in *E* you see the raising of the Crab's first point in *C*, in the same moment you will observe the setting of the last Goat's point in *A*. You will also observe that the points *C*, *E*, *A* are aligned. Therefore, this line will be the diameter of the Zodiac, namely of the ecliptic. On the other hand, after a revolution (namely after six months) the Goat's first point will rise in *B* and the Crab's last point will set in *D*. Thus, the points *B*, *E*, *D* will be aligned. Since the straight lines *BED* and *CEA* are diameters of the ecliptic and mutually cut in *E*, then *E* is the centre of ecliptic. But if a circle—the horizon in this case—cuts a great circle of a sphere—the ecliptic, in this case—in two diametrically opposed points, then it is a great circle, as well. Therefore, the Earth is approximately in the centre of the world. In fact, the Sun is in the centre of the world, but since the distance Sun-Earth is negligible with respect to the distance between the centre of the world and the sky of the fixed stars, the Earth can be considered as the world's centre for all the apparent phenomena that are not connected with the solar system.

In the *Commentariolus* Copernicus had faced the problem of the ratio among the three most important distances of the universe: the Earth's radius, the ecliptic's radius and the sky's radius. He had solved it through the following postulate:

> The ratio of distance between the sun and the earth to the height of the sphere of the fixed stars is so much smaller than the ratio of the semidiameter of the earth to the distance of the sun that the distance between the sun and the earth is imperceptible compared to the great height of the sphere of the fixed stars.[7]

This means that, if we indicate the distance earth-firmament with H, the distance Sun-Earth with D and the radius' Earth with r, it is $\frac{D}{H}:\frac{r}{D}=$ imperceptible.

In the *De Revolutionibus*, as we have seen, Copernicus proved that the ratio $\frac{r}{H}$ is negligible.

In the light of these geometrical demonstrations, we now understand the deeper meaning of Copernicus' remark about the motions of the container and the contained: it would be surprising if the immense universe rather than a relatively little body such as the Earth could rotate in 24 h. Thus, there is no objection to thinking that the Earth is not the centre of the universe. Even if it is not the centre, it is so close to the centre—which, according to Copernicus, is in the proximity of the Sun, though the Sun is not exactly the geometrical centre of the universe—that, with respect to the immensity of the sky, the phenomena concerning the fixed stars (not the planets) are more or less the same whether seen from the Earth or from the Sun.

In the seventh chapter Copernicus refers to the objections of the ancients against the Earth's movement. In particular, he reports Aristotle's ideas on the natural motions and on gravity: all the heavy bodies fall, according to their nature, towards the Earth's centre, which is, therefore, the centre of the universe. The celestial bodies do not fall because they are not heavy. Their natural motion is circular (*Physica*, II, 1 and V, 2; *De Caelo*, I, 2). Ptolemy added: (1) if the Earth rotated in 24 h, its motion would be so fast that the Earth's body would be destroyed and all bodies on it would fly away; (2) if the Earth rotated, the bodies would not fall on the perpendicular.

Copernicus answers these objections in the 8th chapter. Apparently, he does not reject the Aristotelian doctrine of natural motions, since he claims that, if the Earth rotates, it does so according to its natural motion, which excludes the violent effects envisaged by Ptolemy's objection. With regard to this topic, Copernicus adds a further remark which pursues a different line of argumentation, asking why Ptolemy was not afraid that the rotating sky might dissolve. Since the sky moves, according to Ptolemy's logic, it would continuously move away from its centre and its velocity would increase ad infinitum. This is absurd and certainly dangerous for the stability

---

[7] Copernicus in Swerdlow (1973, p. 436). Cf. Copernicus (1884, p. 186): "Minorem esse comparationem distantiarum Solis et terrae ad altitudinem firmamenti, quam semidimetientis terrae ad distantiam Solis, adeo ut sit ad summitatem firmamenti insensibilis". Swerdlow (1973) is the reference text for the *Commentariolus* because of its precise translation and detailed and clear commentary. Another valuable translation of the *Commentariolus* was offered by Rosen (see Copernicus, 1959).

of the sky. But there is something more: according to Aristotle, the infinite cannot move (*De caelo*, I, 5 and 7; *Physica* III, 4). Therefore, the sky should be at rest. Let us clarify every step of Copernicus' argument: (a) suppose the sky is finite, but big, and that it rotates; (b) given its immensity, its speed is extremely high; (c) the circular motion tends to expel the objects from the centre so that the sky will expand; (d) while expanding, the speed will continuously increase; (e) *the sky will become infinite* (here Copernicus is not completely clear, we think intentionally, on the distinction between actual and potential infinite); (f) since the infinite cannot move, the sky must be at rest. The consequence is that, even starting from the hypothesis of a moving sky, we are driven to conclude that it is at rest. This type of reasoning, if we accept all the steps, exhibits the logical pattern of the so called *consequentia mirabilis*, which is a tautological scheme expressed like this: $(\neg A \to A) \to A$, where A is, in this case, the proposition "the sky is at rest". From an intuitive point of view, the scheme shows that if even the assertion that the sky rotates entails that 'the sky is at rest', then it is surely true that the sky is at rest.

We have pointed out that Copernicus did not have a dynamical conception of the universe, but the situation was different with regard to the Earth. Though Copernicus was not a physicist, he had grasped some aspects of the participation of a body in the motion of the inertial system to which it belongs. This is what Copernicus writes in *De Revolutionibus* I, 8:

> For when a ship is floating calmly along, the sailors see its motion mirrored in everything outside, while on the other hand they suppose that they are stationary, together with everything on board. In the same way, the motion of the earth can unquestionably produce the impression that the entire universe is rotating.

And again:

> [...] We would only say that not merely the earth and the watery element joined with it have this motion, but also no small part of the air and whatever is linked in the same way to the earth [i.e. in modern terms they belong to the same inertial system].[8]

Therefore, the stone falls perpendicularly because it participates in the movement of the Earth and of the air.

This is worth focusing on: the belief in a heliostatic system does not necessarily imply a dynamics for the solar system, but it certainly implies a modification of the terrestrial dynamics. Copernicus is consistent in this development.

It is certain that if he had considered his system as a hypothesis, he would not have included so many details to prove the validity of his theory. There is a further

---

[8] Copernicus (1978, p. 16). Cf. Copernicus (1873, p. 22): "Quoniam fluitante sub tranquillitate navigio, cuncta quae extrinsecus sunt, ad motus illius imaginem moveri cernuntur a navigantibus, ac vicissim se quiescere putant cum omnibus, quae secum sunt. Ita nimirum in motu terrae potest contingere, ut totus circuire mundus existimetur. [...] nisi quod non solum terra cum aequo elemento sibi coniuncto sic moveatur, sed non modica quoque pars aëris, et quaecumque eodem modo terrae cognationem habent." Copernicus seems to foreshadow a famous passage of Galileo's *Dialogue*: see Galilei (1632), in Galilei EN, vol. VII, Day 2, p. 212ff. Cf. Galilei (1967, p. 186ff). Obviouly, in a strict sense, the Earth is not an inertial reference frame, but for the local motions on its surface considered by Copernicus and Galileo, it can be thought of as inertial at all effects.

observation that clearly shows that Copernicus believed in the structural truth of his system, but was at the same time aware that the instruments then available could not provide the fine tuning of his theory. While dealing with the preliminary notions concerning the apparent inequalities of the solar movement (book 3, chapter XV), he claims that the problem of the irregularities can be solved by means of an eccentric model or a concentric plus epicyclic model. This was well known and was the basis of Ptolemy's astronomy as well. If the irregularities are complicated (as they are as far as the planets are concerned), more complex models than in the case of the Sun are necessary, but, in principle, the situation does not change.

According to Copernicus:

> From all these analyses it is clear that the same apparent nonuniformity always occurs either through an epicycle on a concentric or through an eccentric equal to the concentric. There is no difference between them provided that the distance between their centers is equal to the epicycle's radius.
>
> *Hence it is not easy to decide which of them exists in the heavens.*[9]

This means, of course, that one of the two exists in the skies!

Copernicus thus deemed his system to be the real structure of the world, not merely a mathematical hypothesis.

In order to fully understand how difficult the creation of new scientific paradigms and their fine tuning are in the evolution of scientific ideas, it is worth focusing on two of the topics that distinguish Copernicus's astronomical thought: (a) the third motion of the Earth; (b) the so called "trepidation of the equinoxes".

(a) As regards the assumption that the Earth rotated around the Sun, Copernicus believed that two phenomena had to be explained: the alternation of seasons and the precession of the equinoxes. He assumed that, during the Earth's rotation around the Sun, the terrestrial axis would move along the orbit and would not remain parallel to itself; as a consequence, in one year it would describe in the sky a conic surface (see Fig. 2.2).[10]

In accordance with this assumption, as Copernicus remarked in I, 9, the alternation of day and night would exist, but the duration of the days would be always equal to that of the nights and no alternation of seasons would exist. Then, to explain these phenomena, Copernicus devised a third motion, besides the daily and the annual

---

[9] Copernicus (1978, p. 156). Cf. Copernicus (1873, p. 207). "E quibus omnibus patet eamdem semper apparentiae inaequalitatem produci, sive per epicyclium in homocentro, sive per eccentrum circulum aequalem homocentro, nullatenusque invicem differre, dummodo distantia centrorum aequalis fuerit ei, quae ex centro epicyclii. Utrum igitur eorum existat in caelo, non est facile discernere". (Our italics).

[10] Copernicus' thesis implies that the Earth's axis and the axis of the ecliptic were coplanar (in this case, the intersection of the two axes would represent the vertex of the cone); if the two axes were skew (an alternative possibility that Copernicus did not take into account), the Earth axis would describe in one year the surface of a one-sheeted hyperboloid of revolution; in this case, it is possible to determine the neck circle of the hyperboloid.

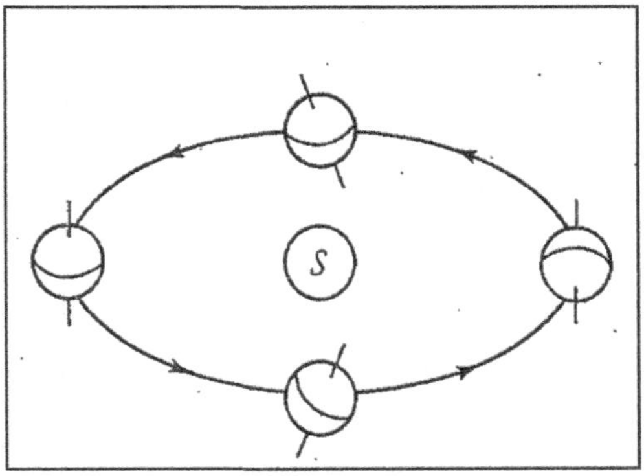

Fig. 2.2 The movement of the Earth's axis according to Copernicus if only the annual motion of the Earth existed

rotations of the Earth, which he named "declination" or "inclination"[11] and ascribed it to the terrestrial axis. As a consequence of this third motion, the axis of the Earth would remain parallel to itself during the annual revolution. He wrote:

> The third motion in inclination is consequently required. This also is a yearly revolution, but it occurs in the reverse order of the signs [namely clockwise], that is, in the direction opposite to that of the motion of the center [the counterclockwise annual revolution of the Earth around the Sun]. These two motions are opposite in direction and nearly equal in period. The result is that the earth's axis and [celestial] equator, the largest of the parallels of latitude on it, face almost the same portion of the heavens, just as if they remained motionless. Meanwhile the sun seems to move through the obliquity of the ecliptic with the motion of the earth's center, as though this were the center of the universe. Only remember that, in relation to the sphere of the fixed stars, the distance between the sun and the earth vanishes from our sight forthwith.[12]

In order to allow the Earth's axis to maintain its position parallel with respect to the fixed stars (so that the observed length of days and nights as well as the alternation of seasons were possible), he supposed that the axis was animated by a motion whose direction is opposite to that of the annual motion (see Fig. 2.3).

The period of the two movements is *almost* equal. Almost, but not exactly equal because Copernicus thought that the third motion should also explain the precession of the equinoxes. He combined appropriately the annual motion of the Earth around the Sun and the motion of declination so as to obtain the exact period of the precession, at least of the mean precession. Since Copernicus believed that the precession motion

[11] The Latin term used by Copernicus is *declinatio*. Some authors translate this term as "declination", others as "inclination".

[12] Copernicus (1978, p. 23). Cf. Copernicus (1873, p. 31): "Sequitur ergo tertius declinationis motus annua quoque revolutione, sed in praecedentia, hoc est contra motum centri reflectens. Sicque ambobus invicem aequalibus fere et obviis mutuo evenit, ut axis terrae, et in ipso maximus parallelorum aequinoctialis in eamdem fere mundi partem spectent, perinde ac si immobiles permanerent. Sol interim moveri cernitur per obliquitatem signiferi, eo motu, quo centrum terrae, nec aliter quam si ipsum esset centrum mundi, dummodo memineris solis et terrae distantiam visus nostros iam excessisse in stellarum fixarum sphaera".

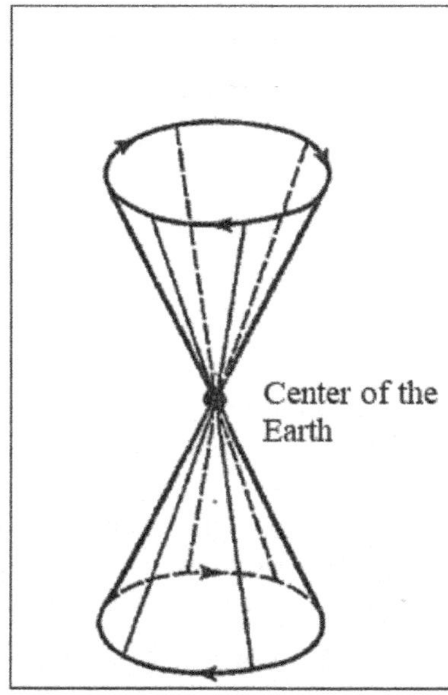

Fig. 2.3 The clockwise motion of declination that explains the length of days and nights and the alternation of seasons

was not regular, the introduction of the third motion of the Earth's axis was not sufficient to explain its irregularities, as we will see below.

(b) The situation is as follows: on the basis of his predecessors' and his own observations, Ptolemy estimated that the movement of precession was 1° every 100 years. Arabic astronomers, who had developed better observations than those of Ptolemy, determined the correct value of 1° every 72 years. However, since they thought that Ptolemy's observations could not be wrong, they concluded that the movement of precession is variable and that it was accelerating. They also invented complicated models to explain this supposed variable movement, called "trepidation",[13] a movement which, in fact, does not exist and whose supposed existence was the result of the principle of authority (i.e. of the reverence of the Arabic astronomers for Ptolemy and the unquestionability of the observational data reported in his *Almagestus*). The idea that Ptolemy's observations were correct was also shared by European astronomers and Copernicus was not an exception.

The first twelve chapters of the third book of the *De Revolutionibus* are dedicated to the precession of the equinoxes and solstices and most of these chapters concern the supposed irregularity of this phenomenon. What Copernicus wrote in III, 1 is a perspicuous summary of the history of astronomy, of his ideas, and his purposes with regard to the precession of the equinoxes:

[13] On the supposed phenomenon of the trepidation, see Cohen (1967), Comes (2001), Goldstein (1965), Mancha (1998), Nothaft (2017), Rajep (1996), Samsó (1994), Swerdlow (1980).

> Now Hipparchus of Rhodes, a man of marvellous acumen, was the first to notice that these years are different from each other. When he was scrutinizing the length of the year more intently, he found that as measured with reference to the fixed stars [sidereal year], it was longer than when measured with reference to the equinoxes or solstices [tropical year]. Hence he thought that the fixed stars too had a motion in the order of the zodiacal signs, but a very slow motion which could not be perceived immediately [Ptolemy, *Syntaxis,* III, 1]. Now however, with the passage of time it has become absolutely clear. Because of it we see that at present the signs and stars rise and set quite differently from the indications of the ancients, and that the twelve signs of the zodiac have shifted a considerable distance away from those constellations of the fixed stars which originally agreed with them in name and position.
>
> Moreover, the motion is found to be nonuniform. The desire to account for this nonuniformity has elicited various beliefs. In the opinion of some people, the universe, being in suspension, has a certain oscillation, a motion such as we find in the latitudes of the planets [VI, 2]; [...] Others have indeed held that the sphere of the fixed stars moves forward, but with unequal strides; yet they have laid down no definite pattern. Besides, another marvel of nature supervened: the obliquity of the ecliptic does not appear as great to us as it did before Ptolemy, as I said above.[14]

In the first paragraph of this quotation Copernicus refers to Hipparchus' (of Nicaea, not of Rhodes) discovery of the precession, while in the second paragraph he claimed: (1) there is no doubt that the movement of precession is variable; (2) the available explanations of such a variability are not satisfactory; (3) the obliquity of the ecliptic is also nonuniform.

Therefore, among his purposes, there was also that to explain the irregularity of the precession and the variable obliquity of the ecliptic. In the long second chapter, Copernicus refers to a series of observations of Greek and Arabic astronomers who should confirm the variability of such phenomena. In the third chapter he offers model to explain them. Copernicus claims that various observations prove the ecliptic to be a fixed circle. Since the phenomenon of the equinoxes depends on the mutual inclination of ecliptic and celestial equator, the only possibility is to admit a movement of the terrestrial axis, which changes its inclination between ecliptic and equator (see Fig. 2.4).

Copernicus argued that: be E the pole of the ecliptic. He indicated by A, C, B, D respectively the first point of the constellations of the Goat, Crab, Ram and Balance. He drew the circle AEC. The points B and D represent the intersection between

---

[14] Copernicus (1978, p. 119). Cf. Copernicus (1873, pp. 157–158): "Hipparchus autem Rhodius, vir mirae sagacitatis, primus animadvertit haec invicem distare, qui, dum anni magnitudinem attentius observaret, maiorem invenit eum ad stellas fixas comparatum quam ad aequinoctia sive solstitia. Unde existimavit stellis quoque fixis aliquem inesse motum in consequentia, sed lentulum adeo nec statim perceptibilem. At iam tractu temporis factus est evidentissimus, quo longe iam alium ortum et occasum signorum et stellarum cernimus ab antiquorum praescriptio, ac dodecatemoria signorum circuli a stellarum haerentium signis magno satis intervallo a se invicem recessisse, quae primitus nominibus simul ac positione congruebant. Ipse praeterea motus inaequalis reperitur, cuius diversitatis causam reddere volentes diversas attulerunt sententias. Alii libramentum esse quoddam mundi pendentis, qualem et in planetis motum invenimus circa latitudines eorum [...]. Alii progredi quidem stellarum fixarum sphaeram opinati sunt, sed passibus inaequalibus, nullum tamen certum modum definierunt. Accessit insuper aliud naturae miraculum, quod obliquitas signiferi non tanta nobis appareat, quae ante Ptolemaeum, ut supra diximus".

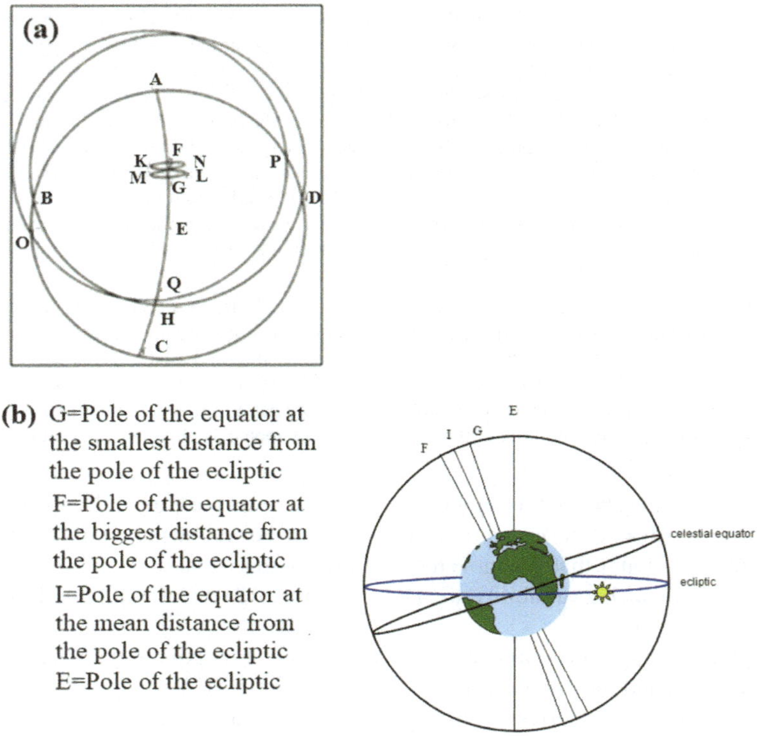

Fig. 2.4 **a** Copernicus' diagram to explain the libration of the Earth's axis that causes the supposed trepidation of the equinoxes and the variable inclination of the ecliptic. *Source* Copernicus (1978, p. 124). **b** Copernicus' original diagram offered a view from above of the supposed phaenomenon, while this view is from a point situated on the nodes' line ecliptic-equator

the ecliptic and the equator. Hence, Copernicus drew the equator *BHD*. Since the inclination of the Earth's axis on the ecliptic is variable, the poles of the equator, obviously, are, as well. Be *G*, *I*, *F* three positions of the equator's pole, which moves on the circle *AEC*. The point *G* represents the smallest distance between pole of the ecliptic and pole of the equator, the point *F* the biggest distance and the point *I* the mean distance. The position *BHD* of the equator, whose pole is *I*, was called "mean equator", while *B* and *D* are the "mean equinoxes". They move clockwise circularly and uniformly around *E*. Now Copernicus imagined two motions of the Earth's axis: through the first motion the axis' pole oscillates between the points which represent its smallest and biggest distances (resp. *G* an *F*) from the ecliptic's pole *E*. Copernicus named this movement "motion of the anomaly". Furthermore, he supposed the existence of a transversal motion, whose speed is the double of the former. Through this motion the Earth's axis oscillates between points as *M*–*L*, *K*–*N*. He called this motion "anomaly of the equinoxes". Both motions were named "oscillating librations". Combining appropriately these motions, it is possible to obtain the rosette *FKILGMINF*. Through this rosette, Copernicus explained both the supposed trepidation of the equinoxes as well as the supposed variable obliquity of the ecliptic.[15]

In the chapters IV–XII of the third book, Copernicus develops all the details of the model we have described: he proved how the rosette could be obtained by a combination of circular motions, gave a precise mathematical demonstration that the model we have described can explain the trepidation of the equinoxes and the variability of the ecliptic's inclination, and so on. He also tabulated the mean motion of the equinoxes and the motion of the equinoxes' anomaly.

Copernicus' solution to the problems of the equinoxes' precession is significant as it enables us to understand aspects of the evolution of science and of the model of universe that he was constructing.

With regard to the first problem, there were two reasonable possibilities to explain the supposed different speeds of the precession movement: (1) the observations of the Greek astronomers were inexact. Their instruments were not completely adequate to detect the speed of the precession. The observations of the Arabs were more accurate. Therefore, there is no trepidation of the equinoxes. There was simply an observative error in the Greek sources; (2) both the observations reported by Ptolemy and those of the Arabic scientists are regarded as accurate: in this case, the equinoxes trepidate.

Copernicus, the man who dared to challenge and overthrow Ptolemy's system, did not think that Ptolemy, whom he considered the greatest astronomer of all time, as well as other great astronomers as Hipparchus, might have carried out wrong observations. Thus, he concluded that the equinoxes trepidate. In this case, Copernicus shared the opinion of most astronomers (be they Arabic or European) and proposed a new and ingenious theory to explain the trepidation. Once again, he proved to be the most prominent scientist of his epoch because of the precision and clarity of his theory of trepidation. This is an evident instance of how intricate and complex the evolution of science is. Copernicus was not led to believe in the trepidation by philosophical or metascientific convictions. He merely assumed that Ptolemy's observations were correct. This conviction was in its turn based on a psychological assumption: great scientists like the Greek astronomers could not make so evident observative mistakes. From the point of view of the logic of science, Copernicus's conviction is equivalent to an axiom of his theory, an axiom from which the consequences are drawn correctly, but the axiom itself is wrong. Thus, the scientific-psychological presuppositions of a theory have to be considered.

The third motion of the Earth, the motion of declination or inclination also is worth noting. Since Copernicus admitted the movement of the Earth, he faced a correct scientific problem: why doesn't the Earth's axis describe a cone in the firmament? This is a reasonable question that shows, once again, Copernicus' sagacity. Obviously, he had no dynamical explanation of the fact that the Earth's axis remains parallel to itself. The solution he devised is worth of our admiration, though it refers to a false problem. The situation is not the same as that of the libratory motion of

[15] For all the details of the explanatory mechanism devised by Copernicus see the chapters III–XII of the third book of the *De Revolutionibus*.

the Earth's axis. While the libratory motion was invented as a consequence of an error concerning the evaluation of the sources of astronomical data, in the case of the third motion Copernicus had to face a problem related to the movement of the Earth around the Sun, a problem that did not exist for the geocentric astronomy. Given that no dynamics of the solar system was available at his epoch, he drew the only geometrically consistent conclusion with regard to the annual axis' positions. Since phenomena did not correspond to such positions, he introduced a third motion, which also allowed him to explain the equinoxes's precession. Copernicus's procedure was correct from the point of view of a scientific logic. However, knowledge of physics in Copernicus's age was by far insufficient to explain such complex phenomena as those he studied. This is an evidence, on the one hand, of the historicity of science, whereas on the other hand it shows the advancements that are implicit in any scientific theory. When scientific knowledge reached a satisfactory dynamical explanation of the phenomenon of precession thanks to Copernicus's astronomical revolution and Kepler's improvement of his theory, it became clear that Copernicus's third motion did not exist. As Federigo Enriques claimed, science is a process of continuous approximations towards increasingly better explanations of phenomena, though scientists are conscious that the final theory and explanation will never be achieved.[16] Under this perspective, the third motion of the Earth can be considered an historical step towards the acquisition of a scientific knowledge of the cosmos, and, above all, towards a scientific approach to cosmological problems.

The next topic concerns the model of the universe that Copernicus had in mind: the theory of trepidation is a further evidence that he was trying to construct a real model of the universe, not a mere mathematical scheme "to save the phenomena". He was not satisfied to admit that the Earth's axis remained parallel to itself with no further explanation; he thought that it was necessary to explain this phenomenon as it was inconsistent with the initial hypotheses on which his theory was based. As will be considered in the next section, he intended to construct a system that could exhibit the real physical structure of the universe. On the other hand, Copernicus's whole theory of celestial movements remains a kinematical theory. There is no indication that he was thinking of the physical causes that produce the movements. In this sense, his theory is comparable with Ptolemy's. The fundamental difference is that Copernicus thought to have discovered the real system of the cosmos, in spite of the fact that he did not search the forces responsible for the rotations of the planets around the Sun. Thus, his model is kinematic and physical-structural. It is not dynamical.

2.1.2 Copernicus' Theory of Planetary Motions

So far we have seen the fundamental novelties introduced by Copernicus. We will summarise the main features of his system at the end of Sect. 2.2, after addressing the

[16] Among Enriques's most important publications in which his ideas on the scientific theories as successive approximations to the truth are expressed, see Enriques (1906, 1912, 1936).

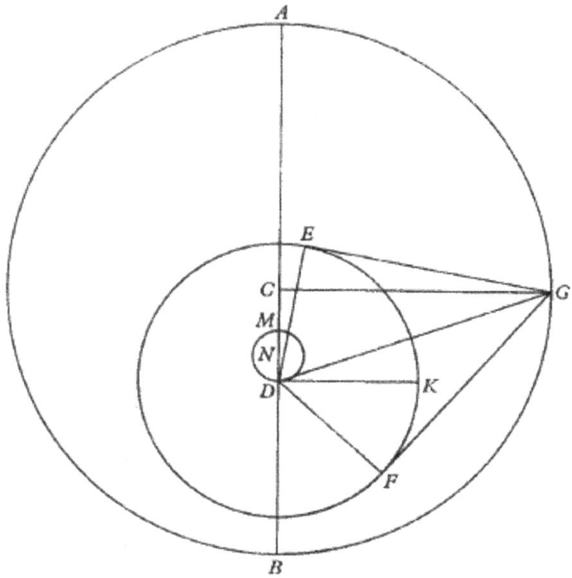

Fig. 2.5 Copernicus's diagram to explain the motion of the centre of Venus' orbit. *Source* Copernicus (1978, I, 22, p. 273)

philosophical implications of his thought. There is a long diatribe among the scholars who consider Copernicus as a "revolutionary" and those who consider him as a sort of new Ptolemy. According to the latter, he moved the centre of the system from the Earth to the Sun, but his general view of the cosmos and his methods of inquiry were, in substance, more similar to those of Ptolemy than to those of Kepler. We agree with the first group. However, in order to offer a correct image of Copernicus' work, it is appropriate to stress that the specific technique through which he constructed the planetary orbits is mostly conceived of in a "Ptolemaic style". We mean that, relying upon observations, he developed a model of planetary movements through a series of geometrical-kinematical devices. They explained the observations and also allowed him to compile good astronomical tables, but are clearly invented a posteriori, and are not derived from a general theory of celestial movements. Each planet needs specific and different explanatory devices and the only unitary feature of Copernicus' theory is the principle that all celestial motions must be strictly circular. To give an example, we will summarize Copernicus' theory of Venus, whose movement is the easiest one to explain.[17]

The basic idea of Copernicus to explain the motion of Venus is to reduce it to a movement around an eccentric circle (see Fig. 2.5).

Thus, in Fig. 2.5 be *D* the geometrical centre of Venus' orbit. The idea that Venus's movement might be regular around the point *D* does not square with the observations. Copernicus referred to two observations of Ptolemy dated resp. 18 February 134 A. D. and 18 February 140 A. D. In both of which the elongation of Venus is maximal. Be

[17] For a complete exposition of Copernicus's mathematical techniques we refer to the fundamental Swerdlow and Neugebauer (1984).

C the geometrical centre of the Earth's orbit, namely the mean Sun, and *G* the position of the Earth in the dates reported by Ptolemy. If, considering the mean movement of the Earth, Venus's movement around *D* were uniform, the elongation *EGC* observed on the 18th February 134 should have been equal to that *CGF* observed six years later. However, this was not the case because the angle *EGC* is 43° 35′, while the angle *CGF* is 48° 20′. The conclusion is that Venus's movement around *D* is not regular. To offer a good model of Venus's motion, Copernicus reasoned like this: it is easy to calculate the angle *CGD*, which is 2° 30′. Since *DCG* is a right angle, in the triangle *CGD* it is possible to calculate the ratio of the sides *CD/CG*. Be *a* such a value.[18] Relying upon a previous calculation,[19] based on the observation of Venus's maximal elongation when the Earth is not in *G*, but in the farthest and in the nearest points of its orbit with respect to *D* (namely in *A* and in *B* resp.), such a ratio should have been *a*/2. Copernicus, then, divided *CD* in *M* into two equal parts so that $DM = MC = a/2$. The segment *DM* represents the whole amount of the departure and approaching of *D* from *C*. Thus, if you bisect *DM* in *N*, you get the mean point of the movement of the point *D*. To conclude, Copernicus claimed that the motion of Venus can be explained through "the eccentric of an epicycle". If around the point *N* a little circle with radius *DN* is drawn, the centre of Venus's orbit moves on this little circle (the epicycle). The law of its movement can be calculated in the following way: when the Earth is in *A* or in *B*, the instantaneous centre of Venus's movement is always in the positions closest to the mean Sun, namely in *m*. Each time that the Earth is in *G* or in the symmetric point of *G* with respect to the apses line *AB*, the instantaneous centre of Venus's motion passes through *D*. Therefore, the centre of Venus's motion rotates around *N* in the same verse as the Earth's rotation around the Sun and with a double speed with respect to the mean movement of the Earth. Obviously, in the meantime, Venus rotates around its instantaneous centre. Hence, the motion of Venus is composed of two uniform circular movements: the movement of the planet around its instantaneous centre and the movement of such a centre around the point *N*. By means of these geometrical-kinematical devices, which are conceived in the style of the Greek astronomy, Copernicus stated that he was able to explain both the mean and the apparent motion of Venus coherently with the observed positions of the planet: "Through this hypothesis for Venus, its uniform and apparent motions agree with every kind of situation, as will soon be clear[20]". As a matter of fact, the following chapters 23 and 24 are exactly dedicated to show that the proposed model fits with the observations.

This is the model for Venus's motion. In V, 35 and 36 Copernicus added his model for the retrograde motions of all planets, but what we have expounded should be enough for the reader to get a clear idea of the devices developed by Copernicus.

---

[18] Copernicus offers the value of this ratio after having conventionally set the length of the radius *CG* of the Earth's orbit to be equal to 10,000 (Copernicus, 1978, V, 22, p. 272). For our purposes, it is not necessary to specify the value of such a ratio and it is more convenient to indicate it by *a*.

[19] This calculation is developed in V, 21.

[20] Copernicus (1978, p. 272). Cf. Copernicus (1873, p. 369): "Per talem enim circa Venerem hypothesim omnimodis exemplis consentiunt aequalitas at apparentia, ut mox apparebit".

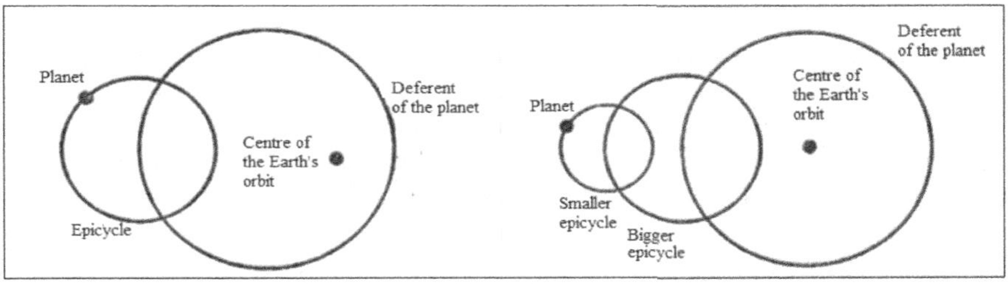

Fig. 2.6 On the left: the planetary model expounded in *De Revolutionibus*. On the right: the planetary model expounded in *Commentariolus*. The figure is ours

It should be recalled that, besides the model eccentric-epicycle expounded in *De Revolutionibus*, Copernicus conceived a different geometrical model for the motion of the planets in another work, the *Commentariolus* (Copernicus, 1878): this model is kinematically equivalent to the one presented in his major work.

This model is concentric-epicycle-epicycle: for any planet, the centre of its deferent is the centre of the Earth's orbit. There is a first bigger epicycle whose centre rotates on the deferent. The planet rotates around the centre of a second, smaller epicycle whose centre rotates on the circumference of the first epicycle (see Fig. 2.6).

Commentariolus was not published in Copernicus' lifetime and the first edition dates back to 1878. At the end of the nineteenth century and at the beginning of the 20th, eminent scholars such as Curtze, Dreyer (1953 [1906]), Berry (1898) thought that the *Commentariolus* was a sort of summary of *De Revolutionibus* written after the major work. On the contrary, nowadays scholars agree that this minor work was written before *De Revolutionibus* because of two reasons: (1) in *Commentariolus* Copernicus explicitly wrote that the geometrical demonstrations were not expounded in this text because he reserved them to a more complete work; (2) it is unlikely that Copernicus presented a model of planetary motion in *De Revolutionibus*—a work to which he devoted more than twenty years and that was published at the end of his life—and that, after the composition of his masterpiece, he rethought of the entire planetary model reaching a different conclusion.

2.2 Epistemological and Philosophical Implications of Copernicus' Thought

It is commonly said that Copernicus' astronomical revolution is an example of how reason can reach a scientific truth despite the opposite evidence of our senses. Galileo, one of the most prominent advocates of Copernicanism, praised in his *Dialogue concerning the Two Chief World Systems* the "sublime intellect" of Copernicus because "with reason as his guide he resolutely continued to affirm what sensible experience seemed to contradict".[21] This is, however, only partially true. The optical relativity of motion is an ordinary sense experience that had been taken into account and studied by ancient and medieval thinkers such as Aristotle, Witelo and Oresme.[22] As we have reported in a previous citation concerning the motion of a ship in the sea, Copernicus appealed to the optical relativity of motion to show the inadequacy of the reasons why the ancients said that the Earth is at rest in the middle of the world (Copernicus, 1978, I, 8, p. 16). Copernicus' inverted point of view is not a triumph of reason over the deceptive limits of sensation, but an intellectual assessment of sensory evidence which may be interpreted in two contrary ways as they are attested by the sense experience itself. Therefore, we are allowed to conclude that the Copernican theory, like every successful scientific theory, is an example of the cooperation and coordination between sensorial evidence and intellectual judgement.

It may be interesting to ask which factors prompted Copernicus to embrace his revolutionary point of view. Some aspects of his historical context might have been influential. The great transformations of that age (the invention of the printing, the new geographic explorations, the Renaissance culture with its revival of ancient texts and philosophies) helped to create an intellectual climate which favoured significant innovations. Copernicus himself cited the ancient precedents of the idea of the motion of the Earth and praised the Platonic and Hermetic tradition of the philosophy of light, which emphasized the primacy of the Sun:

> At rest, however, in the middle of everything is the sun. For in this most beautiful temple, who would place this lamp in another or better position than that from which it can light up the whole thing at the same time? For, the sun is not inappropriately called by some people the lantern of the universe, its mind by others, and its ruler by still others. [Hermes] the Thrice Greatest labels it a visible god, and Sophocles' Electra, the all-seeing. Thus indeed, as though seated on a royal throne, the sun governs the family of planets revolving around it.[23]

However, this intellectual background, though a contributing factor, is obviously insufficient to explain how Copernicus managed to advance a new detailed astronomical theory which took the place of the Ptolemaic system. Whoever browses through Copernicus' treatise, especially if one goes beyond the first eleven chapters of the first book, realizes the obvious truth that Copernicus was first and foremost an astronomer, i.e. a geometer; he was not a natural philosopher devoted to the mission of reviving the ancient wisdom of Pythagoreism, Platonism and Hermeticism. In fact, passages such as the aforementioned one are extremely rare in his work and they should be interpreted merely as the expression of a metaphysical 'world-view' which

---

[21] Galileo (1967, p. 339); cf. Galilei EN, VII, p. 367).

[22] With regard to this topic see, e.g., Koyré (1957, p. 279, n. 14).

[23] Copernicus (1978, I, 10, p. 22). Cf. Copernicus (1873, p. 30): "In medio vero omnium residet sol. Quis enim in hoc pulcherrimo templo lampadem hanc in alio vel meliori loco poneret, quam unde totum simul possit illuminare? Siquidem non inepte quidam lucernam mundi, alii mentem, alii rectorem vocant. Trimegistus visibilem deum, Sophoclis Electra intuentem omnia. Ita profecto tamquam in solio regali sol residens circumagentem gubernat astrorum familiam". Edward Rosen rightly remarked (Copernicus, 1978, p. 359) that the importance of this passage should not be overemphasized. Rosen hinted critically at Koyré (see Koyré 1973, p. 65).

could be usefully associated to the new astronomical hypothesis and could legitimize it in the eyes of the scholars of his age; but in order to understand the deep philosophical relevance of Copernicus' enterprise, we must look at some epistemological issues which are connected, as we will see, to a more fundamental metaphysical assumption than the Platonic and Neoplatonic tradition of the metaphysics of light.

In the preface and dedication of his work to Pope Paul III Copernicus explains how he was led to "consider a different system of deducing the motions of the universe's spheres". What prompted Copernicus to "venture to conceive any motion of the Earth"—and so to think "against the traditional opinion of astronomers and almost against common sense"—was an assessment of the deeply unsatisfactory condition of astronomy, which was ridden by the uncertainties of the mathematicians about the movements of the Sun, of the Moon, of the planets and, above all, about "the structure of the universe and the true symmetry of its parts". Copernicus remarked upon the disagreements of the astronomers who did not employ the same principles in their demonstrations, since they made use either of homocentric circles, or of eccentric circles and epicycles. With all their assumptions they "were unable to obtain any incontrovertible result in absolute agreement with the phenomena" and, even worse, "they introduced a good many ideas which apparently contradict the first principles of uniform motion" (Copernicus, 1978, p. 4). Copernicus started his search for a new explanation of heavenly phenomena because the existing status of his discipline was characterized by a lack of certitude, which he could not accept in a mathematical theory. He was also convinced that no irregularity should be admitted in the physical consideration of the world: a sure scheme "for the movements of the world machine" must be established on the assumption that "it has been created for our sake by the best and most systematic Artisan of all" (ibid.). This was a traditional metaphysical assumption—the idea of the divine Artificer of the orderly mechanism of the cosmos—which justified the search for a new astronomical hypothesis that had to be consistent and capable of explaining the observed phenomena. It is indeed a metaphysical assumption but with a clear and basic epistemological implication: astronomy is a mathematical discipline and the physical order of the world is mathematically structured.

The assumption of the mobility of the Earth—that is the reversion of the point of view of previous astronomy—is according to Copernicus the way to explain the observations "more compactly and more appropriately" (Copernicus, 1978, p. 307), as he wrote in the Introduction of Book Six. In particular, in the Introduction of Book Five, he highlighted that the solution offered to the irregularities of the planetary motions shows that the new hypothesis has a superior explanatory power and therefore corresponds to the harmonious structure of the world:

> Now I tackle the motions of the five planets. The order and size of their spheres are connected with remarkable agreement and precise symmetry by the earth's motion, as I indicated generally in Book I [ch. 9], when I showed that the centers of these spheres are not near the earth, but rather near the sun.[24]

[24] Copernicus (1978, p. 227). Cf. Copernicus (1873, p. 307): "Aggredimur modo quinque errantium stellarum motus, quorum orbium ordinem et magnitudines ipsa terrae mobilitas consensu mirabili

The new hypothesis, according to Copernicus and his followers, has the epistemic advantages of being more systematic, more consistent, more economical than the old one, and these features correspond to the idea of simplicity and rigorous regularity which Copernicus attributed to nature as the creation of the most perfect artificer.

An observation is here necessary: there is a long debate in the history of science whether the number of epicycles is bigger in the Ptolemaic system or in the Copernican system. If the number of epicycles is taken as the sole criterion of simplicity, one has to conclude that the two systems exhibit—more or less—the same degree of simplicity. However, as Richard Hall showed against Kuhn's opinion, the Copernican system—even taking into account the knowledge available in the sixteenth century—was simpler and preferable to the Ptolemaic one because: (1) in the Ptolemaic system the explanations of the limited elongations of Mercury and Venus as well as of the rotation period of the centre of Mercury's and Venus' epicycles on the deferent—which is exactly one year—are completely arbitrary. In contrast to this, in the Copernican system they have a clear and simple explanation: they depend on the fact that Mercury and Venus rotate around the Sun, not around the Earth; (2) a more simple solution of the problem of the retrograde motion (superior planets retrograde only when they are in opposition) is given. Since in Ptolemy, the retrograde motion depends on the motion of the planet on the epicycle, whereas the position of the planet depends on its motion along the deferent, a relation has to be postulated between these two motions and the position of the planet and the Earth with respect to the Sun. These relations are entirely arbitrary, whereas, assuming the rotation around the Sun, they are more easily explained; (3) with regard to the frequency of the retrogressions, it is also possible to show that Copernicus' treatment is easier and more natural than Ptolemy's (Hall, 1970). This means that the arbitrary elements in the Copernican system are by far less numerous and less significant than in the Ptolemaic one, though even Copernicus had to use epicyclic models to explain the "fine tuning" of many phenomena.

The advantages of the Copernican astronomy over the Ptolemaic system, highlighted by Hall (see above: 1 and 2), had been already noted by Kepler in his *Mysterium*: for, in the Ptolemaic system there is no reason why the elongations of Venus and Mercury ought to be limited and why the period of the epicycle's centre of these two planets on the deferent is exactly one year, whereas the Copernican system offers easy solutions to these problems. Furthermore, it is inexplicable in the Ptolemaic system why the planets retrograde, while the Sun and the Moon do not. Kepler wrote:

> For in the first place one might ask of Ptolemy how it comes about that the three eccentrics of the Sun, Venus, and Mercury have equal times of revolution? For the answer is, that they themselves do not really revolve, but instead of them the Earth does on its own. 2. Why do the five planets make retrogressions, whereas the luminous stars do not? The answer is first, in the Sun's case, that it is because it is at rest; and the result is that the motion of the Earth,

ac certa symmetria connectit, ut in primo libro summatim recensuimus, dum ostenderemus, quod orbes ipsi non circa terram, sed magis circa solem centra sua haberent".

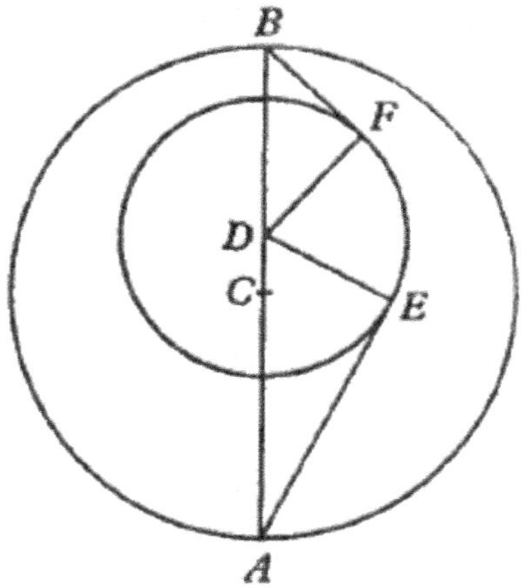

Fig. 2.7 Copernicus' diagram to determine the ratio between Earth's and Venus' orbital radiuses. *Source* Copernicus (1978, p. 271)

> which is always in the same direction, seems straighforwardly and uninterruptedly to belong to the Sun itself, though in the opposite direction with respect to the heaven.[25]

Kepler added a further observation: why are the epicycles of Venus and Mars enormous with respect to their deferent while the epicycles of Mercury, Jupiter and Saturn are much smaller? The answer is that, since Venus and Mars are close to the Earth, the variation of their distances from the Earth has a greater ratio with respect to their deferent's diameter than in the case of the other three planets. Big epicycles are therefore necessary to account for their positions with respect to the Earth, which is easy to explain if one thinks that all the planets orbitate around the Sun rather than around the Earth.

With regard to the ratios among the planets' distances, they are completely arbitrary in the Ptolemaic system because there is no constraint, while in the Copernican system a precise ratio is deducible. To explain this point, we refer to what Copernicus wrote about Venus; his argumentation can be extended, with modifications, to the other planets. In the chapter XXI, fifth book of his masterpiece, Copernicus addressed the problem of finding the ratio between Earth's and Venus' orbital diameter ("Quae sit ratio diametrorum orbis Terrae et Veneris", Copernicus, 1873, pp. 366–367). He argued as follows (see Fig. 2.7).

Be AB the Earth's orbit whose diameter Copernicus indicated by ACB and whose centre is C. Be D the centre of Venus' orbit, which is eccentric with respect to AB. Be A the point of the Earth's orbit which is the farthest from D (apogee) and B the point which is the closest to D (perigee). Draw the two tangents BF and AE from B and A to the Venus' orbit. Since the elongation $D\widehat{A}E$ is known from the observations (it is

---

[25] Kepler (1981, p. 17). Cf. KGW I, p. 18: "Nam primo a Ptolemaeo quaeri potuit, qui fiat, quod Eccentrici tres Solis, Veneris et Mercurij habeant equales revolutiones? Respondetur enim, quod non vere revolvuntur ipsi, sed pro ipsis unica terra. 2. Quare quinque Planetae fiunt retrogradi, Luminaria non item? Respondetur primo de Sole, quia is quiescit: unde fit, ut motus terrae, qui semper directus est, ipsi Soli mere et imperturbate inesse videatur, tantum per partem oppositam coeli".

44° 48′), and $A\widehat{E}D$ is a right angle, the ratios among the sides of the triangle *ADE* are known. Therefore, it is possible to calculate the ratio between *AD* and *DE*, which is the radius of Venus' orbit. Analogously, since the angle $D\widehat{B}F$ is 47° 20′, the ratio *BD* to *DF* is known. It is now simple to calculate the ratio between *AC* (i.e. the radius of Earth's orbit) and *DF* or *DE* (i.e. the radius of Venus' orbit). Copernicus calculated that, posed $AC = 10{,}000$, it is $DE = 7193$, which is a very good approximation.[26]

A further and last point sheds light on Copernicus' approach to the Ptolemaic system: his criticism of Ptolemy's equant or *punctum equans*. As is well known, to explain all the features of the retrograde motion of the planets, Ptolemy introduced a point in respect of which the motion of the planet is uniform. Such a point coincides neither with the geometrical centre of the planet's orbit around the Earth nor with the position of the Earth within the deferent. Both in the *Commentariolus* and in the fifth book of the *De Revolutionibus*, while dealing with Mercury's motion, Copernicus criticized the *punctum equans* because it violated the principle of explaining planetary motions through the composition of circular uniform motions around the centre of the deferent. To Copernicus the *punctum equans* appeared only an ad hoc device to explain the retrograde motion.[27] Copernicus' criticism of the *punctum equans* reveals his firm conviction of the perfection of the circular geometric shape. At the same time, it shows his epistemological acuity, since the *punctum equans* appeared to be a fictitious expedient inconsistent with a general assumption of the Ptolemaic theory itself.

With Copernicus for the first time in the history of science a real cosmological model of the solar system is presented. By the phrase "a real cosmological model" we mean a model in which: (1) the ratio of the distances among the bodies belonging to the system are exactly determined; (2) all the phenomena, as well as the supposed different motions of the Earth's axis, and the supposed trepidations of the equinoxes are explained; (3) the oddities of the enormous epicycles of Venus and Mars are removed; (4) the limited elongations of Venus and Mercury are clarified; (5) the explanations of the planetary movements are based on simpler and more unitary principles than in Ptolemy, though the explanations of all the details need a number of epicycles not smaller than in Ptolemy; (6) all the possible objections against the movement of the Earth are removed. Because of all these characteristics Copernicus' model of the universe is both structural and kinematical, and is more precise and unitary than any other model devised before him. In this respect, Copernicus is a milestone in the history of cosmology. But there are further interesting elements. Let us consider Copernicus' idea that all the movements of the planets should be of the circular uniform kind. There is no doubt that, under a certain standpoint, this is a step back with respect to Ptolemy, who used freely the equant, thus departing, as a matter of fact, from the rigid principle of the explanations based on the composition of uniform circular motions. However, in a different perspective, it shows that Copernicus tried to describe a cosmological structure which had to be consistent not

---

[26] The major half axis of the Earth's orbit is around 1.5×10^8 km and that of Venus' orbit is around 1.08×10^8 km.

[27] For this question see Rabin (2004), revised 2019, where a rich bibliography is also referred to.

only with observations, but also coherent in its own concepts since it was built with uniform mathematical methods. This is the main reason why Copernicus adopted the principle of uniform circular motions.

A second point concerns the dynamic aspect of a cosmological model. There is no doubt that Copernicus' cosmological system is not dynamic. However, as we have seen in the previous section, he had ideas on the Earth's physics that are surely of a dynamic kind, though they are not integrated in a well defined dynamic picture of the universe.[28] The most important passage where Copernicus showed his new view is the one relative to the participation of a body to the movement of the system to which the body belongs. Obviously, this does not mean that Copernicus developed the concept of inertial system, even less that he thought of something like the modern concept of force to explain local movements, but he certainly guessed that a new physics was required: a physics, based on principles coherent with his new view of the solar system, and different both from the Aristotelian ones and from the physics of the impetus.

The elements of a Pythagoric and Platonic metaphysics that can be detected in the first book of the *De Revolutionibus* confirm that Copernicus' astronomical thought is characterized by a structural geometric realism. Copernicus opens his treatise with a short exposition of the ancient idea that the spherical shape is the most perfect and that it is the form of divine bodies (Bk. I, chap. I). In the second and third chapters he proceeds to prove that the Earth too is spherical: "it is perfectly round as the philosophers hold".[29] The next step (chap. 4) is the reassertion of the ancient tenet that the motion of the heavenly bodies is circular in accordance with their spherical shape:

> I shall now recall to mind that the motion of the heavenly bodies is circular, since the motion appropriate to a sphere is rotation in a circle. By this very act the sphere expresses its form as the simplest body, wherein neither beginning nor end can be found, nor can the one be distinguished from the other, while the sphere itself traverses the same points to return upon itself.[30]

Finally, Copernicus addresses the question if the circular motion pertains to the Earth too, since the Earth is a globe like the other heavenly bodies. The leading principle of Copernicus' astronomical physics is that "the form entails the motion"[31]: the geometrical shape is naturally associated to a kind of motion of the same form.

---

[28] On the first book of Copernicus' *De Revolutionibus* see De Pace (2009). The author addresses not only the issue of the physical reality of Copernicus' cosmos, but also the problem of his dynamical conception.

[29] Copernicus (1978, pp. 8–10).

[30] Copernicus (1978, p. 10). Cf. Copernicus (1873, p. 14): "Post haec memorabimus corporum caelestium motum esse circularem. Mobilitas enim sphaerae est in circulum volvi, ipso actu formam suam exprimentis, in simplicissimo corpore, ubi non est reperire principium, nec finem, nec unum ab altero secernere, dum per eadem in se ipsam movetur".

[31] Copernicus (1978, p. 11).

We may even say that the geometrical structure of the bodies is the cause of their motions, as Koyré remarked.[32]

The exposition of Copernicus' new astronomical system by Rheticus in his *Narratio Prima* is another evidence of how the metaphysical background of Copernicus' science implied a realistic view of the geometrical structure he had devised. When Rheticus compares "the ancient hypotheses" to the "hypotheses of my teacher [sc. Copernicus]", he says that he "feels more inclined" to the latter because it enabled him to understand fully the truth of the maxim, attributed to Plato, that "God ever geometrizes". Copernicus' revival of astronomy has finally lifted the fog and the sky is now clear, exclaims Rheticus.[33]

Copernicus' realism is confirmed also by the references to the metaphysics of light that Rheticus evokes to illustrate the relevance and novelty of his teacher's system. The rule of the Sun in the universe is deduced by the geometrical truth "that all the celestial phenomena conform to the mean motion of the sun and that the entire harmony of the celestial motions is established and preserved under its control". Copernicus was persuaded that, because of this ruling function of the Sun, it was convenient to pose the Sun at the centre of the heavenly motions, and Rheticus draws two similitudes to justify this idea: in human affairs "the emperor needs not himself hurry from city to city in order to perform the duty imposed on him by God"; analogously, in the human body "the heart does not move to the head or feet or other parts of the body to sustain a living creature" (p. 139). These metaphysical opinions favored and 'accompanied' Copernicus' astronomical revolution, and should not be confused with its conceptual core; nonetheless, if Rheticus' testimony is reliable, they are evidence that Copernicus was convinced that he had uncovered the "absolute system" of the "order and motions of the heavenly spheres".[34]

References

Berry, A. (1898). *A short history of astronomy*. London: Murray.

Birkenmajer, L. (1924). *Stromata Copernicana. Studja, poszukiwania i materiały biograficzne [Stromata Copernicana. Studies, pieces of research and biographic material]*. Kraków: Nakł. Polskiej Akademii Umiejętności.

Bortz, F. (2013). *The sun-centered Universe of Nicolaus Copernicus*. New York: Rosen Publishing.

Cohen, G. D. (1967). *The book of tradition by Abraham Ibn Daud*. Philadelphia: Jewish Publication Society.

Comes, M. (2001). Ibn al-Hā'im's Trepidation Model. *Suhayl, 2*, 291–408

Copernicus, N. (1543, 1978). *On the revolutions*. Baltimore: The John Hopkins University Press.

---

[32] Koyré (1973, p. 58): "Copernicus seems to believe that a spherical shape, the most perfect geometrical form, and the one that all natural bodies endeavour to assume because of this very perfection, is not only more suitable for motion (as is universally admitted), but is sufficient cause thereof, and naturally engenders *the most perfect and most natural* of motions, namely *circular* motion".

[33] Rheticus (1959, p. 167f).

[34] Ibid., p. 138f.

Copernicus, N. (1873). *De Revolutionibus Orbium Caelestium libri VI*. Thorn: Societas Copernicana.
Copernicus, N. (1878). De hypothesibus motuum coelestium a se constitutis Commentariolus. In M. Curtze (Ed.), *Mitteilungen des Coppernicus-Vereins für Wissenschaft un Kunst zu Thorn* (Vol. I, pp. 5–17).
Copernicus, N. (1884). *De hypothesibus motuum caelestium a se constitutis Commentariolus*. In L. Prowe & N. Coppernicus (Eds.), *Werke* (Vol. II, pp. 184–202). Berlin: Weidmannsche Buchhandlung.
Copernicus, N. (1959). *Three Copernican treatises: The* Commentariolus *of Copernicus, the letter against Werner, the* Narratio Prima *of Rheticus*. Translated with introduction and notes by E. Rosen. New York: Dover.
Copernicus, N. (1978). *On the revolutions*. In F. Dobrzycki (Ed.), Translation and commentary by E. Rosen. Warszawa and London: Polish Scientific Publishers and Macmillan Press.
Copernicus, N. (1979). *Opere*, a cura di F. Barone. Torino: UTET.
De Pace, A. (2009). *Niccolò Copernico e la fondazione del cosmo eliocentrico*. Milano: Mondadori.
Dreyer, J. L. E. (1953). *History of planetary systems from Thales to Kepler*. New York: Dover.
Enriques, F. (1906). *Problemi della scienza*. Bologna: Zanichelli.
Enriques, F. (1912). *Scienza e razionalismo*. Bologna: Zanichelli.
Enriques, F. (1936). *Il significato della storia del pensiero scientifico*. Bologna: Zanichelli.
Freely, J. (2014). *Celestial revolutionary. Copernicus, the man and his universe*. London-New York: I.B. Tauris.
Galilei, G. (1632). *Dialogo sopra i due massimi sistemi del mondo tolemaico e copernicano* (pp. 21–520). EN, VII.
Galilei, G. (1890–1909). *Edizione Nazionale delle Opere*. Direttore A. Favaro, (20 vols). Firenze: Barbèra. Abbreviated as EN.
Galilei, G. (1967). *Dialogue concerning the two chief world systems: Ptolemaic and Copernican* (2nd ed.).Translated and with notes by S. Drake. Foreword by A. Einstein. University of California Press.
Gingerich, O. (2004). *The book nobody read. Chasing the revolutions of Nicolaus Copernicus*. London: William Heinemann.
Goddu, A. (2010). *Copernicus and the Aristotelian tradition*. Leiden, Netherlands: Brill.
Goldstein, B. R. (1965). On the theory of trepidation according to Thābit b. Qurra and al-Zarqāllu and its implications for homocentric planetary theory, *Centaurus, 10*(4), 232–247.
Hall, R. (1970). Kuhn and the Copernican Revolution. *British Journal for the Philosophy of Science, 21*(2), 196–197.
Hoyle, F. (1973). *Nicolaus Copernicus. An essay on his life and work*. New York: Harper and Row.
Kepler, J. (1981). *Mysterium Cosmographicum (The Secret of the Universe)*. Translation by A. M. Duncan. Introduction and commentary by E. J. Aiton. Preface by I. Bernard Cohen. New York: Abaris Book.
Kepler, J. (1937–2012). *Gesammelte werke*. In W. Van Dyck & M. Caspar, et al. (Eds). Revised April 2013. (10 vols). München: Deutsche Forschungsgemeinschaft und Bayerische Akademie der Wissenschaften. Beck'sche Verlagsbuchhandlung. Abbreviated as KGW.
Koyré, A. (1973). *The astronomical revolution: Copernicus—Kepler—Borelli* (1st ed., 1961). London: Cornell University Press.
Koyré, A. (1957). *From the closed world to the infinite universe*. John Hopkins Press.
Kuhn, T. (1957). *The Copernican Revolution: Planetary astronomy in the development of western thought*. Harvard University Press.
Mancha, J. L. (1998). On Ibn al-Kammād's table for trepidation. *Archive for History of Exact Sciences, 52*, 1–11.
Omodeo, P. D. (2014). *Copernicus in the cultural debates of the Renaissance: Reception, legacy, transformation*. Leiden: Brill.
Nothaft, E. (2017). Criticism of trepidation models and advocacy of uniform precession in medieval Latin astronomy. *Archive for History of Exact Sciences, 71*, 211–244.

Rabin, S. J. (2004, revised 2019). Nicolaus Copernicus. In E. N. Zalta (Ed.), *Stanford Encyclopedia of Philosophy*. https://plato.stanford.edu/archives/fall2019/entries/copernicus/

Rajep, F. J. (1996). Al-Battānī, cosmology, and the early history of trepidation in Islam. In J. Casulleras & J. Samsó (Eds.), *From Baghdad to Barcelona: Studies in the Islamic exact sciences in honour of Prof. Juan Vernet* (pp. 267–298). Barcelona: Universitat de Barcelona.

Repcheck, J. (2007). *Copernicus' secret: How the scientific revolution began*. New York: Simon & Schuster.

Rheticus, G. J. (1540, 2015). *Narratio prima or first account of the books on the revolutions by Nicolaus Copernicus*. With an introduction by J. Włodarczyk. Warsaw: University of Warsaw Press.

Rheticus, G. J. (1959). *Narratio Prima*. In *Three Copernican treatises*. Translated with introduction and notes by E. Rosen. Revised with an annotated Copernicus bibliography 1939–1958 (2nd ed.). New York: Dover Books.

Rosen, E. (1984). *Copernicus and the scientific revolution*. Robert Krieger Publishing Company.

Rosen, E. (1995). *Copernicus and his successors*. London: Hambledon Press.

Rybka, E. (1964). *Four hundred years of the Copernican heritage*. Kraków: Jagellonian University of Krakow.

Samsó, J. (1994). Trepidation in al-Andalus in the 11th Century. In *Idem, Islamic astronomy and medieval Spain*. Aldershot: Ashgate-Variorum.

Somervill, B. (2005). *Nicolaus Copernicus: Father of modern astronomy*. Minneapolis: Compass Point Books.

Swerdlow, N. M. (1973). The derivation and first draft of Copernicus's planetary theory: A translation of the *Commentariolus* with commentary. *Proceedings of the American Philosophical Society, 117*(6), 423–512.

Swerdlow, N. M. (1980). Long-period Motions of the Earth in *De revolutionibus*. *Centaurus, 24*(1), 212–245.

Swerdlow, N. M., & Neugebauer, O. (1984). *Mathematical astronomy in Copernicus's De Revolutionibus. In two parts*. New York-Berlin-Heidelberg-Tokyo: Springer.

Westman, R. (2011). *The Copernican question: Prognostication, skepticism, and celestial order*. Los Angeles: University of California Press.

Chapter 3
Kepler: The Cosmographer *Par Excellence*

Before Newton, Johannes Kepler and Galileo Galilei were the two most important scientists of the modern age and their contribution was essential in enabling Newton to achieve his final synthesis. Galileo was the creator of the modern idea of physics, but, as an astronomer and a cosmologist Kepler was superior.[1] The latter's cultural background was not restricted to a single discipline. Although this was the case for several scientists of the early modern age, it is particularly true for Kepler. He had a thorough knowledge of Renaissance culture, which was, in part, tied to a rediscovering of Pythagoreanism and of Platonism. Renaissance thought was also connected to what we nowadays call "natural sciences", since one of its various purposes was to understand the structure and the functional laws of the universe. This aim was pursued following certain paths and relying upon some hypotheses which turned out to be fruitless, but which were not a priori absurd. Kepler also achieved a good expertise of music theory, which played an important role in his late cosmology. Besides his deep knowledge of traditional culture, Kepler was a fervent Copernican from his youth; he then became one of the greatest mathematicians of his age and was a highly creative thinker. His research in astronomy and cosmology can be divided into three phases marked by his major works: (1) *Mysterium Cosmographicum*[2]: in this work, Kepler advances a series of arguments in favour of the Copernican system: however, his main scope was to propose a cosmological model that might be defined as Pythagorean, but whose features are entirely original; (2) *Astronomia Nova*[3]: this is Kepler's most important work because in it he formulates his first two laws on the ellipticity of the planetary orbits. He also tried to offer dynamical explanations for

[1] Before the birth of modern cosmology in the twentieth century, Johannes Kepler was surely the scientist who most deserves the title of cosmologist.

[2] The first edition of the *Mysterium* was published in 1596, the second edition enriched with several notes in 1621: see KGW, I and VIII, Kepler (1981).

[3] The *Astronomia Nova* was published in 1609: see KGW, III, Kepler (1992).

P. Bussotti and B. Lotti, *Cosmology in the Early Modern Age: A Web of Ideas*,
Logic, Epistemology, and the Unity of Science 56,
https://doi.org/10.1007/978-3-031-12195-1_3

the movements of the planets; (3) *Harmonice Mundi*[4]: to summarize a book whose structure and arguments are highly complicated we can say that here Kepler argues that cosmology is also tied to musical harmonies. This is a Pythagorean view, but Kepler tried to define it through a precise mathematical treatment. Kepler's Third Law is enunciated in the *Harmonice*, which also contains Kepler's late cosmology. He also wrote a fourth masterpiece: the *Epitome Astronomiae Copernicanae*, composed of seven books.[5] Numerous concepts related to dynamics had already been expounded in *Astronomia Nova* and were later reworked and clarified in *Epitome Astronomiae Copernicanae*. Given Kepler's role as a cosmographer, his ideas are particularly worth perusing.

3.1 Kepler and the Unity of Science: His Attempt to Unify All the Aspects of Cosmology

We will focus on the three primary aspects of Kepler's cosmology: (1) the structural aspect as it is presented in the *Mysterium Cosmographicum*[6]; (2) the kinematical aspect, explained in the *Astronomia Nova* in connection to the dynamical one; (3) the dynamical aspect expounded by Kepler in the *Astronomia Nova* and improved in the *Epitome*. Finally, we will also examine how he introduced the third law and how the three aspects of Kepler's cosmology are interrelated.

3.1.1 The Structure of the World in Mysterium Cosmographicum

Kepler imagined that the structure of the solar system was like this (Fig. 3.1).

The basic concept is that the spheres of the planets are inscribed in and circumscribed by the five Platonic polyhedra. Each sphere is inscribed in a polyhedron and circumscribed by another, different polyhedron. More precisely, starting from the sphere of Saturn, which is the most external element, the sequence of the spheres and polyhedra circumscribing each other is the following: the sphere of Saturn—cube—the sphere of Jupiter—tetrahedron—the sphere of Mars—dodecahedron—the sphere of the Earth—icosahedron—the sphere of Venus—octahedron—the sphere of Mercury, which is the most internal one, proximate to the Sun. Kepler thus provided

[4] The *Harmonice Mundi* was published in 1619: see KGW, VI, Kepler (1997).

[5] The books I–III were printed in 1617, book IV in 1620 and books V–VII in 1621 (see KGW, VII).

[6] A classical work on the *Mysterium* is Field (1988). There are interesting observations on Kepler's text by Eric Aiton in Duncan's English translation of the *Mysterium* (Kepler, 1981) as well as by Alain Segond in his French translation (Kepler, 1984). For a detailed explanation of the geometrical aspects as well as for references to more modern literature, see Pisano and Bussotti (2012, pp. 121–135).

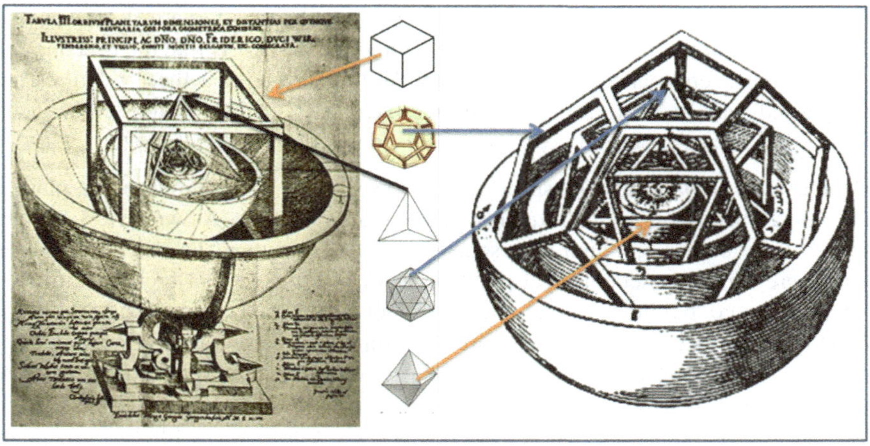

Fig. 3.1 Kepler's system of the word as explained in the *Mysterium Cosmographicum. Source* https://bluedragonfly10.wordpress.com/2012/04/

a model of the solar system. Outside the solar system, there is the sphere of the fixed stars.

Kepler's cosmology is one of the most perspicuous examples of the interrelations among metaphysical, religious, mathematical, and physical considerations.

First of all, Kepler's model is structural since he tries to determine the true, *ideal* structure of the universe. The spheres and the polyhedra were not conceived by Kepler as physical bodies—as was, instead, the case with the spheres in the classical Aristotelian tradition. In Kepler's view, no such body exists in the sky.[7] Rather, between the orbits of the planets it is ideally possible to insert those solids that are conceived as mathematical entities. Kepler argues that three main problems exist in cosmology: (1) to understand why the planets are six in number; (2) to explain the dimensions of planetary orbits; (3) to determine the planetary movements according to the Copernican doctrine. All these issues suggest a view of the universe which is characterized by the following assumptions and arguments.

Religious arguments: The universe has a perfect correspondence with the Holy Trinity. While looking for the splendid harmony of those cosmic elements which are at rest, Kepler compared the Sun to the Father, the fixed stars to the Son and the intermediate space between the Sun and the fixed stars to the Holy Spirit (KGW, I, pp. 24–25). He did not offer any reason for the use of this analogy. The Sun represents the Father because it is the source of motion and the motion of the Sun is the equivalent of the divine creation. The Sun moves the planets in the space delimited by the fixed stars like the Father creates in the Son. Empty intermediate space can be seen as the medium on which the Sun exerts its force. In this sense it is compared to the virtue of the spirit by means of which the Father creates (KGW XIII, p. 35).

[7] An interesting book dealing with "the dissolution of celestial spheres" is Donahue (1981).

The universe, having a spherical form, is the physical image of an ideal *geometrical figure*. The centre of the sphere is like the Father, the surface is like the Son and the void between centre and surface is like the Holy Spirit so that these relations are established: Father-centre-Sun, Son-surface-fixed stars, Holy Spirit-space between centre and surface-space between Sun and fixed stars.

Metaphysical arguments: The space around the Sun has a structure too: the planets are ordered according to the sequence of the Platonic solids. As we have seen, the figure of the universe, which represents the Holy Trinity and its manifestation, is the sphere. From a metaphysical point of view, it is the perfect archetype of that which is curved, but in the universe there are not only curves, but also straight lines. Which are the three-dimensional figures that represent perfection as far as straight lines are concerned? Here, it is impossible to indicate a sole figure, because there are five figures that represent the archetypical perfect model of what is straight. These are the Platonic polyhedra. Each one is perfect in its way, although a scale of perfection can be traced, as will be considered. Therefore, it is only natural that the planets move along great circles of spheres and that such spheres are separated by the regular Platonic polyhedra. Furthermore, these polyhedra are exactly 5 in number and there are 6 planets. This means that the number of regular convex polyhedra and the number of the planets match perfectly. This answers the first of the three aforementioned questions. The cosmos refers to perfect archetypical models. Since there are 5 perfect polyhedra, there must be 6 planets. The ideal archetype corresponds to and justifies the apparent empirical structure (KGW, I, p. 26).

Metaphysical-mathematical arguments: chapters III–VIII of the *Mysterium Cosmographicum* are surprising from a modern point of view: Kepler attempted to explain the reason why a particular solid is interposed between the spheres of two particular planets. In order to do so, he evaluated the mathematical properties of Platonic solids according to his notion of geometrical perfection.[8] Kepler distinguished the regular polyhedra into two classes: the first class (primary solids) includes cube, tetrahedron, and dodecahedron; the second class (secondary solids) includes octahedron and icosahedron. The primary solids are regarded as more perfect than the secondary ones. This depends on seven of their features, four of which are connected with the form of the solid, while the fifth and sixth ones concern respectively the way in which the figure can rotate and the equilibrium-position of the solid. The seventh one refers to the angles. Let us examine closer these seven properties (ibid., p. 29):

(1) and (2): the three primary solids have mutually different faces, while the secondary solids have equilateral triangles as faces;
(3) the angle at the vertex of the primaries is "simple", that is only three planes concur to form it. Of course, 3 is the smallest number of planes sufficient to form a solid angle; as to the secondary solids, there are four and five planes respectively;

[8] Obviously, Kepler's judgement is arbitrary and subjective, but in his view it corresponded to perfection-criteria inherent in the universe.

(4) Kepler stresses that the primaries have absolutely original features, while the secondaries derive theirs from the primaries. He enumerates all these features (we refrain from going into detail here).
Property (5) concerns the way in which a regular solid can rotate around an axis passing through its centre and through two of its vertices or through the centres of two opposite faces This property clearly shows that Kepler guessed several features concerning the rotational symmetries of regular polyhedra and their dual character (we refer to the duality cube-octahedron; dodecahedron-icosahedron; the tetrahedron is in a separate rotational class).

The nature of this fifth property differs from the plain properties ascribed by Kepler to the Platonic polyhedra [as those (1)–(4)]. He possibly discovered with property (5) profound mathematical features of these figures, which are much less evident than the previous four properties. This fascinating aspect of Kepler's cosmology shows how the logic of scientific discovery is not unilinear: Kepler possibly discovered the rotational symmetries of the Platonic polyhedra while he was elaborating his cosmological ideas. We contend that this is a fundamental and profound discovery, which found its full explanation and setting only between the end of the 19th and the beginning of the twentieth century, with the group theory within abstract algebra. It is remarkable that Kepler allegedly discovered this property—though only from a geometrical point of view—at the end of the sixteenth century. However, from Kepler's perspective, the rotational features of the regular polyhedra are no more significant than properties (1)–(4), which, from our point of view, are trivial and represent only elementary aspects of mathematics. Therefore, property (5) is fundamental for the history of mathematical ideas, but it is not more important than the trivial properties (1)–(4) in Kepler's way of thinking. According to him, all these properties represent on an equal footing the harmonic and perfection criteria that he thought he had discovered in the Platonic polyhedra.

Let us now analyse in detail property (5): Kepler claimed that the primary polyhedra can move appropriately[9] only through a diameter traced across the centres of two opposite faces, while the secondaries through a diameter traced across the opposite vertices. Field and Stephenson report Kepler's assertion on the rotational symmetries of the regular polyhedra (Field, 1988, p. 55; Stephenson, 1987, p. 80): the cube has three axes of rotation, which pass across the centres of two mutually opposite faces. The rotations of $\frac{\pi}{2}$, π, $\frac{3}{2}\pi$ are symmetries of the cube. Globally there are 9 symmetries of this kind. Among the rotations around the four diagonals, only those of $\frac{2}{3}\pi$, $\frac{4}{3}\pi$ are symmetries, hence we obtain eight rotations. As to the six straight lines that join the middle points of two opposite edges, only the rotation of π are symmetries, hence six rotations. Adding the identical rotation, there are 24 rotations of the cube, which are symmetries (Fig. 3.2).

Therefore, the cube is more symmetric with respect to an axis passing through the centres of the faces because it has 9 rotations of this kind, while the other rotational symmetries of the cube through each of the other axes are less than 9. In the light

[9] The term used by Kepler is "concinne" (KGW I, p. 29). We explain in the running text how to interpret the meaning of this term.

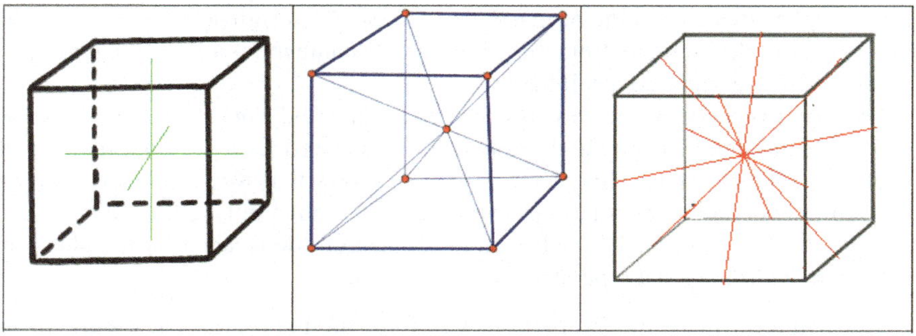

Fig. 3.2 The axes of rotational symmetries of a cube as probably devised by Kepler. On the left: the three segments joining the central points of the cube's faces. Around these segments the rotations of 90°, 180° and 270° are symmetries of the cube. In the centre: the four diagonals of the cube. Around them, the rotations of 120° and 240° are symmetries of the cube. On the right: the six segments joining the central points of the cube's edges. Around them, the rotation of 180° is a symmetry of the cube. The figures are ours

of Field's and Stephenson's convincing interpretation, it can be said that the cube moves more appropriately around these axes than around the axes of the two other kinds. The octahedron is the dual figure of the cube. This means that the axes of rotation, which join the centres of the faces in the cube, become the axes passing through the three couples of opposite vertices of the octahedron. Hence, there are 9 symmetries of this kind. Through an analogous reasoning, it is possible to determine the other 8 and 6 rotations of the octahedron which are symmetries. Thus, this solid is more symmetric with respect to the three axes of rotation that pass through the three couples of opposite vertices, as Kepler claims. *Mutatis mutandis*, the reasoning for the dodecahedron and the icosahedron is analogous. The tetrahedron escapes this picture because it can be defined as being more symmetric with respect to the four straight lines passing through the vertices and perpendicular to the respective opposite faces because 8 of the 12 rotational symmetries of the tetrahedron are of this kind. According to Kepler, property (5) is an indication of the greater perfection of the primary solids.[10]

(6) The shapes of the solids look visually and aesthetically more harmonious if the primaries stand on one of their faces and if the secondaries are suspended by a vertex.
(7) The number of the primary solids, 3, is perfect; while that of the secondary solids, 2, is not.

[10] Our interpretation of property (5) is probable, but not certain, in so far as Kepler stated explicitly that the different regular polyhedra "move more appropriately" around different axes, but did not explain why. We suggest that he had an insight of the most important characteristics of the groups of rotations of the Platonic polyhedra. With regard to the rotational symmetries explained by Kepler in the *Mysterium* see also Pisano and Bussotti (2012, p. 128).

This concept of geometrical perfection justifies the positions of the solids in the structure of the world. For example: since the Earth is the noblest of the planets, it necessarily separates the primary solids from the secondaries. The primary polyhedra are associated with the three external planets, while the secondaries are associated with the two internal ones. Since the fixed stars are the most important part of the universe after the Sun and the Earth, the most perfect solid has to stay as close as possible to the fixed stars. This is the reason why the cube (the most perfect solid) is also the most external one in Kepler's structure of the world, and so on. Therefore, according to Kepler, a new metaphysics based on mathematical properties plays a crucial role in the structure of the universe. As we have seen, Kepler does not refer generically to a mathematical natural order, but connects precise geometrical characteristics with physical elements and structures of the world. Mathematics tends to substitute, in Kepler's way of thinking, a non-quantitative metaphysical ontology. The use of mathematics in Kepler is quite precise (apart from the naïve reference to 3 as the perfect number), although, to our eyes, it is arbitrary, as is the definition of perfection itself according to the seven features we have enumerated.[11]

Mathematical-empirical arguments: Kepler believed that his system really reproduced the structure of the world. Therefore, it was necessary that the mathematical calculations were coherent with the real distance-ratios between the planets and the Sun, at least with the distance-ratios computable in the Copernican system. He operated in the following manner: (1) for every Platonic solid, he calculated the ratio between the radiuses of the inscribed and circumscribed sphere; (2) relying upon the data available in Copernicus's *De Revolutionibus*, Kepler calculated the ratios between the distances of the planets from the Sun; (3) he calculated the ratios of such distances, when his own polyhedral hypothesis was inserted within the Copernican theory. He remarked that the discrepancies were, in most cases, irrelevant. This was, in his opinion, a further confirmation that Copernican theory was correct. In Chapter 18 of *Mysterium* Kepler speculated on the relations between the theory and empirical data setting forth this argument: as to the distances among the planets and the distances between each planet and Sun, the empirical data were not quite precise; they were compatible both with the Copernican theory without polyhedra and with his own theory with polyhedra. Relying on his calculations, these data were, anyway, better suited to the theory with polyhedra, although the uncertainty deriving from the approximate nature of the data was, of course, unavoidable. He distinguished

---

[11] The relations between geometrical and theological aspects in the peculiar cosmological structure of the *Mysterium* (i.e. the series of spheres and Platonic solids) are encapsulated in this statement by Regier (2013, pp. 152–153): "In the second chapter of the *Mysterium*, creation is portrayed as an act of divine expression, wherein God communicated his being through geometry. Mainly, God wished to describe his status over his creatures' by the incommensurability of the curve and the line. In Kepler's account, the creation of matter was necessary to achieve a medium that might express quantity—and quantity (as in extension or dimension) was essential for the realization of geometric form. Reading between the lines, a priori reasoning must incorporate geometric forms that have a status determined by their nearness to the sphere, that is, by their simplicity and their equalities". As Regier points out, this topic is also addressed by Hon and Goldstein (2008, pp. 170–176).

between astronomers and cosmographers. For the cosmographers (that is, for those scientists aiming to understand the structure of the solar system), the accuracy of data and the agreement between data and theory were sufficient to justify the creation of a model. In a sense, this situation, though far more simplified, is comparable to the modern one: the cosmographers have a series of data which, in many cases, are not precise and whose interpretation is not easy. They nonetheless create models of the universe. Things were easier for Kepler because he had to choose only one of two models: Copernican theory with or without polyhedra.

The problem arises for the astronomers who, less interested in foundational principles and theoretical structures, look for some mechanisms that would allow them to obtain ephemerides that are as accurate as possible. Which theory shall we draw the ephemerides from? From the Copernican theory with or without the polyhedral hypothesis? Here Kepler claimed, in substance, that the instruments available at that time were insufficient to form an accurate idea of the real distances and that, such a discussion would have been pointless. In conclusion, Kepler believed that his theory was acceptable from a cosmological point of view and that astronomers had no reasons to prefer a Copernican theory without polyhedra.

Physical-dynamical arguments. The *Mysterium* is not a text of physics. There is no precise theory of the forces that make the planets rotate around the Sun. (Kepler tried to develop such a theory in his successive works.) However, he knew that this problem had to be addressed. In Chapters XX–XXII of *Mysterium*, Kepler tries to offer a mathematical characterization of the force responsible for the movements of the planets around the Sun, which is a clear attempt to offer a dynamical treatment of forces. The attempt is naïve and only briefly outlined, but it is remarkable that Kepler felt the need to develop it. He started from the idea that the velocity of the planets in their orbits is inversely proportional to their distances from the Sun. The force emanates from the Sun and is due to the Sun's *anima motrix*. It decreases in proportion to the distance from the Sun. Kepler's force produces *velocities* and not *accelerations*, unlike Newton's force. With regard to its nature, the idea of the Sun's *anima motrix* derives from the Renaissance Pythagorean-Platonic tradition. In the following works Kepler will progressively abandon this vitalistic conception in favour of a more mechanical approach. However, the idea that forces produce velocities recurs throughout his writings. This is also connected to Kepler's idea of *natural inertia*. On this point he followed the Aristotelian tradition: the natural "inertial" state of a body is the state of rest. This is one of the reasons why a force has to produce velocity. Though hindered by these erroneous ideas, Kepler, in *Astronomia Nova* and *Epitome Astronomiae Copernicanae* developed a physics of forces (see below C).

3.1.2 Kinematical Aspects of Kepler's Cosmology in Connection with His Dynamical Ideas

The main purpose of *Astronomia Nova* is to determine the shape of Mars' orbit, which means to determine its parameters. In particular, Kepler focused on the determination of: (1) longitude of the nodes; (2) inclination of Mars' orbit with respect to the ecliptic; (3) position of the apses line; (4) eccentricity of the orbit. At the beginning of his investigations, which lasted from 1602 to 1609, Kepler supposed, in the wake of Ptolemy and Copernicus, that the orbit of Mars was an eccentric circle. He exploited the enormous amount of observations left by Tycho Brahe and relied on his own mathematical capabilities and patience in repeating long and exhausting series of calculations. First of all, Kepler was able to determine the longitude of the ascending node which is 46 and 1/3°.[12] He also proved that the plane of Mars' orbit is inclined by $1^\circ\ 50'$ on the plane the ecliptic, that Mars' orbital plane is constant and that it passes through the Sun.[13] The most difficult problems concern items (3) and (4). To determine the position of the apses line and the eccentricity of Mars' orbit, Kepler resorted to an eccentric model with *punctum aequans*. The whole construction we are presenting here was named by Kepler *hypothesis vicaria*. In Fig. 3.3, points D, G, F, E represent four of the Mars' positions observed by Tycho, in particular they are the oppositions observed by Tycho respectively in 1587, 1591, 1593, 1595. Point A represents the Sun, point B the geometrical centre of the orbit and point C is the *punctum aequans*. The straight line HI is the apses line.

By means of a complex argumentation that we do not examine here,[14] Kepler obtained: (a) the longitude of the aphelion (year 1587) that was $28^\circ\ 48'\ 55''$ of the Lion constellation; (b) posed equal to 1 the segment HD, the length of the segments $BC = 0.07232$ and $BA = 0.11332$.

This model represented accurately the longitudes of all the twelve oppositions reported by Tycho, but failed to represent correctly the latitudes. Then, Kepler hoped to save his model by posing $CB = BA = 0.09282$, the so-called bisection of eccentricity, an old device used by Ptolemy in the context of geocentric astronomy. This supposition improved the model because Kepler was able to represent correctly the positions of Mars at 90° from the apses, but a crucial problem remained: the positions at 45°, 135°, etc. from the apses were represented with an error of $8'$. Thanks to the improvements in the precision of astronomical observations fostered by Tycho Brahe, Kepler regarded this error as incompatible with the standard of accuracy of his own theory. Thus, the model did not work, unless one admitted that the *punctum*

---

[12] See the XII Chapter, entitled *Investigatio Nodorum Martis*, of the second part of *Astronomia Nova* (KGW III, pp. 130–133). To reach this result Kepler made use of Tycho's observations of the years 1589, 1590, 1592, 1593, 1594, 1595.

[13] These results were achieved in two chapters of the second part of the *Astronomia Nova*, the XIII and the XIV, entitled respectively *Investigatio Inclinationis Planorum Eclipticae et Orbitae Martis* (ibidem, pp. 133–140) and *Plana Eccentricorum Sunt Ἀταλαντα* (ibidem, pp. 140–142).

[14] For these details a fundamental text is still the chapter on Kepler in Dreyer's *History of planetary systems from Thales to Kepler* (1953, original 1906).

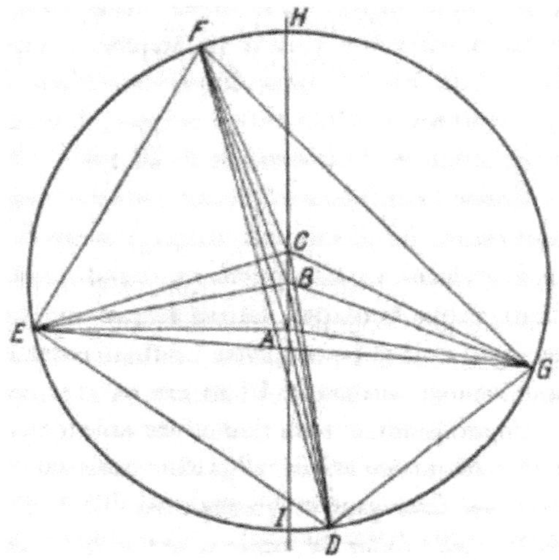

Fig. 3.3 Kepler's figure to illustrate his *hypothesis vicaria* (KGW, III, p. 153). The figure is inserted in the fundamental Chapter 16 of *Astronomia Nova*'s second part, entitled *Methodus Inquirendi Hypothesin pro Inequalitate Prima Solvendi* (ibid., pp. 151–169)

aequans oscillates back and forth on the apses line, but—as Kepler wrote—"I do not see how this can be reconciled with natural reasons" ("Quod quomodo cum rationibus naturalibus conciliari possit, non video") (KGW, III, p. 182).

Since no physical reason could be consistent with the libration up and down of the *punctum aequans*, Kepler abandoned entirely the idea that Mars' orbit was an eccentric circle.

From this moment onwards Kepler began an authentic *tour de force*, which is impossible to describe here, and concluded that the orbit of Mars is an ellipse with the Sun located in one of its foci. As first hypothesis, Kepler supposed the path to be an oval and, after several uncertainties on the form of this oval, he realized that it was an ellipse. Kepler describes the ellipticity of the orbit in the fundamental Chapter LIX of the *Astronomia Nova*'s fourth part.[15]

The area law is a topic of remarkable conceptual interest: as will be considered in Sect. 3.1.3. Kepler was convinced that the planets move around the Sun because of a force emanating from our star whose intensity decreases with the increasing of the distance from the Sun. This implies that a planet moves quicker at the perihelion and slower at the aphelion. Obviously, this dynamic view is not yet the area law, but it is certainly a good heuristic idea to reach such a law. Kepler conceived the idea of a solar force moving the planets before he reached the conclusion of the ellipticity of the planets' path: as a matter of fact, the idea of a solar force already existed *in nuce* in the *Mysterium*. Indeed, the area law was formulated in the *Astronomia Nova* and was originally conceived by Kepler for eccentric circular orbits Therefore, the

[15] For Kepler's complex argumentations through which he arrived at conceiving the ellipticity of the orbits, see Aiton (1978), Baigre (1990), Donahue (1993), Donahue (1996), Stephenson (1987), Whiteside (1974), Wilson (1968). Caspar's notes at KGW III, pp. 455–484 are a precious guide to the technical aspects of Kepler's astronomy as expounded in the *Astronomia Nova*. For Kepler's deduction of the ellipticity of the orbits see Koyré (1973, [1961], II, II).

area law should be more appropriately called "first Kepler law" rather than "second Kepler law". Importantly, it is noteworthy that Kepler obtained the area law relying upon two mistakes which mutually compensate.[16]

Commentary: though the main details of Kepler's dynamics will be illustrated in the next subsection, the relations between kinematics and dynamics in Kepler's theory are worth discussing here. His dynamical conceptions are wrong, whereas his kinematical laws are correct. We may therefore wonder how Kepler's dynamical ideas influenced his kinematical achievements. A satisfying answer requires the distinction between heuristics and scientific deduction. As to the heuristic standpoint, the connection between Kepler's dynamical ideas and his results on the form and laws of planetary paths is indisputable: (1) Kepler was convinced that no physical cause might have determined the libration of the *punctum aequans* on the apses line. This dynamical conviction had fundamental consequences on his kinematics. If Kepler had not believed in a dynamical requirement, in principle he could have accepted the movement up and down of the equant[17]; (2) Kepler thought that a solar force moves the planets with a speed inversely proportional to the distance from the Sun. This dynamical conviction led him to deduce the area law. Kepler realized an important dynamical fact: the velocities of the planets are an inverse function of the distance Sun-planet. He erroneously thought of a simple inverse relation, but this idea had a seed of truth. Analogously, the idea that a force emanating from the Sun determined the motion of the planet was right, though his general views on the nature of such force were wrong. Kepler's dynamical convictions were a heuristic guide towards his kinematics.

From the point of view of the scientific deduction, things are more complicated: with regard to the form of the orbit, Kepler's dynamical convictions led him to abandon the belief in the circularity of the orbits, but played no role in determining their ellipticity with the Sun located in one of the two foci. This result was obtained through a series of mathematical hypotheses and calculations, through a specific

[16] With regard to Kepler and the area law see Davis (2003), where the author argues that, though the proof in *Astronomia Nova* is unsatisfying, the one offered by Kepler in the *Epitome* is correct. Other works containing remarkable insights on the area law are: Aiton (1973), Davis (1992a), Russell (1964) Stephenson (1987). Stephenson (pp. 94–103) offers an accurate account of how Kepler derived the area law. The first scientist who provided a rigorous proof of the area law was Newton in the second section of *Principia*'s first book. The area law is the first proposition of Newton's masterpiece. He also proved (*Principia*, Book I, Section III, Proposition XVI, Corollary 5) that, given an elliptical orbit in which the centripetal force is located in one of the two foci, the speed of the body in each point *P* is inversely as the perpendicular drawn from the focus to the tangent at *P*.

[17] The movement of the equant was admitted by Ptolemy, as Koyré recalls: Koyré (1973, [1961], p. 402), note 23 to the chapter II—entitled *First Attack upon the Theory of Mars*—of the section on Kepler. With regard to the relation between dynamical and kinematical aspects in Kepler a fundamental reference text is Stephenson (1987). See also Petroni (1989), who holds that the influence of Kepler's dynamics upon his kinematics was negligible. He does not agree with Koyré's opinion, which is, instead, shared by Bernard Cohen (see Cohen, 1975). Our opinion will be expressed in the running text.

observation[18] and a profound analysis of the empirical data. Instead, for the area law, the dynamical convictions played an important role because the first part of the reasoning[19] relies exactly on two dynamical convictions: (a) the forces produce velocities: (b) these velocities are directly proportional to the forces, thence, according to Kepler, they decrease as the distances from the body emanating the force.

Our interpretation can be summarised like this: Kepler was guided by his dynamical convictions in so far as they were useful to free his mind from those old conceptions that were not consistent with the data. These convictions helped him to abandon the classical idea that planetary movements had to be explained through a system of circles. On the other hand, Kepler was so brilliant that, when he had to infer from Brahe's observational data and from his own calculations the new exact form of planetary orbits, he did not deduce the ellipticity of the orbits from his dynamical ideas. If he had tried to do so, this would have been impossible.

This is not to say that this intellectual process was either conscious or voluntary. However, the developments of *Astronomia Nova* are susceptible to this interpretation. This enables us to understand both the leading views of this great scientist—which characterised the whole of his production—and his flexibility to exploit a set of ideas when it was useful for him, without binding himself to it.

3.1.3 Kepler's Concept of Force

According to Kepler, gravity did not determine the motion of the planets around the Sun; however, in order to illustrate Kepler's cosmology an outline of his doctrine of gravity is necessary for the following reasons: (a) Kepler devised a new conception of gravity which made Aristotelian physics and, hence, Aristotelian cosmology, unacceptable; (b) Kepler explained a terrestrial phenomenon—the tides—referring it to the action of the Moon; (c) Kepler's concept of gravity is still, in part, qualitative, and this prevented him from reaching a universal conception of gravity.[20]

Kepler presented his *doctrina de gravitate* in the Introduction to *Astronomia Nova* (KGW, III, pp. 24–28) starting with a definition of gravity:

---

[18] The whole reasoning that drove Kepler to guess that the shape of Mars's orbit is an ellipse and not a more complicated oval—as he had thought for many months—starts with his observation of the particular value assumed by the secant of the maximum Mars' optical equation (Chapter 56 of *Astronomia Nova*). From this observation, through a series of arduous—and not always easily reconstructable—deductions Kepler arrived at the ellipticity of Mars' orbit. This topic is skilfully expounded by Stephenson (1987, pp. 107–110). We think that Kepler noticed that the secant's value is typical of the ellipse because he worked on the problem of Mars' orbit for months and months. The connection between the secant's value and the ellipse was undoubtedly facilitated by Kepler's almost maniacal research on the possible orbit of Mars.

[19] Here Kepler assumed that the velocity of a planet is inversely as its distance from the Sun.

[20] With regard to this section, see Pisano and Bussotti (2018). As to the concept of gravity in Kepler, see Elena (1983), the fundamental Goldbeck (1896), Ihmig (1990), Pisano and Bussotti (2018, pp. 304–311).

> Gravity is a mutual corporeal disposition among kindred bodies to unite or join together [...] (The magnetic faculty is another example of this kind).[21]

This statement shows the qualitative character of Kepler's gravity, for two bodies attract mutually only if they are cognate. Therefore, gravity is not a universal disposition of all bodies, but acts only in the case that two bodies are cognate. An interpretative problem thus arises: when is it possible to consider two bodies as *cognata*? Kepler did not provide any precise answer, but we can infer from his reasoning that two bodies are *cognata* if they have the same material composition. Since in Kepler's time no chemical theory had been developed, the material composition of a body was conceived in very vague terms. Kepler wrote that a stone and the whole Earth exert a mutual attraction because they are *cognata* bodies (KGW, III, pp. 25–26). He was certainly aware that the material composing the Earth is not exactly the same as that composing the stone. This means that, at all appearances, Kepler used the word *cognata* in an intuitive sense, to mean that the Earth and the stone are composed more or less of similar materials, while, for example, the matter composing the Sun is different. In conclusion, for Kepler gravity does not act among bodies insofar as they are bodies, but only if they have the same composition.

Kepler also specified that if the Moon were not maintained in its orbit around the Earth by a "vi animali", the two bodies would collapse one on the other. This is very interesting because he guessed that some other physical entity is necessary to avoid that gravity joins the Earth and the Moon. Instead, Newton understood that no additional force is necessary to avoid such a gravitational "collapse", but that an appropriate inertial velocity of the Moon with respect to the Earth is the necessary and sufficient condition for the Moon to describe its orbit around our planet. Kepler did not have a correct view of inertia, and so he guessed that without the existence of some other physical action, the Moon and the Earth would be joined by gravity.

A further fundamental question concerns the way in which gravity acts among cognate bodies. This is what Kepler wrote:

> If the moon and the earth were not each held back in its own circuit by an animate force or something else equivalent to it, the earth would ascend towards the moon by one fifty-fourth part of the interval, and the moon would descend towards the earth about fifty-three parts of the interval and there they would be joined together, provided, that is, that their substance of each is of the same density.[22]

What is the meaning of the assertion that, if the distance Earth-Moon were divided into 54 parts, the Earth would ascend along one of these parts and the Moon along 53 of them? The answer is not to be found in the *Astronomia Nova*, but, as Donahue points out, in a letter to Fabricius of 11 October 1605, where Kepler wrote:

---

[21] Kepler (1992, p. 55). Cf. KGW, III, p. 25: "Gravitas est affectio corporea, mutua inter cognata corpora ad unitionem seu conjunctionem (quo rerum ordine est et facultas Magnetica)".

[22] Kepler (1992, p. 55). Cf. KGW III, p. 25: "Si Luna et Terra non retinerentur vi animali, aut alia aliqua aequipollenti, quaelibet in suo circuitu; Terra ascenderet ad Lunam quinquagesimaquarta parte intervalli, Luna descenderet ad Terram quinquaginta tribus circiter partibus intervalli: ibique jungerentur: posito tamen, quod substantia utriusque sit unius et ejusdem densitatis".

> [...] not only does a stone approach the earth, but the earth also approaches the stone, and they divide the space between them in the inverse ratio of their weights.[23]

And what is the meaning of the reference to density? Again, we don't find any answer in the *Astronomia Nova*, but in what we read in the first part of the fourth book of the *Epitome*:

> If the volume of Saturn, namely 10, were multiplied by its density, namely 5, you would obtain the same "mass", that is 50, as that you would obtain multiplying the volume of Jupiter, 5, by its density, 10.[24]

Without reporting Kepler's complicated reasons to motivate his choice of the values for the volume and the density of Saturn and Jupiter, it is to be noted that he defined the concept of *copia materiae* as the product of the volume by density. A few pages later, in the *Epitome* Kepler identified *copia materiae* with weight: "pondus seu copia materiae" (KGW, VII, p. 306). Therefore, Kepler's view of gravity can be summarised like this: gravity acts only between cognate bodies. Under this condition, the gravitational action of a body is proportional to its *copia materiae.* Kepler gave a definition of *copia materiae* which is more similar to that of our concept of mass than to our concept of weight, though he considered equivalent *copia materiae* and *pondus.* This is not surprising since the distinction between weight and mass will be clarified only by Newton and it would have been unattainable without the Newtonian concept of force.

Thus, to some extent, Kepler's conception of gravity is extremely advanced. Each body has its gravitational power which depends on its *copia materiae.* Therefore, in contrast to what Aristotle and the Aristotelians believed, the Earth is not the only centre of gravity. Each object—a stone, another planet, the Moon and so on—is endowed with its own gravitational power. A further evidence of the connection between gravity and *copia materiae* is Kepler's statement in the *Astronomia Nova* that no mathematical point, even if it were the centre of the world, can attract (KGW III, p. 24). This statement is clearly addressed against the Aristotelians: even in the pre-Copernican hypothesis that the Earth were the centre of the world, it would attract not for its position as the centre of the world, but as a massive object. Kepler asserted that no object is absolutely light or absolutely heavy, but it is such only in comparison with other objects (ibid., p. 27).

On the other hand, the qualitative aspect of gravity remains essential since Kepler held that only cognate bodies mutually attract. It is not by chance that he compared gravity with magnetism.

An interesting feature of Kepler's doctrine of gravity are his ideas about tides. He identified the origin of tides in the gravitational attraction exerted by the Moon on the Earth. We read:

---

[23] Kepler (1992, p. 55, note 7). Cf. KGW, XV, p. 241: "non tantum lapis ad Terram eat, sed etiam Terra ad lapidem, diuidantque spacium interjectum in euersa proportione ponderum".

[24] Our translation. Cf. KGW VII, p. 283: "Multiplicata enim mole Saturni 10, in densitatem 5, prodiret copia materiae 50, tantundem scilicet, quantum, si molem lovis 5 in densitatem ejus 10 multiplicasses".

> If the earth should cease to attract its waters to itself, all the sea water would be lifted up, and would flow onto the body of the moon. The sphere of influence of the attractive power of the moon is extended all the way to the earth, and in the torrid zone calls the waters forth, particularly when it comes to be overhead in one or another of its passages. This is imperceptible in enclosed seas, but noticeable where the beds of the ocean are widest and there is much free space for waters' reciprocation.[25]

Here we have a clear explanation—though in qualitative terms—of the tides phenomenon. The Moon attracts each part of the Earth. The solid parts are attracted, but, due to their compactness, the Moon's gravitational attraction does not exert a visible influence on them, while it is evident on the seas. If the Earth did not exert a far stronger attraction on its waters, they would fly towards the Moon. Kepler thus gave a cosmological explanation of the tides phenomenon: he thought that it was not explicable referring it only to the motion of the Earth, but that it should be related to a celestial object such as the Moon.

Galileo was extremely critic of Kepler's ideas on tides: he was convinced that Kepler resorted to mysterious entities, to influxes among bodies, which are not supported by any scientific evidence, but are the heritage of an animistic and vitalistic conception of the cosmos which should be abandoned. We read in the *Dialogo*:

> But among all the great men who have philosophized about this remarkable effect, I am more astonished at Kepler than at any other. Despite his open and acute mind, and though he has at his fingertips the motions attributed to the earth, he has nevertheless lent his ear and his assent to the moon's dominion over the waters, to occult properties, and to such puerilities.[26]

Was Galileo's opinion justified? First of all, as we have seen, Kepler's doctrine of gravity relied upon the concept of weight or *copia materiae*, which is a quantitative concept. It is nothing mysterious. The idea of *cognatio*, though not expressed in a perspicuous manner, clearly refers to the material composition of a body. In this respect, Galileo's accusations are not justified. However, he was probably unsatisfied with the way in which Kepler's gravity should act. Galileo thought that Kepler regarded gravity as an immediate action at a distance, comparable to the magic and vitalistic attractions of the Renaissance thinkers against whom he vindicated his

---

[25] Kepler (1992, p. 56). Cf. KGW III, p. 26: "Si Terra cessaret attrahere ad se aquas suas; aquae marinae omnes elevarentur, et in corpus Lunae influerent. Orbis virtutis tractoriae, quae est in Luna, porrigitur usque ad Terras, et prolectat aquas sub Zonam Torridam, quippe in occursum suum quacunque in verticem loci incidit, insensibiliter in maribus inclusis, sensibiliter ibi ubi sunt latissimi alvei Oceani, aquisque spaciosa reciprocationis libertas." As pointed out in Pisano and Bussotti (2018, p. 309), Kepler's discovery of the origin of the tides was favored by his profound knowledge of Renaissance literature, in this case of Patrizi's *Nova de universis philosophia* (see also Goldbeck, 1896, pp. 15–19), for Patrizi referred to the opinions of the ancients on the tides. The ancient opinions were, in many cases, more advanced than those of Kepler's contemporaries. Actually, Kepler quoted Patrizi in *Astronomia Nova*, though in a different context (KGW III, p. 62).

[26] Galileo (1967, p. 462). Cf. EN VII, p. 486: "Ma tra tutti gli uomini grandi che sopra tal mirabile effetto di natura [le maree] hanno filosofato, piú mi meraviglio del Keplero che di altri, il quale, d'ingegno libero ed acuto, e che aveva in mano i moti attribuiti alla Terra, abbia poi dato orecchio ed assenso a predominii della Luna sopra l'acqua, ed a proprietà occulte, e simili fanciullezze". It is not easy to guess which of Kepler's works Galileo read and how thorough his knowledge was. To be precise, Kepler did not speak of *copia materiae* in the *Astronomia Nova*, but in the *Epitome*.

rational-mechanistic natural philosophy. In order to realize how Kepler understood the action of gravity, it is worth focusing on the *Epitome*, where he illustrated the idea that the planets can be moved by a *species immateriata* emanating from the Sun[27] He claimed that the gravitational action of the Moon on the Earth, which is responsible for the tides, is produced by something similar to the Sun's *species immateriata*, though—this has to be stressed—the Sun's *species immateriata*, and the *species* responsible for gravity are not identifiable. We read:

> Therefore, if someone doubts whether the magnetic, namely, the terrestrial faculties, are also in the sky, and whether the earth, which is a heavy body, might be moved from a place to another by the Sun's *species immateriata*, then he should look at the Moon, which is cognate to the Earth, and observe that it rotates without any support of a solid orb. Actually, the species of the bodies which move each other reciprocally are adequate to deduce the motion. This is manifest in the Moon itself, which through an emitted species moves our seas.[28]

Therefore, Kepler did not think that gravity acted immediately at a distance. Rather, he thought that it was spread through a *species*, of which he gave no specification, but which was certainly a component—though *immateriatus*—of his physical world. In other terms: he believed in the existence of a medium through which gravity might be spread.

As regards the gravity's range of action, Kepler clearly explained that the Earth's gravity power extends far beyond the Moon indefinitely:

> For it follows that if the moon's power of attraction extends to the earth, the earth's power of attraction will be much more likely to extend to the moon and far beyond [...].[29]

Therefore, one might be led to think that, according to Kepler, the *virtus tractoria* of gravity could explain the movements of the planet around the Sun. But this is not the case, for two fundamental reasons: (a) the *virtus tractoria* is only able to attract a body towards another body, not to produce a curvilinear movement as that of the planets around the Sun; (b) the Sun and the planets are not cognate bodies and, as a consequence, the Sun's gravity has no effect on the planets and vice versa. In Kepler's view, gravity is thus a local force whose action radius is limited to distances as those Earth-Moon, which are much smaller than the distances Sun-planets, the ones significant to analyse the global dynamics of the solar system.

Kepler resorted to actions different from gravity to explain the movement of the planets around the Sun. He identified four forces whose combined action determines such movement:

1. the mean motion around the Sun;

---

[27] In the next pages we will discuss in details the action and the nature of such *species immateriata*.

[28] Our translation. Cf. KGW VII, p. 306: "Deinde si quis dubitat, an magneticae, h. e. terrestres facultates in coelo sint, et an terra, grave corpus, de loco in locum transponi possit a specie immateriata Solis: is Lunam intueatur, quam Terrae cognatam, videt circumire nullo substrato sollido orbe. Valere vero ad inferendum motum species corporum mutuo commeantes, patet in eadem Luna, quae per emissam speciem penes nos movet maria".

[29] Kepler (1992, p. 57). Cf. KGW III, p. 27: "Sequitur enim, si virtus tractoria Lunae porrigitur in Terras usque, multo magis virtutem tractoriam Telluris porrigi in Lunam et longe altius [...]".

2. the elliptical orbit, which he interpreted as a perturbation of the mean movement;
3. the movement of the planetary apses-line;
4. the movement in latitude.

Since force 1 is the most intense and the most significant for planetary motion, we will focus on it while also hinting at the nature of force 2.[30]

In *Astronomia Nova*, Kepler described four features of the *virtus motrix* which determines the mean motion of the planets (Pisano & Bussotti, 2018, p. 314):

(1) it is located in the centre of the system and its intensity decreases as the distance from such centre increases (KGW, III, p. 237, lines 14–24);
(2) the Sun is in the centre of the system and rotates around its axis (KGW, III, p. 243, lines 22–25);
(3) The *virtus motrix* depends on the rotation of the Sun. It is spread through a *species immateriata* (KGW, III, pp. 240–242, 350) radiating from the Sun.
(4) The *virtus* is *promotoria*[31] (induces movements), not *tractoria* (attracts).

Let us see how Kepler justifies these characteristics. With regard to the item (1) Kepler wrote:

> […] To the extent that a planet is farther from the point which is taken as the centre of the world, it is less strongly urged to move about that point. […]. Just so, in this instance, the intension and remission of motion is always in the same ratio as the approach and recession from the centre of the world.[32]

It follows that the *virtus motrix* decreases inversely as the distance from the centre as can be inferred from the above quoted passage. The speed decreases, as well, inversely as the distance from the centre of the world. Thus, in Kepler's view, the solar virtue and the planet's speed are directly proportional and both of them are inversely as the distance from the Sun. The action of the solar *virtus motrix* is conveyed by the *species immateriata* which radiates from the Sun. Each point of the Sun's surface emanates a ray of the *species immateriata.* The Sun rotates around its axis. When a ray of the *species immateriata* touches the planet, it moves the planet in the same direction of the Sun's rotation.

The *species immateriata* is not material by definition and yet it moves the planets. Kepler claims that this is possible because the *species immateriata* is "not free from geometrical laws" ("non liberam esse a legibus Geometricis", Kepler, 1992, p. 383,

[30] For an account of the forces (3) and (4) we refer to Pisano and Bussotti (2018, pp. 330–332), where the reader can also find bibliographical indications.

[31] This adjective is not used in *Astronomia Nova*, but in the letter to Fabricius mentioned above in note 23. In this book our aim is not to analyse all the difficult technical details and interpretations of the solar virtue, but to present its main features in order to show the connection between the dynamical and the kinematical aspects of Kepler's cosmology. For more specific details on the solar virtue, see: Davis (1992b), Granada (2009), Guidi Itokazu (2006, 2007), Holton (1956), Hoyer (1979), Krafft (1991), Rabin (2005), Pisano and Bussotti (2018), Stephenson (1987).

[32] Kepler (1992, p. 376). Cf. KGW, III, p. 236: "[…] quo longius abest Planeta a puncto illo, quod pro centro mundi assumitur, hoc debilius illum incitari circa illud punctum. […] Ut hic intentio et remissio motus, cum accessu et recessu a centro mundi, in proportione perpetuo coincidit".

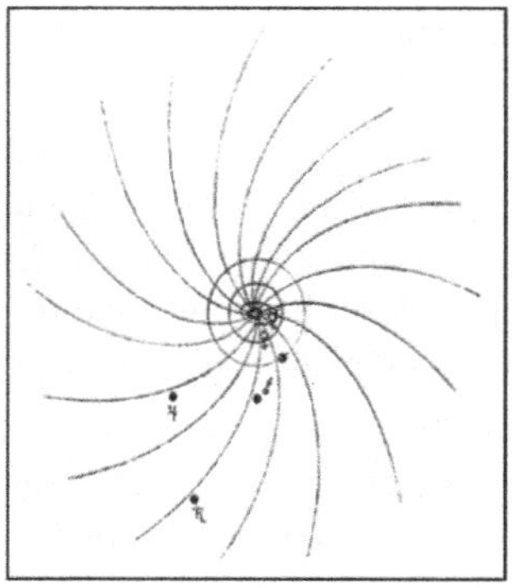

Fig. 3.4 A description of Kepler's solar *virtus motrix*

KGW, III, p. 241). This assertion is remarkable: for Kepler the *species immateriata* was not something spiritual. Had Kepler wished to indicate a spiritual essence acting on bodies, he could have used the term *immaterialis* rather than the unusual *immateriata*. Kepler's species is immaterial, but it is not spiritual: it is a component of the physical world that, as the other components, is subject to the laws of geometry. Therefore, the *species immateriata* is a part of Kepler's physical theory that has nothing to do with something animistic or spiritual. Light is another example of something that is neither material nor spiritual. It is not by chance that, as we will see, Kepler developed a comparison (which, obviously, is not an identity) between light and *species immateriata* (KGW, III, p. 240). With regard to the single rays of the *species immateriata*, Kepler claimed that their intensity is constant, namely it does not decrease with the distance from the Sun. However, since the number of rays is the same for every arc of circumferences concentric to the Sun—as it is shown in Fig. 3.4—the rays' density decreases the farther a circumference is from the Sun (KGW, III, chapter 33, p. 239). Thus, the farther a planet from the Sun, the slower its motion, since, considering two planets P and P', with P internal to P', and the same part S of their orbit (see Fig. 3.5), the *species immateriata*'s rays are denser in the orbit of P than in that of P'. Thus, P move faster than P'.

The "mass" also has an influence on the mean motion because the more massive a planet is, the bigger its resistance to the action of the solar *virtus* is, due to the natural inertia of the "masses".[33]

If the solar *virtus motrix* were the only force to act, the paths of a planet would turn out to be a circumference with the Sun in the centre. However, the real orbit is an ellipse. Kepler introduced another force in order to explain the elliptical form of the orbit (which he called "libratory force")[34]. He thought that both the Sun and the

[33] We used the expression "mass" rather than mass to signify that Kepler in *Epitome* spoke of *copia materiae* as equal to the volume multiplied by density (KGW, VII, p. 283), but then identified this quantity with weight (ibid., p. 306).

[34] The action of this force is explained in the *Astronomia Nova* (KGW III, pp. 348–364) but more clearly in the *Epitome Astronomiae Copernicanae* (KGW VII, pp. 337–342). For more specific

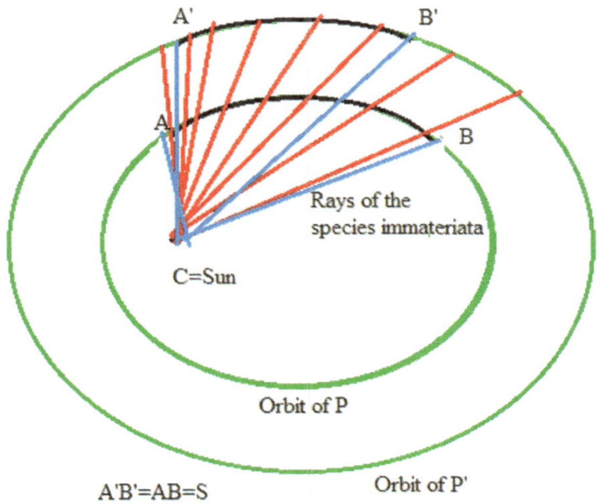

Fig. 3.5 This diagram illustrates Kepler's idea that, though the single rays of the *species immateriata* do no lose their power while increasing their distance from the Sun, the power of the species weakens the further a planet is from the Sun. Since the lengths of the two arches AB and $A'B'$ is the same, but the orbit of P is more internal than that of P', the angle ACB is bigger than $A'CB'$ and thus the density of the rays is higher in AB than in $A'B'$. The figure is ours

planets are magnets: the Sun is an enormous magnet, of which one pole is on the surface while the other is internal. As for the planets, they have an attractive pole (which Kepler called *amicus*, the friendly pole, indicated by the arrow in Fig. 3.6a) and a repulsive, unfriendly one, called *discors*, indicated by the tail in Fig. 3.6a. Let us consider the planet in position *A* (Fig. 3.6a).

The magnetic poles of the planet, indicated by the point and the tail of the arrow, are equidistant from the Sun. In this position there is no magnetic attraction. But when the planet is in the positions *B*, *C*, *D*, the friendly pole is turned towards the Sun and thus the Sun attracts the planet. After having reached and passed the perihelion in *E*, the *discors* pole (KGW, VII, p. 337) of the planet is turned towards the Sun, hence the magnetic interaction pushes the planet distant from the Sun. This is the general dynamic interaction which explains why the orbit is not circular.

However, it is necessary to take into account:

(A) The *virtus motrix* is far stronger than the attraction–repulsion between the Sun and the planetary magnetic axis (see, e.g., in the *Epitome*, KGW, VII, pp. 337–338).
(B) The magnetic axis of the planet has a sort of magnetic inertia and hence tends to maintain its position.

---

details on how this force acts see Caspar's *Nachbericht* (KGW VII, pp. 598–600), Davis (2003), Pisano and Bussotti (2018, pp. 322–330), Stephenson (1987, pp. 154–172).

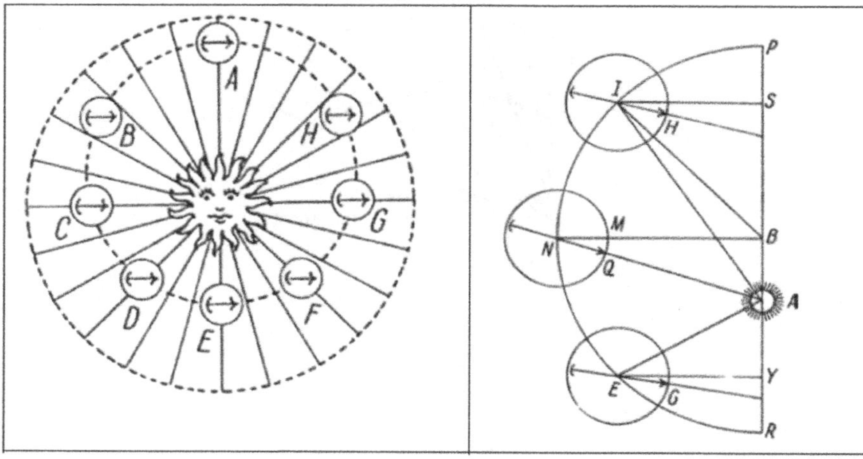

Fig. 3.6 **a** The general explanation of the action of the force responsible for the ellipticity of planetary paths (KGW, VII, p. 337). **b** Specification of the previous diagram. When the centre of the Sun, the centre of the planet and the *amicus* pole are aligned (position *AQN*), the planet is in the quadratures (ibid., p. 339)

The conclusion is that the tail of the planet's magnetic axis is directed exactly towards the Sun only when the planet is in the quadratures. This situation is explained by Kepler in Fig. 3.6b. Analogous considerations hold when the *discors* pole is oriented towards the Sun.

It should be noted that Kepler also gave a mathematical description of the libratory force because he tried to deduce the ellipticity of the orbits and the area law from such description. Therefore, had his reasoning been correct, he would have obtained the fundamental achievement of deducing kinematics from dynamics. We summarize here his most complete reasoning, expounded in the *Epitome*[35]: first of all, it is to be pointed out that, though, as we said, the magnetic axis of the planet has an inertia, it does not remain exactly parallel to itself. Its position changes slightly due to the magnetic attraction exerted on it by the Sun. However, Kepler first analysed the action of the libratory force supposing that the magnetic axis of the planet remains parallel to itself and, precisely, that it remains perpendicular to the apses line.[36] Hence, in Fig. 3.7, the segments *IH* and *FG* indicate the direction of the magnetic axis for two different positions of the planet.

In the first part of the fifth book of the *Epitome* Kepler reasoned like this: be $A =$ Sun; $I, E =$ different positions of the planet's centre; $EG, IH =$ position of the planet's magnetic axis that remains perpendicular to *RP* during the planet's revolution; *AI*,

[35] For a complete treatment of this topic, we refer to the accurate analysis by Stephenson (1987, pp. 146–172).

[36] As a matter of fact, the situation is more complicated because Kepler introduced the concept of "magnetic fibers". However, for our present aims, it is enough to refer to the magnetic axis. For Kepler's concept of magnetic fibers, see Stephenson's book mentioned in the previous note.

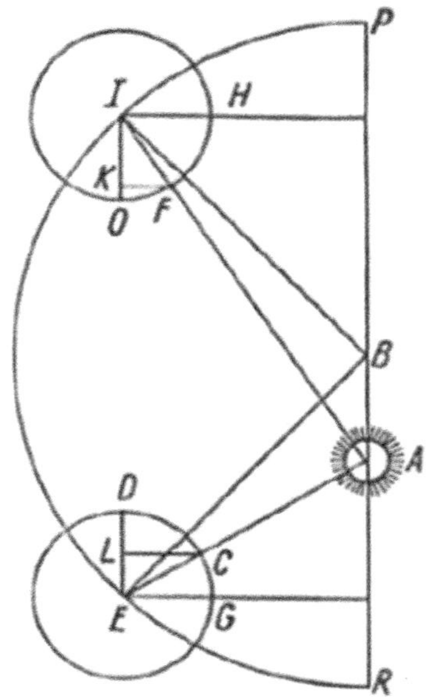

Fig. 3.7 The segments *IH* and *EG* represent the magnetic axis of the planet in the simplified hypothesis that it remains parallel to the apses line *PR* while the planet rotates around the Sun (figure from KGW VII, p. 369)

AE = radius vector Sun–planet. Let *ED* and *IO* be drawn parallel to *RP*. If the perpendiculars to these lines are drawn from *F* and *C*, they cut *ED* and *IO* in *L* and *K*, respectively. When the planet is in *P* and in *R*, its magnetic axis is perpendicular to the radius vector Sun-planet and hence, there is no magnetic attraction.

Kepler, then, developed his analysis taking into account that, in the movement of the planet from *P* to *R*, its radius-vector is not perpendicular to its magnetic axis and, hence, the Sun exerts a magnetic attraction. Kepler calculated its intensity in this way: if *IH* and *EG* represent the global attractive force of the Sun, the ratio between this force and its part acting on the planet in *I* is as *IH* (= *IF* = *EG*) to *KF* and that acting on *E* is as *EG*(= *IH*) to *LC*. The true or coequated anomaly is the angle *IAP*, which is congruent to *AIK*. The libratory force, which is as *KF*, is, therefore, as the sinus of the true anomaly. This means that the intensity of the libratory force has it maximum absolute value when the centre of the planet is in *Q* (see Fig. 3.8a).

In the fourth book of the *Epitome*, Kepler supposed that the libratory force were proportional to the sinus of the eccentric anomaly (see Fig. 3.8b) *IBP* and, hence, the maximal intensity would have been reached when the planet is in Q' rather than in *Q*. Kepler then abandoned this first hypothesis for physical reasons, because the geometrical centre *B* of the orbit plays no physical role.

However, Kepler realised that a relation between the eccentric anomaly and the libratory force would have been very useful, because in this case, as it is not difficult to prove, he could have derived the exact equation of an ellipse for the path of the planet and his kinematical theory would have relied upon his dynamical one. Kepler thus introduced the idea that the magnetic axis of any planet does not remain exactly parallel to itself and perpendicular to the apses line, but during the path's planet,

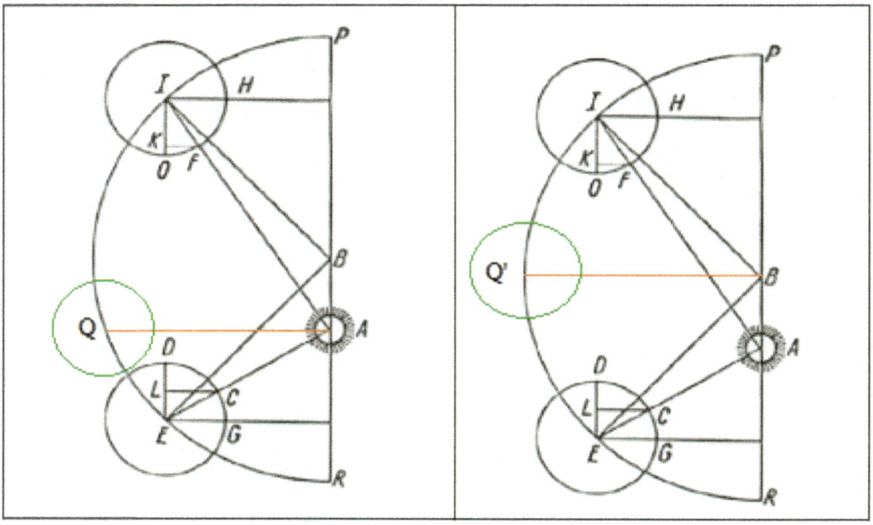

Fig. 3.8 **a** The maximal libratory force if it were proportional to the sinus of the true anomaly. **b** The maximal libratory force if it were proportional to the sinus of the eccentric anomaly

deviates slightly due to the Sun's magnetic attraction on it. If the amount of deviation holds appropriate values, it is possible to prove (Stephenson, 1987, pp. 151–153) that the libratory force is proportional to the sinus of the eccentric anomaly in each point of the planetary orbit.

Now Kepler has established that the libratory force in each point is as the sinus of the eccentric anomaly. The following step is to calculate the amount of the libratory force along the whole planet's trajectory. Kepler developed this calculation through a series of brilliant reasonings.[37] Finally, he reached the following results:

1. The whole amount of the libration is proportional to the versed sine of the eccentric anomaly.
2. He was able to construct the orbit.
3. He was able to show that the constant of proportionality between the amount of libration and the eccentric anomaly is exactly the eccentricity of an ellipse. The orbit itself is, thence, an ellipse.

Kepler developed furtherly his reasoning until he reached the area law and what we now call Kepler's equation

$$\alpha = \beta + e \sin \beta$$

[37] See KGW VII, pp. 370–375; Stephenson (1987, pp. 154–157), Pisano and Bussotti (2018, pp. 325–327).

where α is the mean anomaly, which is also the measure of the time necessary for a planet to reach a point P of eccentric anomaly β starting from the aphelion, and e is the eccentricity of the elliptical orbit.[38]

If Kepler's deduction of the orbits' ellipticity and of the area law from the libratory action had been correct, then he would have obtained a dynamical derivation of his first two laws, which would have been a result of immense importance.

Commentary: Kepler was the first scientist who tried to explain the solar system in terms of physical causality, that is he was the first who attempted to determine the set of actions that might explain the kinematics of the solar system. This is to his great credit. In Sect. 3.1.2, we have analysed the connections between the dynamical and kinematical aspects. This granted, Kepler's dynamics is unsatisfactory under several respects: first of all, it is not exactly a theory, i.e. it is not a system of truths based on few concepts and axioms from which kinematics is deduced. Rather, given the geometrical and kinematical features of the planetary paths, Kepler seems to construct a posteriori a force which determines them, i.e. a force ad hoc. Therefore, an epistemological analysis of Kepler's procedure shows that, in fact, kinematics is the foundation of dynamics. The whole process of Kepler's thought might be summarised like this: (1) there is a series of supposed movements which no physical cause can determine, like the movements of the equant in his *hypothesis vicaria*; (2) as a consequence, the models based on the equant are discarded and, relying on the observations and on a long series of mathematical deductions and calculations, the right kinematical laws are deduced; (3) a series of forces compatible with the new kinematics are constructed. However, and this is the fundamental point, if these forces had been posed into exact mathematical terms, the kinematics of item (2) would not have been obtained.

In particular: the Sun's *virtus motrix* is not mathematized, but Kepler specified that:

(a) it is the main force responsible for the motion of the planets;
(b) it is proportional to the inverse distance Sun-planet;
(c) the libratory force is a perturbation of the *virtus motrix*.

This means that the global force deriving from the combination of the *virtus motrix* and of the libratory force is not exactly like the inverse distance Sun-planet; however, it is as an inverse power of the distance which deviates by a little amount from 1 and that, certainly, cannot be 2. It should be recalled that Newton proved that an elliptical orbit (and more generally an orbit that is a conic) in which the attracting body is in one of the foci can be obtained if and only if the inverse square law holds.[39]

---

[38] For the complete reasoning see KGW VII, Book V, Part II, Chapters IV–V, pp. 390–396. See also Caspar's *Nachbericht* (KGW VII, pp. 598–600) and Stephenson (1987, pp. 154–172).

[39] In *Principia*'s Book I, Section III, Proposition XI, Problem VI Newton solved the direct problem of central forces for an ellipse because he proved that, given an elliptical orbit where the centripetal force is in one of the foci, the inverse square law holds. In the following Proposition XII, he proved the same property for a hyperbola and in Proposition XIII for a parabola in which the force is located in the parabola's focus. In the fundamental Proposition XLI, Problem XXVIII (Book I, Section VIII), Newton solved the inverse problem of centripetal forces: given the force, to find the

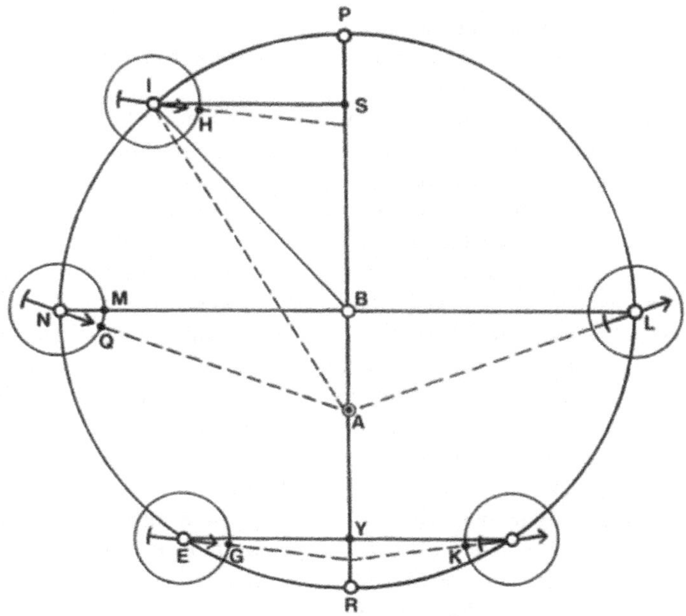

Fig. 3.9 The explanation devised by Kepler of the functioning of the librational force. (Stephenson, 1987, p. 148)

Therefore, Kepler could not obtain an ellipse as an orbit according to the forces he had introduced. One might thus wonder how Kepler deduced the ellipticity of the orbits and the area law in the *Epitome*. First of all, Stephenson observes that Kepler made a serious mistake in the deduction of a mathematical property of the orbits from the physical properties of the librational force (see Fig. 3.9):

> Kepler's theory of the deflection of the magnetic fibers[40] was thus burdened with a serious contradiction. The cause he postulated for the deflection was the tug of the sun on the attracted end of the magnetic fibers, and its push at the repelled end. The tugging and pushing should have twisted the fibers in a clockwise direction in Figure 30 [3.9], until the motion of the planet around the sun carried the fibers to a position where they pointed "below" the sun, that is toward perihelion; thereafter the fibers should have been twisted back counterclockwise. The mathematical requirement that the *effect* of the deflection should be a force proportional to the sine of the eccentric anomaly was incompatible with this supposed physical cause; for this mathematical condition implied that the deflection equaled the optical equation, so that the fibers pointed directly at the sun when the eccentric anomaly was 90°. (Stephenson, 1987, pp. 153–154)

---

trajectories and the times of the motions in the trajectories. As a particular case, it follows that if the force is as the inverse square distance between the moving point and the point in which the force is applied, then the trajectory is a conic and the force is in one of the foci.

[40] For our purposes, the expression "magnetic fibers" can be replaced with "magnetic axis of the planet".

This mistake affects Kepler's theory of libration and makes his deduction incorrect. Besides, another significant flaw can be detected: in his deduction of the orbits' ellipticity and of the area law, Kepler analysed only the libratory force. It is as if the main force responsible for the planetary motion, namely the solar motive virtue, had completely disappeared. There is no reference, in mathematical terms, to the combined effects of the two forces. The solar virtue is subject to no mathematical determination—if you exclude the idea that it is inversely as the distance—and, therefore, Kepler did not offer any mathematical treatment of the combined effects of the solar virtue and of the libratory force. It seems that, at a certain point in Kepler's treatment, the effects of the *virtus motrix*, which is the main force, were left in the background and all the most important features of the planetary paths depended on the libratory force. It would have been desirable that Kepler had shown the combined effects of the *virtus motrix* –which is responsible for the mean motion of the planets—with the libratory force, which is responsible for the ellipticity of the orbits.

It is for these cogent reasons that Kepler's dynamics is not and cannot be the scientific-mathematical foundation of his kinematics. As noted above, Kepler started from correct kinematical achievements and then tried to construct an ad hoc dynamics from which his kinematics might have been hopefully derived. Though he failed in his purpose, his idea to set a dynamical system for the universe is grandiose and shows the depth of his conceptions.

In the analysis of Kepler's dynamical conceptions there are two other points to consider:

(1) A correct idea of inertia is missing. Kepler conceived of natural inertia, according to which a body stays at rest with reference to a sort of absolute space unless an external action intervenes. This means that a force is needed to produce a uniform rectilinear motion. In this respect, Galileo's idea of inertia is far more advanced than Kepler's.

(2) The quantities used by Kepler to determine the libratrory force are classical geometrical quantities, i.e. the trigonometric functions; there is no reference to the concept of mass that we consider the element separating a geometrical-kinematical treatment of physics from a dynamical one. This means that—even if Kepler's deduction of the orbit's ellipticity and of the area law had been correct—this would have represented, so to say, a successful geometrical-kinematical treatment of dynamics, not a dynamical treatment of dynamics.

All in all, what was missing is the idea of a sole cause which, jointly with some initial conditions, determines the movement of the planets. A further aspect typical of Kepler's mentality may be here noted: his will to explain everything. In Newton's theory, the orbits of the planets are due to the Sun's centripetal force and to the initial uniform velocity with which the planet entered the gravitational field of the Sun; Newton did not claim to explain the origin of this initial velocity, whereas Kepler tried to explain everything.

In regard to the relations between the structural-cosmological level and the kinematic one, they seem, at a first glance, rather tenuous: the theory expounded by Kepler in *Mysterium Cosmographicum* seems to have no link with his further achievements, considering that it was referred to circular orbits. However, this theory is taken up again and is substantially enriched in *Harmonice Mundi* (1619). Furthermore, in the second edition of *Mysterium Cosmographicum* (1621), Kepler claimed that, basically, he maintained the opinions expressed in the first edition of this work (1596). There is, thus, a line of continuity between *Mysterium* and *Harmonice*. Since the so-called Kepler's third law is presented in the *Harmonice Mundi*, which is a cosmological work, we will examine how such a law was introduced. Kepler formulated his third law in these terms:

> But it is absolutely certain and exact that *the proportion between the periodic times of any two planets is precisely the sesquialterate* [namely 3/2] *proportion of their mean distances, that is, of the actual spheres.*[41]

We do not intend to analyse the details of this formulation, but to point out that this law was deduced empirically, as was also the case with the ellipticity of the orbits. Kepler relied upon Tycho's observations, which he believed to be correct in defining the distances among the celestial spheres.[42] At the same time, if Kepler had not been guided by his idea of the harmony of the universe, according to which a certain ratio between the periodical times and the mean distances of a planet from the Sun should exist, he would have never started the research which led to the third law. This law was considered by Kepler as the fundamental step towards the definitive demonstration of the geometrical-musical harmonies of the universe, which, in fact, is the topic of the *Harmonice*. Kepler's words reveal his train of thoughts:

> Again, therefore, a part of my *Secret of the Universe,* put in suspense 22 years ago because it was not yet clear, is to be completed here, and brought in at this point. For when the true distances between the spheres were found, through the observations of Brahe, by continuous toil for a very long time, at last, the genuine proportion of the periodic times to the proportion of the spheres
>
> *only at long last did she look back at him as he lay motionless.*
>
> *But she looked back and after a long time she came*
>
> and if you want the exact moment in time, it was conceived mentally on the 8th March in this year one thousand six hundred and eighteen, but submitted to calculation in an unlucky way, and therefore rejected as false, and finally returning on the 15th of May and adopting a new line of attack, stormed the darkness of my mind. So strong was the support from the combination of my labor of seventeen years on the observations of Brahe and the present study, which conspired together, that at first I believed I was dreaming, and assuming my conclusion among my basic premises.[43]

---

[41] Kepler (1997, p. 411). Cf. KGW VI, chapter V, p. 302: "Sed res est certissima exactissimaque quod *proportio quae est inter binorum quorumcunque Planetarum tempora periodica, sit praecise sesquialtera proportionis mediarum distantiarum,* id est *Orbium ipsorum*" (italics in the text). On the *Harmonice* and on the role of Kepler's third law in his system see: Field (2009), Gingerich (1975), Haase (1998), Knobloch (1992, 1995), Stephenson (1994), Vijaya (2019).

[42] Kepler maintained the denomination "spheres", though he obviously knew that they were not spheres in the strict sense of the term.

[43] Kepler (1997, p. 411). Cf. KGW VI, chapter V, p. 302: "Rursum igitur hic aliqua pars mei Mysterij Cosmographici, suspensa ante 22. annos, quia nondum liquebat, absolvenda, et huc inferenda est. Inventis enim veris Orbium intervallis, per observationes BRAHEI, plurimi temporis

This passage shows that Kepler's cosmological ideas were fundamental for his discovery of the third law. Without these erroneous ideas, the (correct) kinematical law would not have been conceived. At the same time, Kepler searched and found the empirical confirmations of this law. In conclusion: the idea of a universal harmony induced Kepler to find a law that connected the mean distance of a planet in its orbit around the Sun with its periodical time; once this law was discovered and empirically verified, it was used by Kepler to furtherly reinforce the musical harmonic theory. This is the same process by which Kepler arrived at the polyhedra theory in the *Mysterium*. Therefore, Kepler's cosmological views influenced the determination of his third law, though, obviously, it would have been possible to think of a relation between distances and periodical times without the idea of a general universal harmony. Nonetheless, it must be recognized that it is hard to think how Kepler might have discovered such a complicate law only through empirical observations, if he had not been guided by a strong leading idea. There is thus an actual link between the three levels of a cosmological model in Kepler's theory in the sense that his wrong harmonic and dynamic ideas played a role—at least a heuristic, but fundamental one—in determining his correct kinematical laws. From this point of view, the three laws belong to a unified theory. However, the unity of Kepler's theory was incomplete because he did not deduce his kinematical laws from his dynamical presuppositions. This would have been impossible since such presuppositions were erroneous. With our previous remarks we did not intend to criticize Kepler's conceptions, but, rather, to highlight their genesis and greatness, while also discussing their scientific limits.

3.1.4 Virtus Motrix *and Light in Kepler's Theory: A Comparison*

We have defined Kepler as the "cosmographer par excellence" since he tried to offer a unitary explanation of all the most important cosmic phenomena. Light was among the most studied cosmic phenomena of Kepler's time: what is its nature and origin? How and according to which laws is it spread? Kepler dedicated two of his masterpieces to light: the *Paralipomena ad Vitellionem* and the *Dioptrice* (resp.

labore continuo; tandem, tandem, genuina proportio Temporum periodicorum ad proportionem Orbium - - *sera quidem respexit inertem, Respexit tamen et longo post tempore venit*; eaque si temporis articulos petis, 8.Mart. hujus anni millesimi sexcentesimi decimi octavi animo concepta, sed infoeliciter ad calculos vocata, eoque pro falsa rejecta, denique 15. Maji reversa, novo capto impetu, expugnavit Mentis meae tenebras; tanta comprobatione et laboris mei septendecennalis in Observationibus Braheanis, et meditationis hujus, in unum conspirantium; ut somniare me, et praesumere quaesitum inter principia, primo crederem" (italics in the text). The lines "*sera quidem respexit inertem, Respexit tamen et longo pòst tempore venit*" are drawn from Virgilius' *Eclogues*, I, 27–29. The complete sentence is "Libertas, quae sera tamen respexit inertem, / candidior postquam tondenti barba cadebat, / respexit tamen et longo post tempore venit /[...]". The meaning is that, when Kepler was already old, the inspiration (*libertas*) finally arrived after a long time and enabled him to complete his cosmological theory with the addition of his third law.

Kepler, 1604, 1611). What matters for our cosmological exposition is a comparison between the properties of light and of the *virtus motrix*. Like *virtus motrix*:

(1) light is composed of a matter that is different from *virtus motrix*'s *species immateriata*, but is equally imponderable and is subject to geometrical determinations (Simon, 2019, p. 55);
(2) light is emitted by any single point of the emanating surface (Kepler, 1604, p. 20);
(3) light propagates at infinity through rays. Each single ray does not loose its intensity while receding from the source because light, being imponderable, is subject to no resistance in its path (ibid., pp. 20 and 21);
(4) light propagates instantaneously; its speed is infinite (ibid., p. 21).

On the other hand:

(1) light's spread is prevented by the opaque bodies, whereas this is not the case for the *virtus motrix*. A ray of *species immateriata* suffers no impediment if a body is interposed between the Sun and the planet;
(2) the intensity of light and of *virtus motrix* decrease in space with different ratios and this is a very important distinction, as we will see. Let us consider a point *P* emanating light. It emits rays of light in any direction. Each single ray does not lose its intensity withdrawing from the source. Light is spread according to concentric spherical surfaces or, to use Kepler's term, *orbiculariter*. This means that the density of the rays decreases from the source *P* in the same exact proportion as the spherical surface increases. But, given a sphere, the spherical surface is like the square of the sphere's radius. Hence, light's intensity decreases as the square of the distance from the source. We have seen that the intensity of the *virtus motrix* also decreases, though, as it is the case with light, the single rays do not lose their power with their distance from the source. However, according to Kepler, the *virtus motrix* decreases as the distance from the source (the Sun) not as the square of the distance. How is this possible? Kepler offered an answer with these sibylline words, which have given rise to several interpretations:

> Light flows out in straight lines spherically (*orbiculariter*), the moving virtue in straight lines also but circularly (*circulariter*); that is, it presses in only one direction of the heaven, from west to east, not backwards nor toward the poles [...].[44]

This means that, assuming as direction North–South that of the Sun's poles, the *virtus* propagates parallel to the plane of the solar equator. Namely: each point of the Sun does not spread the *species immateriata* in any direction of space, but only in a plane which is parallel to the ecliptic. In this manner, a series of mutually parallel planes (each parallel to the Sun's equator) exists, in which the *virtus motrix* is spread

[44] Translation from Stephenson (1987, p. 70). Cf. KGW III, p. 240: "Rursum lux rectis effluit *orbiculariter*, virtus movens rectis quidem sed *circulariter*; hoc est in unam tantum plagam mundi ab occasu in ortum nititur, non contra, non ad polos." (Our italics).

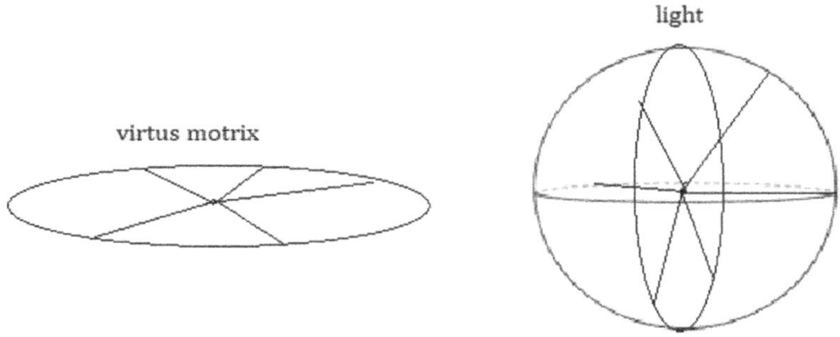

Fig. 3.10 On the left: according to Kepler, a point of the Sun's surface emanates *circulariter* (that is, "in circle") the rays of the *species immateriata* composing the *virtus motrix*. This means that the intensity of the *virtus motrix* decreases as the distance from the source. Instead, light (on the right) is spread from a source *orbicularite*r, namely "in a sphere". Therefore, light's intensity decreases as the inverse square distance from the source. The figure is ours

(see Fig. 3.10).[45] Therefore, its intensity decreases as the length of a circumference increases, and not as the surface of a sphere increases (as it is the case of light). But the length of a circumference increases as the radius. Hence the *virtus motrix* decreases as the inverse of the radius, namely as the distance Sun-planet, not as the square of such distance.

Commentary. Here we face a remarkable issue from an epistemological standpoint: Kepler offered a simple and clear explanation for the spread of light. There is no possible interpretative ambiguity. This explanation led him easily to establish the inverse square law for the decrement of light intensity, while receding from the source. Why did not he apply the same theory to the *virtus motrix*? Why did he construct a complicate device, which is rather obscure and which appears extremely artificial? Given a source, unless some impediments exist, it is natural enough to think that such a source radiates uniformly in all directions of space. Why should there be any preferred directions? From our post-Newtonian perspective one might claim that, if Kepler had followed the seemingly easiest way of reasoning, he would have reached the inverse square law before Newton, though without all the deductive steps connoting Newton's theory. From a historical standpoint, we must try to understand why he did not follow this path. The answers cannot be but conjectural. We have seen that, according to Kepler, the *virtus motrix* has to produce the velocity itself of the planets, and not only the change in the direction of their velocity, that is the planets'

[45] The exact way in which the *virtus motrix* is spread is a very problematic question. Until Stephenson (1987), scholars thought that it was spread only in proximity of the Ecliptic. Stephenson (ibid., pp. 67–75) proved this to be false. He proposed a different spread-mechanism which, though very ingenious, was criticized by Guidi Itokazu (2006, 2007). Pisano and Bussotti (2018) proposes a solution of which we have here offered the conclusions, namely that the *virtus motrix* is spread in circles belonging to planes parallel to the Ecliptic's. For a complete historiographical account of the interpretations of this problematic and technical step of Kepler's theory see Pisano and Bussotti (2018, pp. 318–322).

centripetal acceleration. This assumption implies that Kepler lacks a clear concept of inertia. Furthermore, he was convinced that the speeds of the planets decreased as the distance from the Sun. Therefore, the *virtus motrix*'s intensity itself has to decrease as the distance. As a consequence, to preserve the coherence of his whole dynamical system, Kepler had to contrive an explanation which allowed the *virtus motrix*'s intensity to decrease linearly as the distance from the Sun. It is remarkable that in chapter 36 of the *Astronomia Nova* Kepler wrote that, if the analogy with light had been perfect, then the *virtus motrix* would have decreased as the square of the distance from the source; he also contemplated the possibility that light's intensity might decrease as the cube of the distance from the source. However, eventually he invented the idea that *virtus motrix* spreads *circulariter* and not *orbiculariter* just to preserve the simple inverse proportionality between the distance from the source and the decrement of the *virtus motrix*'s intensity, which was necessary for the coherence of his dynamical system. This confirms that Kepler was constructing a true scientific theory with its conceptual interconnections and that, in this respect, he was a truly modern scientist. His dynamical presuppositions were, in many cases, wrong, but this is a different question; his dynamical conceptions were still unripe because physics was in its initial steps.

Even if, from an ontological point of view, Kepler's universe was populated of immaterial entities—such as light, solar motive virtue, magnetic virtue, and gravity—these entities were not conceived on the basis of vague vitalistic notions, but, rather, they were subject to precise mathematical-geometrical laws. Kepler elaborated a theory in which both matter and interactions obey to laws that are exactly quantifiable. The causes producing light phenomena, such as reflexion, refraction, etc., or determining the Sun's attraction on the planets, are not final causes and have nothing to do with will and souls. Rather, they are mechanical causes and Kepler tried to determine their dynamical properties. This does not at all exclude that Kepler believed the celestial bodies to be living beings endowed with a soul.

3.2 Kepler: The Role of the A Priori and the A Posteriori in the Making of Science

Adopting and developing the Copernican hypothesis, Kepler shared with Copernicus the belief that the physical world has a mathematical order; he was also convinced that Copernicus had not discovered the real mathematical structure of the cosmos and that it was his task to reveal it. Finally, he believed that he had disclosed the rational secret structure of the universe (which for him was still equivalent to the solar system plus the sphere of the fixed stars) by means of the Platonic solids and of the harmonic musical proportions of the heavenly spheres. Kepler was not only an astronomer but also a natural philosopher in the Renaissance style, and was more deeply rooted than Copernicus in the Pythagoric and Platonic traditions: for instance, in chap. 24 of his book *De Stella Nova* (1606) he attributed the origin of the 'new star' (which appeared in the sky 8–9 October 1604, known today as Kepler's supernova) to the Soul of the World, "the architectonic natural faculty" that by its fecund heat generates every sort of being both on Earth and in the ethereal heavens. Kepler wrote:

> For, it seems that everything fat, everything flammable and everything mutable in the luminous matter necessarily proceeds from the same architectonic natural faculty, that propagates native heat in its body and by that heat is prepared to modify and assimilate that matter to which it applies itself.[46]

In *De Stella Nova* references to Patrizi, and especially to Scaliger, are as numerous as those to Aristotle, Ptolemy, Copernicus or Tycho and comparisons between the stars and the living terrestrial beings are very frequent; for example, Kepler draws an analogy between the way the blood-life is pushed in the arteries by the heart and the way in which light is spread by the stars. The Sun is seen as the heart; light is seen as blood-life (ibid., pp. 241–242). In the *Stella Nova* topics derived from the Renaissance tradition are as numerous as those which come from recent astronomical discoveries and theories. This is understandable since Kepler embraced vitalistic conceptions to some extent, but we have also to realize that those conceptions cannot be dismissed merely as superstitions and mistaken ideas. They were evidence of the attempt to understand the unity of nature, though within a general frame that will not turn out to be scientifically productive and that Kepler himself will mostly abandon in his more mature works.[47] For example, the idea that the universe is a dynamical entity where several changes and new phenomena occur—in contrast to the Aristotelian theory of the heavenly sphere, always equal to itself—is shared both by the Renaissance thought and by the new astronomers. With regard to the alterations occurring in the universe, it is significant that Kepler thought the stars to be objects susceptible of change and not eternally immutable. Vitalism extended to the stars implies a transformative view of the heavens:

> Scaliger thinks, and I vehemently approve, *that in the body of any star a natural power of spreading light, not without movement, exists* [...]. The vital spirit is something corporeal: light, which is a very pure quality, does not need this rough impulse. However, the motion in a star has to be considered as something analogous, though in this case the motion is not a local one, rather a modification of the star with regard to that quality by which the star is a bright body. Actually, I propose a natural example of this alteration in the living carbons, which, because of this motion, existing, so to say, with alternating power, have obtained the name of "living coals".[48]

---

[46] Our translation. Cf. KGW I, p. 268: "Etenim pinguedo omnis, et omne inflammabile, omne in materiam lucidam mutabile, videtur opus esse cujsdam architectonicae naturalis facultatis, quae nativi caloris in suo corpore sit propagatrix, eoque calore instruatur, ad alterandam et sibi assimilandam eam materiam, in quam incumbit".

[47] On the other hand, we will see that Kepler disregarded completely the approach of those thinkers like Fludd, who also belonged to the Renaissance tradition and who developed speculative and fanciful views without any analysis of the phenomena.

[48] Our translation. Cf. KGW I, pp. 241–242: "Placet SCALIGERO, et ipse vehementer approbo, *naturalem esse in sideris corpore vim luminis sese spargentis non sine motu* [...] Spiritus enim vitalis corporeum quippiam est: lumen, sincerissima qualitas, crassa ista impulsione non indiget. Sed tamen analogus aliquis intelligatur in stella motus, non localis, sed alterationis, secundum eam

Even if Kepler suggests a solution to the question of the efficient cause of the new star that belongs to the tradition of Renaissance vitalism, the very attitude that drives him to search for a cause of this extraordinary celestial phenomenon is evidence of his scientific inquiring mindset. Kepler wants to investigate the natural cause of the production of the new star, as he is not satisfied with the miraculous explanations of theologians. In this pursuit, he does not allow himself to be hindered by his friend Johann Georg Brengger, who warned him that it was impossible to assign the natural cause of the new star unless a new celestial physics was first devised. Kepler decides to advance his own thesis in order to start a scientific debate on the topic.[49]

Kepler's most significant innovation with respect to the traditional philosophy of nature was that he abandoned the notion of the circularity of the orbits and thus was able to define the three laws of planetary motions which determine the true mathematical relations among the bodies in the solar system. The discovery of the elliptical shape of the orbits shows that Kepler's belief in the regular geometry of the cosmos was not bound to traditional assumptions; he believed that a geometrical theory had to be accurately tested with observational data and that the physical truth could be reached only by combining these two sources.

On the one hand Kepler seems to have been an apriorist speculative thinker, while on the other hand he was innovative and paid scrupulous attention to empirical data. When he asked Tycho for his opinion on his *Mysterium cosmographicum*, Tycho replied in 1598 with a letter in which he praised Kepler's "original and ingenious idea [of relating], as you have done, the distances and periods of the planets to the symmetrical properties of the regular solids". Tycho, however, urged Kepler to take into account "the true values for the eccentricities of the planets" and, in a following letter, stated openly that "the harmony and proportions of this arrangement must be [sought for] a posteriori, where the motions and the circumstance of the motions have been definitely established, and must not be determined a priori, as you and Maestlin would do; even then they are very difficult to find" (Koyré, 1973, pp. 160–162). If Tycho's reproaches indicate Kepler's apriorism,[50] his correspondence with the German astronomer Johann Albert Fabricius in 1607 tells us a different story. Reacting to Kepler's announcement of the discovery of the elliptical trajectories of the planets, Fabricius defended the traditional circular orbits and accused him of destroying the uniformity of motion. Kepler replied that circular motions "do not agree with the phenomena", that Tychonian astronomy had shown that "there is no sphere, no epycicle", that in his new conception the constant uniformity of motion was preserved in a new manner, and that his view of the shape of the orbits was not a conjecture, but "was founded on geometric proof". He concluded emphatically: "The simplicity of Nature must not be judged by our imagination" (Koyré, 1973,

---

qualitatem, qua stella lucidum est corpus. Imo vero vivum propono hujus alterationis exemplum, in vivis carbonibus; qui ab hoc ipso motu, et quasi vigore alternante, nomen vivorum sunt adepti." (Italics in the text).

[49] Cf. Kepler (1606, p. 267): "Primus igitur ego sententiam dicam: ut habeant caeteri materiam dicendi tanto copiosiorem".

[50] On the relation in Kepler between apriorism and observations as a necessary test to validate a theory see Di Liscia (2007, 2009), Field (1982), Regier (2013).

p. 263). Kepler's search for the mathematical structure of the world never meant that he cultivated mystical numerology or Pythagoric speculations to the detriment of empirical observations.

Further remarks may help us to understand in greater depth the wavering relation in Kepler's activity between a priori geometrical insights and interpretations of observational data. The discovery of the elliptical shape of the orbits, achieved after a careful study of the observations of Mars, has been traditionally regarded as evidence of the attention Kepler paid to the empirical data collected under Tycho's guidance; more recently, a Keplerian scholar, William Donahue, has maintained that in *Astronomia Nova* Kepler fabricated figures or adjusted observations to support his hypotheses on the two laws of planetary motions (Donahue, 1988). This whole issue is controversial and highly technical, and we should not forget the important truth that Kepler's numerical errors "did not affect the final answer" (Field, 1995, p. 348). A further remark concerns the hypotheses of the Platonic solids and of the harmony of the spheres: Kepler made them roughly compatible with the available data, but we may wonder if in both cases he tolerated a degree of discrepancy that was perhaps higher than the one he did not allow himself to admit in his study of the orbit of Mars, when he discarded the circular orbits in favour of the ellipse. Finally, it must be mentioned that Kepler stuck to the hypothesis of the Platonic solids even after he had discovered the elliptical orbits, without substituting the inscribed or circumscribed circles with elliptical shapes.

Kepler's intellectual attitude as a combination of traditional philosophy of nature and innovative science was not unusual in a pre-Cartesian age (see William Gilbert for another example) and may be considered ambiguous only from our point of view. We may get a clear example of Kepler's mixed attitude from his reaction to the positions of other prominent intellectuals of his time. As a consequence of his astronomical calculations Kepler enlarged the dimensions of the cosmos to such a point that "the ball of the Universe ... extends like a soap-bubble before bursting among [his] successors" (Koyré, 1973, p. 360); nonetheless, he vehemently refused the cosmological infinitism advanced by Giordano Bruno and supported by William Gilbert. Citing them in *De Stella Nova*, chap. 21, Kepler describes the cosmological infinitists as a "sect of philosophers who [...] do not deduce from senses the beginning of reasoning and do not accommodate the causes of things to the experiments; on the contrary, they out of a sudden, and almost as if they are driven by a certain enthusiasm, conceive and depict inside the walls of their brain the constitution of the World"[51]; in doing so, they accommodate the appearances of ordinary experiences to their axioms. The view of the 'unhappy' Giordano Bruno—who imagined an infinite universe and numberless solar systems since there are as many worlds as stars—carries with it an incomprehensible and hidden horror because no limit, no middle point and, therefore, no certain place are conceivable in this immensity. For Kepler it

[51] Our translation. Cf. Kepler (1606), in KGW I, p. 251: "Sed est alia philosophantium secta, eorum, qui [...] non initium ratiocinationis ex sensibus deducunt, neque causas rerum ad experimenta accommodant: sed qui ex abrupto, et quasi quodam Enthusiasmo, concipiunt et depingunt intra sui cerebri parietes, aliquam de Mundi constitutione opinionem".

is certain that the world is finite and that its dimensions may be determined through astronomical calculations. Every other view is a metaphysical superstructure and infinitism is a "folly of philosophers". Kepler wrote:

> And so the unhappy Giordano Bruno defends that [infinitistic doctrine] and claims his thesis not obscurely, as a doubting person would do. William Gilbert, too, in his book *De Magnete*, which is excellent for other aspects, showed, however, a religious disposition insofar as he thought that the infinite God's power cannot be better understood but admitting the universe to be infinite as to its dimension. But Bruno makes the universe so infinite that there are as many worlds as fixed stars and makes this our region where the mobile bodies exist one of the innumerable worlds, which differs from the other surrounding worlds in almost no feature. Thus, if someone were in a star of the Canis constellation [...] the world would appear to him the same as it does to us who observe the fixed stars from our world. Therefore, according to these philosophers, this new star would be a new world. The mere thought of it carries with itself I do not know what horror of the occult catching him who wanders in this immense space, of which the determinations of borders, middle point and any certain place are denied.[52]

What is interesting in Kepler's polemics against Bruno and Gilbert is that it is conducted from a scientific and empirical perspective; if Kepler had been driven by the impulses of a metaphysical and mystical Neoplatonism, he might have accepted, for instance, Cusanus' infinitism; on the contrary, on this issue, Kepler is perfectly aligned with the tradition that links Aristotle with the modern empiricists. At the same time, the empirical and scientific reasons for Kepler's opposition to infinitism are mixed with his religious idea that there is only one world which is the finite image of the divine Trinity (see above, p. 33).

Another episode that revealed Kepler's frame of mind was his polemic against the English Paracelsian alchemist Robert Fludd. In the Appendix to book 5 of the *Harmonice Mundi* Kepler examined Fludd's Hermetic work *Utriusque cosmi maioris scilicet et minoris, metaphysica, physica, atque technica historia* (Fludd 1617–1624). Since Fludd had presented the idea of cosmic music and of harmonic analogies between macrocosm and microcosm, Kepler felt compelled to point out with detailed criticism how he had addressed the same subject in an entirely different manner. The profound difference in their approaches was that "he takes great delight in topics which are hidden in the darkness of riddles, where I strive to bring topics which are wrapped in obscurity out into the light of understanding. The former is familiar to chemists, Hermeticists, and Paracelsians; the latter is considered their own by

[52] Our translation. Cf. KGW I, p. 253: "Itaque defendit illam infelix ille JORDANUS BRUNUS: nec obscure asseruit, specie dubitantis, et GULIELMUS GILBERTUS, libro de Magnete, caetera praeclarissimo, religiosum tamen affectum eo demonstravit, quod existimaret non alia re rectius intelligi infinitam Dei potentiam, quam si infinitum mole conderet mundum. Sed BRUNUS ita infinitum facit mundum, ut quot sunt stellae fixae, tot mundos, et hanc nostram regionem mobilium, unum ex innumerabilibus mundis faciat, nulla fere nota a caeteris circumpositis distinctam: adeoque si quis in stella Canis esset [...], ei faciem eandem mundi inde esse apparituram, quae nobis hinc apparet, ex nostro mundo fixas suspectantibus. Itaque secundum illos, hoc novum sidus, novus aliquis mundus fuerit. Quae sola cogitatio, nescio quid horroris occulti prae se fert; dum errare sese quis deprehendit in hoc immenso; cujus termini, cujus medium, ideoque et certa loca, negantur".

mathematicians".[53] Fludd's teachings on cosmic harmony "are mere symbolism", and their content is "poetical or rhetorical rather than philosophical or mathematical" (Kepler, 1997, p. 505). There could not be a clearer formulation of the distance between Kepler's approach and the obscure and confused numerology of Renaissance Hermeticism.

A further interesting aspect of Kepler's intellectual character is that he understood that the logic of science is different from formal logic. It was well known in formal logic that the truth of the premises cannot be inferred from the truth of the conclusion. Kepler claimed in the *Mysterium Cosmographicum* that in natural knowledge, on the other hand, if a series of premises entails a series of true conclusions, then the premises are also true:

> On this point I have never been able to agree with those who rely on the model of accidental proof, which infers a true conclusion from false premises by the logic of the syllogism. Relying, as I say, on this model they argued that it was possible for the hypotheses of Copernicus to be false and yet for the true phenomena to follow from them as if from authentic postulates. For the model does not fit. The conclusion from false premises is accidental, and the nature of the fallacy betrays itself as soon as it is applied to another related topic—unless you gratuitously allow the exponent of that argument to adopt an infinite number of other false propositions, and never in arguing forwards and backwards to reach consistency.[54]

His reasoning seems to be based on what we would regard today as a statistical inference: it is highly implausible and, as a matter of fact, very unlikely that many true conclusions derive from false premises. Kepler understood that the logic of natural science is not identical with abstract formal logic. He employed this argument against the adversaries of Copernicus who claimed that Copernicus' conclusions might have been true even though his premises were false. This shows that Kepler was a well-accomplished scientist who practised science and reflected on its methods. Therefore his views turn out to be interesting also from an epistemological point of view.

Philosophically, the most significant feature of Kepler's complex personality has been identified by Koyré as a relentless and radical search for causes that runs through all his writings:

> Whilst his predecessors accepted the cosmological state of affairs as the ultimate fact beyond which it was impossible to enquire, and at most believed (as did Copernicus) that they had discovered its specific law of order, Kepler boldly sought the underlying reason. The Cosmos was not formed by chance; it was created by God; and God, assuredly, did not create it *temere,*

[53] Our translation. Cf. KGW VI, p. 374: "Videas etiam, ipsum plurimum delectari rerum aenigmatibus tenebrosis, cum ego res ipsas obscuritate involutas in lucem intellectus proferre nitar. Illud quidem familiare est Chymicis, Hermeticis, Paracelsistis; hoc proprium habent Mathematici".

[54] Kepler (1981, p. 85). Cf. KGW I, p. 15: "Atque hoc loco nunquam assentiri potui illis, qui freti exemplo accidentariae demonstrationis, quae ex falsis praemissis necessitate Syllogistica verum aliquid infert. Qui, inquam, hoc exemplo freti contendebant, fieri posse, ut falsae sint, quae COPERNICO placent hypotheses, et tamen ex illis vera φαινόμενα tanquam ex genuinis principijs sequantur. Exemplum enim non quadrat. Nam ista sequela ex falsis praemissis fortuita est, et quae falsi natura est, primum atque alij rei cognatae accommodatur, seipsam prodit: nisi sponte concedas argumentatori illi, ut infinitas alias falsas propositiones assumat, nec unquam in progressu, regressuque sibijpsi constet".

haphazardly, but on the contrary was guided by rational considerations and followed a perfect architectural plan (Koyré, 1973, p. 122).

This attitude explains not only his repeated attempts at detecting the geometrical archetypes implemented by God in the creation of the world, but also his effort to determine the physical causes of the motions of the planets around the Sun. In his celestial physics the Sun is the origin of the force that causes the planetary motions and it is noteworthy that Kepler was the first to conjecture a dynamic explanation of the motions of the planets without ascribing the origin of this motive force to an incorporeal cause: "No mind is needed for implementing the motion; for the direction in which the Sun rotates was fixed from the very beginning of things; and the constancy of the revolution as well as of the periodic times depends... on the constant ratio of the motive power to the resistance of the material substance" (*Epitome*, quoted by Koyré, 1973, p. 296). To be more precise—as clarified in Sect. 3.1, Kepler thought of four different forces to determine completely the planets' orbits.[55] His dynamical ideas were wrong and the description of the action of these forces is couched in a language which today sounds animistic or anthropomorphic; for instance, when he compares one of them to magnetic attraction and repulsion in these terms: "the Sun has the strength to attract the planet by its friendly part and to repel it by its hostile part…" (*Epitome*, quoted by Koyré, 1973, p. 297). However, if we go beyond the phraseology, what matters is that Kepler designated two components of a quasi-magnetic force in order to explain the orbits' ellipticity. Kepler's attempt to devise the dynamics of the solar system is a clear step in the direction of the scientific treatment of this problem which will culminate in Newtonian dynamics.

We have pointed out that several traits of the vitalistic Renaissance tradition converge in Kepler's thought with a Pythagorean approach and with fundamental features of the new mechanical and mathematized paradigm of science. Was one of these aspects preponderant in Kepler's approach to astronomy?

The relations between the vitalistic and the mechanical viewpoints in Kepler's understanding of the universe are an intriguing and difficult subject. Some historians of science, probably inspired by Gérard Simon, hold that Kepler's way of thinking belongs to a pre-modern "structure of thought". Actually, it seems to us that the concept itself of "structure of thought" is debatable if it is used as a rigid and homogeneous historiographic category.[56] Let us examine the elements that Simon ascribes to Kepler's supposed structure of thought. Some of Kepler's views can surely be considered metaphysical: the idea that the cosmos and cosmic bodies are living beings; the importance of the trinitary archetype; the idea that a mathematized cosmic harmony exists. As for Kepler's views on physics, Simon himself

[55] See Stephenson (1987), Pisano and Bussotti (2018).

[56] Simon has made several valuable contributions to the study of Kepler, the most important of which is his Ph.D. dissertation (Simon, 1976) that was partly published in the celebrated (Simon, 1979). The rest has been published recently (Simon, 2019). On the concept of "structure of thought", see also Simon (1996, 2003). Simon's studies are always refined, profound and informative. We only intend to express doubts on this particular category if interpreted as a rigid historiographical pattern.

points out that Kepler searched for the causes of phenomena, since he found a mere geometrical-kinematical approach to natural philosophy unsatisfactory. Furthermore, Kepler devised a series of hypotheses, formulated in mathematical terms, that he submitted to the test of observation and to experiment. The whole of Kepler's production as well as his polemic with Fludd are clear evidence that he considered mathematics to be a fundamental element for scientific knowledge.

Mathematics as a unifying approach to describe natural phenomena is the foundation of modern physics, whatever ontological conception of mathematical objects one entertains. The relation of cause-effect, sought by Kepler in any possible circumstance, has been for a long time an essential component of theoretical physics. The search for a dynamical cause of kinematical phenomena is a very modern character of Kepler's thought. Moreover, Kepler tried to verify or falsify a huge amount of hypotheses through indefatigable mathematical, experimental and observational work. For these reasons we may say that his approach to natural inquiry is closer in some respects to that of a modern scientist than were those of Galileo or Descartes, both of whom lacked a dynamical approach to physical investigations. Although Descartes showed a 'scientific' attitude in *La Dioptrique* and *Les Météores*, he did not pay attention either to the mathematization or to the experimental verification of his statements in the general exposition of his physics in his *Principia philosophiae*. Kepler, on the other hand, had been always careful to validate his hypotheses and discoveries by means of mathematics, experiment and observation.

Whoever claims (as Simon did) that Kepler's structure of thought is pre-modern must prove that what in Kepler's production is considered scientific today depends on his pre-modern metaphysics. Otherwise, such a thesis amounts merely to the observation that metaphysical and traditional ideas 'accompanied' Kepler's scientific achievements without determining them. Let us, then, analyse some aspects of Kepler's metaphysics, beginning with the analogy between the structure of the cosmos and the Holy Trinity that Kepler outlined both in the *Mysterium* and in the *Paralipomena ad Vitellionem*. In the *Mysterium* Kepler, before conceiving the theory of the Platonic polyhedra, considered a series of different hypotheses about the geometrical-archetypical structure of the universe, but rejected all of them because they were not consistent with the empirical data. It is true, as Simon claims (Simon, 2019, pp. 53–57) that the hypothesis of the Platonic solids is particularly suitable for Kepler's conception of the Holy Trinity, but we are entitled to suppose that Kepler would have rejected it, if it had not been coherent to some degree with the available observations. The trinitary metaphysics plays no role in the origin of this hypothesis, which Kepler formulated after he rejected the others. In any case, the hypothesis of the Platonic solids is not linked to the discovery of the three laws of planetary orbits, which is Kepler's most outstanding scientific achievement.

Furthermore, in the *tour de force* that Kepler undertook to establish the elliptical form of the orbits—that he described in the *Astronomia Nova*—no metaphysical element appears. He dealt with a series of hypotheses which he sifted and rejected because they were inconsistent with Tycho's observations. Certainly, a circular form of the orbits would have been more coherent with the analogy between Trinity and

Cosmos, but Kepler did not hesitate to give up this traditional assumption because his calculations and data were incompatible with it.

Simon claims that Kepler's world is a world of analogies and influences among similar entities. This statement is true, but deserves qualification: we have seen that gravitational attraction acts only between *cognata corpora*; the Sun is not a *cognatum corpus* with respect to the planets; however, it acts on them through the *virtus motrix*. It is, therefore, true that like bodies exert mutual attraction, but it is not necessarily true that unlike bodies do not interact with each other, though the interaction between the Sun and the planets, in Kepler's view, was not of a gravitational kind. This consideration diminishes the importance of *cognatio* in Kepler's physics.

Simon emphasizes the relevance of astrology in Kepler's activity. In this case as well, astrological practice, that was common for astronomers at that time, is not conceptually significant in explaining Kepler's scientific achievement. Besides, Kepler adopted a scientific method even in astrology, a polymorphic and problematic discipline whose characters were not always well defined. He tried to identify basic assertions, to use a deductive method and to control whether his predictions were consistent with his presuppositions.

It is undeniable that Kepler considered cosmic bodies as living beings. However, at least insofar as his dynamical explanations are concerned, he tried to avoid as much as possible resorting to their living nature. He understood that claiming that a planet moves because motion belongs to its living nature is not an explanation at all. He, therefore, developed his physical astronomy to explain the movements of the planets.

As we have seen, a characteristic of Kepler's cosmology, that represents both its charm and its weakness, is the continuous search for harmonies—be they geometrical or musical—that are stated mathematically and that he tried to verify empirically. This aspect is evidence of the 'ambiguity' of Kepler's mindset, but is not a mere expression of speculative preconceptions and shows his bent towards an ascertained knowledge of physical phenomena.

To show how Kepler's mindset differs from a purely metaphysical approach to natural philosophy, we point out that he distinguished meaningful and meaningless questions concerning the structure of the cosmos.[57] As Regier points out, in the second edition of his *Mysterium* (KGW VIII) Kepler stated that the following questions are meaningless:

(1) Why is the Zodiac where it is, when it could have been placed in infinitely many other positions?
(2) Why did God establish the universe in one part of the void rather than another?
(3) Why did God create the universe at such and such a moment rather than at some different moment?
(4) Why do the planets turn from West to East, rather than from East to West?[58]

[57] For the considerations developed in this paragraph, we refer to Regier (2013, pp. 158–160).
[58] Regier (2013, p. 158).

Kepler relies on a limited application of the principle of sufficient reason to prove that these questions are absurd: the position of the Zodiac, of the universe in the void, the moment of Creation and the rotation-sense of the planets are indifferent with respect to both the geometrical-musical structure of the universe and to its physical laws.[59] There is, thence, no possible rational answer to these questions because there is no sufficient reason to prefer an answer among many possible others. Thus, these questions do not belong to any conceivable cosmology. Regier compares Kepler's use of the principle of sufficient reason with Leibniz's. In his epistolary with Clarke, Leibniz used against Newton's absolute space the argument that by swapping East with West, nothing would change in the universe. Kepler, instead, drew from this same counterfactual hypothesis a completely different conclusion: since what is indifferent cannot be used in any scientific argument, the possible existence of a void space independent of matter has to be argued on other bases. In this case he relied on a criterion of plentitude: as existence is preferable to non-existence, and since there is no contradiction in admitting the void, then it is appropriate to assume that void exists (Regier, 2013, pp. 158–159). While this conclusion is drawn from a purely speculative consideration, the premiss that the existence of void cannot be settled by reasoning about an indifferent feature of the universe is evidence of Kepler's epistemological intelligence: Kepler understood that science can only answer questions related to a precise conceptual frame. There are no absolute questions in science independently of a reference theory. This marks a wide difference between Kepler and traditional metaphysical thinkers (among whom, under this respect, Leibniz should also be included) who pose absolute questions beyond any possible context. Thus, if it is doubtless that Kepler was also a metaphysician, his metaphysics did not belong to the absolute ontological-theological doctrine of our philosophical tradition; it was rather a metaphysics based on science, where the empirical and the contextual aspects of physics and astronomy play a fundamental and limiting role.

In conclusion: we have tried to stress the interconnections among the various aspects of Kepler's thought and we have recognised that Kepler's mindset is truly unique. However, we think that in most respects he may be considered a modern scientist and that the concept of a "pre-modern structure of thought" has to be rejected if it is interpreted as a sort of compact frame of mind that led Kepler to some achievements while preventing him from reaching others.

References

Aiton, E. J. (1973). Infinitesimals and the area law. In F. Krafft, K. Meyer, & B. Stickel (Eds.), *Internationales Kepler—symposium* (pp. 285–305). Hildesheim: Gerstenberg.

Aiton, E. J. (1978). Kepler's path to the construction and rejection of his first oval orbit for Mars. *Annals of Science, 35*(2), 173–190.

Baigre, B. B. (1990). The justification of Kepler's ellipse. *Studies in History and Philosophy of Science, 21*(4), 633–664.

[59] KGW VIII, p. 65.

Cohen, B. I. (1975). Kepler's century: Prelude to Newton's. In A. Beer & P. Beer (Eds.), *Kepler's Four Hundred Years, Vistas in Astronomy 18*, 3–36. Oxford, New York, Toronto, Sydney, Braunschweig.
Davis, A. E. L., & Field, J. V. (1995). Kepler translated. Essay review of J. Kepler, *New astronomy*. Translated, introduced and notes by W. H. Donahue. C.U.P. (1992). *The British Journal for the History of Science, 28*, 345–349.
Davis, A. E. L. (1992a). Kepler's "distance law"—Myth, not reality. *Centaurus, 35*(2), 103–120.
Davis, A. E. L. (1992b). Kepler's physical framework for planetary motion. *Centaurus, 35*(2), 165–191.
Davis, A. E. L. (2003). The mathematics of the area law: Kepler's successful proof in *Epitome Astronomiae Copernicanae* (1621). *Archive for History of Exact Sciences, 57*(5), 355–393.
Di Liscia, D. A. (2007). El concepto de causalidad y el desarollo de una teoria cosmólogica en Johannes Kepler. *Anales De La Academia Nacional De Ciencias De Buenos Aires, 34*, 723–753.
Di Liscia, D. A. (2009). Kepler's A Priori Copernicanism in his *Mysterium Cosmographicum*. In M. A. Granada & E. Mehl (Eds.), *New sky, new earth: Copernican revolution in Germany during the Reform (1530–1630)* (pp. 283–317). Paris: Les Belles Lettres.
Donahue, W. H. (1981). *The dissolution of the celestial spheres*. New York: Arno Press.
Donahue, W. H. (1988). Kepler's fabricated figures: Covering up the mess in the *New Astronomy*. *Journal for the History of Astronomy, 19*, 217–237.
Donahue, W. H. (1993). Kepler's first thoughts on oval orbits: Text, translation and commentary. *Journal for the History of Astronomy, 24*(1–2), 71–100.
Donahue, W. H. (1996). Kepler's approach to the oval of 1602, from the Mars notebook. *Journal for the History of Astronomy, 27*(4), 281–295.
Elena, A. (1983). On the different kinds of attractive forces in Kepler. *Archive Internationales D'histoire Des Sciences, 33*, 22–29.
Field, J. V. (2009). Kepler's harmony of the world. In R. Kremer & J. Wlodarczyk (Eds.), *Johannes Kepler. From Tübingen to Zagan* (pp. 11–28). Institute for the History of Science, The Polish Academy of Sciences, The Copernicus Centre for the Interdisciplinary Studies, Instytut Historii Nauki PAN.
Field, J. V. (1982). Kepler's cosmological theories: Their agreement with observations. *The Quarterly Journal of the Royal Astronomical Society, 23*, 556–568.
Field, J. V. (1988). *Kepler's geometrical cosmology*. Chicago: The University of Chicago University Press.
Fludd, R. (1617–1624). *Utriusque cosmi maioris scilicet et minoris, metaphysica, physica, atque technica historia*. Oppenheim and Frankfurt: Hieronymus Gallerus for Johann Theodorus de Bry.
Galilei, G. (1890–1909). *Edizione Nazionale delle Opere*. Direttore A. Favaro, (20 vols). Firenze: Barbèra. Abbreviated as EN.
Galilei, G. (1967). *Dialogue concerning the two chief world systems: Ptolemaic and Copernican* (2nd ed.). Translated and with notes by S. Drake. Foreword by A. Einstein. University of California Press.
Gingerich, O. (1975). The origins of Kepler's third law. In A. Beer & P. Beer (Eds.), *Kepler's Four Hundred Years, Vistas in Astronomy, 18*, 595–601.
Goldbeck, E. (1896). *Keplers Lehre von der Gravitation*. Hildesheim-New York: Olms.
Granada, M. A. (2009). Novelties in the heavens between 1572 and 1604 and Kepler's unified view of nature. *Journal for the History of Astronomy, 40*(4), 393–402.
Guidi Itokazu, A. (2006). A Força que Move os Planetas: Da Noção de *Species Immateriata* na Astronomia de Johannes Kepler. *CHPC, 3*(16/2), 211–231.
Guidi Itokazu, A. (2007). Da Potência Motriz Solar Kepleriana como Emanação Imaterial. *CHPC, 3*(17/2), 303–324.
Haase, R. (1998). *Johannes Keplers Weltharmonik. Der Mensch im Geflecht von Musik, Mathematik und Astronomie*. München: Eugen Diederichs.
Holton, G. (1956). Johannes Kepler's universe: Its physics and metaphysics. *The American Journal of Physics, 24*(5), 340–351.

Hon, G., & Goldstein, B. R. (2008). *From summetria to symmetry: The making of a revolutionary scientific concept*. Dordrecht: Springer.

Hoyer, U. (1979). Kepler's celestial mechanics. *Vistas in Astronomy, 23*(1), 69–74.

Ihmig, K. N. (1990). Trägheit und Massebegriff bei Johannes Kepler. *Philosophia Naturalis, 27*(2), 156–205.

Kepler, J. (1604). *Ad Vitellionem paralipomena quibus astronomiae pars optica traditur*. In KGW, II.

Kepler, J. (1606). *De Stella Nova in Pede Serpentarii*. In KGW, I (pp. 150–292).

Kepler, J. (1611). *Dioptrice*. In KGW IV (pp. 324–414).

Kepler, J. (1618–1621). *Epitome Astronomiae Copernicanae*. In KGW VII.

Kepler, J. (1937–2012). *Gesammelte Werke*. In W. Van Dyck & M. Caspar, et al. (Eds). Revised April 2013. (10 vols). München: Deutsche Forschungsgemeinschaft und Bayerische Akademie der Wissenschaften. Beck'sche Verlagsbuchhandlung. Abbreviated as KGW.

Kepler, J. (1981). *Mysterium Cosmographicum (The Secret of the Universe)*. Translation by A. M. Duncan. Introduction and commentary by E. J. Aiton. Preface by I. Bernard Cohen. New York: Abaris Book.

Kepler, J. (1984). *Le secret du monde*. Introduction, Traduction et Notes de A. Segonds. Paris. Les Belles Lettres.

Kepler, J. (1992). *New Astronomy* (William H. Donahue, Trans.). Cambridge: The Cambridge University Press.

Kepler, J. (1997). *The Harmony of the World*. English translation and introduction by E. J. Aiton, A. M. Duncan, J. V. Field. Philadelphia: American Philosophical Society.

Knobloch, E. (1992). Rapports historiques entre musique, mathématique et cosmologie. In *Quadrivium, Musiques et sciences*, Colloque conçu par D. Lustgarten, C.-H. Joubert, S. Pahaut, M. Salazar (pp. 123–167). Paris: Cité de la musique.

Knobloch, E. (1995). Harmony and cosmos: Mathematics serving a teleological understanding of the World. *Physis, 32*(1), 54–89.

Koyré, A. (1973). *The astronomical revolution: Copernicus—Kepler—Borelli* (1st ed., 1961). Ithaca: Cornell University Press.

Krafft, F. (1991). The new celestial physics of Johannes Kepler. In Unguru (Ed.), *Physics, cosmology and astronomy, 1300–1700: Tension and accommodation. The Boston Studies in the Philosophy of Science* (Vol. 126, pp. 185–227). Dordrecht-Boston-London: Kluwer Academic Publishers.

Petroni, A. M. (1989). *I modelli l'invenzione e la scoperta. Saggio su Keplero, la rivoluzione copernicana e la «New philosophy of science»*. Milano: Franco Angeli.

Pisano, R., & Bussotti, P. (2012). Galileo and Kepler: On *Theoremata circa Centrum Gravitatis Solidorum* and *Mysterium Cosmographicum*. *History Research, 2*(2), 110–145.

Pisano, R., & Bussotti, P. (2018). On the conceptualization of force in Johannes Kepler's *Corpus*: An interplay between physics/mathematics and metaphysics. In R. Pisano, J. Agassi, & D. Drodzova (Eds.), *Hypotheses and perspectives in the history and philosophy of science. Homage to Alexandre Koyré 1892–1964* (pp. 295–345). Cham: Springer.

Rabin, S. J. (2005). Was Kepler's *species immateriata* substantial? *Journal for the History of Astronomy, 36*, 49–56.

Regier, J. (2013). Method and the a priori in Keplerian metaphysics. *Journal of Early Modern Studies, 2*(1), 147–162.

Russell, J. L. (1964). Kepler's laws of planetary motions: 1609–1666. *British Journal for the History of Science, 2*, 1–14.

Simon, G. (1976). *Structures de pensée et objets du savoir chez Kepler* (Vol. 2). Thèse: Philosophie: Paris IV, 1976. Service de reproduction de thèses de l'Université de Lille III, 1979.

Simon, G. (1996). *Sciences et savoirs aux XVIe et XVIIe siècles*. Villeneuve d'Ascq: Presses de Septentrion.

Simon, G. (2019). *Kepler, rénovateur de l'optique*. Paris: Classiques Garnier (Original ed., 1976).

Simon, G. (1979). *Kepler astronome astrologue*. Paris: Gallimard.

Simon, G. (2003). *Archéologie de la vision: l'optique, le corps, la peinture*. Paris: Seuil.

Stephenson, B. (1987). *Kepler's physical astronomy*. New York - Berlin - Heidelberg - London - Paris - Tokyo: Springer.

Stephenson, B. (1994). *The music of the heavens. Kepler's harmonic astronomy*. Princeton: Princeton University Press.

Vijaya, G. K. (2019). Original form of Kepler's third law and its misapplication in propositions XXXII–XXXVII in Newton's *Principia* (Book I). *Heliyon*. On line publication: https://www.cell.com/heliyon/pdf/S2405-8440(18)38696-1.pdf

Whiteside, D. T. (1974). Keplerian planetary eggs, laid and unlaid, 1600–1605. *Journal for the History of Astronomy, 5*, 1–21.

Wilson, C. (1968). Kepler's derivation of the elliptical path. *Isis, 59*, 5–25.

Chapter 4
Galileo and the Spread of the Copernican System

With regard to cosmology, Galileo was not an original thinker. He adhered to the Copernican system of the world and did not elaborate it further: he did not investigate the details of the form of the planetary orbits, he remained faithful to the traditional idea of their circular form and always rejected Kepler's theory of elliptical orbits. He did not investigate the force determining the movements of the planets around the Sun. Nonetheless, Galileo was important for Copernican cosmology because he developed a series of observations and arguments, that, though not decisive in favour of the Copernican system, effectively destroyed the Aristotelian-Ptolemaic system. In particular, as we will see, the discovery of Venus's phases observed by Galileo was not compatible with the Ptolemaic system. Beyond this strictly astronomical issue, Galileo undermined all the philosophical assumptions on which the Ptolemaic system was based, such as the incorruptibility of the skies, the impossibility of the diurnal as well as of the annual Earth's motion, and the idea that no other celestial body apart from Earth might have satellites. There is a further aspect that should be considered when dealing with Galileo's contributions to cosmology: his idea of inertia. This is particularly interesting from a historical and conceptual point of view: we have seen that one of the reasons that prevented Kepler from constructing a working theory of the forces determining the movements of the planets around the Sun was his concept of *inertia naturalis*. Galileo had a more advanced idea of inertia, though not completely coincident with ours, as we shall see. Galileo, however, did not apply his idea to cosmology and, with regard to the cosmological context, he adhered to a more traditional concept of inertia. Anyway, he exploited his new view to show that numerous physical arguments of the Aristotelians concerning the local motion were false and erroneous. Newton realized that some of Galileo's ideas on inertia had to be applied to a cosmological context in order to obtain a working theory of forces. Galileo's concept of inertia is thus a milestone in the history of cosmology, though he was not fully conscious of this.

Like no one before him, Galileo was able to use observations, reasoning, and the art of rhetoric in support of a scientific hypothesis. For the purposes of this book,

P. Bussotti and B. Lotti, *Cosmology in the Early Modern Age: A Web of Ideas*,
Logic, Epistemology, and the Unity of Science 56,
https://doi.org/10.1007/978-3-031-12195-1_4

his two major works are the *Sidereus Nuncius* (Galilei, 1610, in EN, III, 1) and the *Dialogo intorno ai due massimi sistemi del mondo* [*Dialogue Concerning the Two Chief World Systems: Ptolemaic and Copernican*] (1632, see Galilei, 1967). After Galileo's writings it became difficult, though not impossible, to propose serious arguments in favour of a system different from that of Copernicus. In the period between Galileo and Newton, the Copernican system had a competitor, which was not the obsolete Ptolemaic system, but the new one developed by Tycho Brahe in the late 1570s. Several scholars, and in particular Jesuit scientists, adhered to the Tychonic system in an extreme attempt to save the centrality of the Earth's position.[1]

It is worth highlighting the various stages of development of modern science: Galileo constructed, so to say, the walls to protect the Copernican system because a great part of the town, the system itself, had already been built. The reaction of the scientists and of the political-theological milieus that did not want to accept the Copernican system consisted in proposing a revival of the Tychonic system. In the mid seventeenth century, in central and northern Europe nearly all scientists were Copernican, whereas in Italy the debate was still alive. This belated and controversial reception of the Copernican system is a remarkable fact, both historically and conceptually. After Kepler and Galileo, any system different from the Copernican one was extremely implausible, but not impossible, since there were no decisive dynamical reasons that would render the Tychonic system absurd. Most significantly, this shows the interplay of scientific, political, and theological aspects, which is not uncommon in the history of scientific theories.

Let us now examine the aspects of the two great works by Galileo we have mentioned above.

4.1 Galileo's Contributions to Cosmology

This section is divided into three subsections: in the first one, we will analyse three subjects that are directly tied to cosmological issues, namely the main discoveries expounded in the *Sidereus Nuncius*, the phases of Venus and the sunspots. The second one will be dedicated to Galileo's tides theory, which enables us to draw a comparison with Kepler's view. The third subsection, the most important one, concerns Galileo's concept of inertia.

[1] An interesting publication on the spread of the Tychonic system around the half of the seventeenth century is Marcacci (2018). For a good description of the Tychonic system see Schofield (1965, 1981).

4.1.1 *Discoveries in the* Sidereus Nuncius, *Sunspots, Phases of Venus*

In the *Sidereus Nuncius* Galileo referred to the astronomical observations he had carried out between 7 January and 2 March 1610 by means of the telescope, a recently invented instrument. He made three discoveries; (1) The Moon's surface is not at all perfect or composed of ethereal matter. It is irregular and rough. The Moon has mountains, valleys and lowlands, just like the Earth; (2) Jupiter has satellites; (3) The Milky Way is not a nebula, but a set of stars. The booklet was published on 12th March 1610.

Item (1) undermines the Aristotelian tenet that the super-lunar world was made up of incorruptible, unmodifiable, ethereal matter. By means of item (2), one is driven to deny that the Earth holds a privileged position since it is the only body to have a satellite. But this is not the case: Jupiter has satellites too, and more than one. (3) The proof that the Milky Way is not a halo of low dense matter around the Sun, but a set of stars, leads one to reconsider the concept of the sphere of the fixed stars. Galileo remarked that the Milky Way appeared to be a dense conglomeration of stars at a small angular distance from each other. Even if he did not comment on it further, this discovery rendered the idea of the sphere of the fixed stars less plausible (though not impossible) since all the stars were traditionally conceived as set at the same distance from the Earth. Galileo's observation had revealed an irregular disposition of the stars in the allegedly eternal and immutable sphere of the fixed stars.

We will now focus on the reasoning that led Galileo to realize that the lunar surface is similar to the Earth's, with mountains and valleys, since this represents a paradigmatic example of how several factors may contribute to achieving an important discovery. In this case such factors are:

(1) Galileo's idea that the Aristotelian distinction between sublunar and superlunar world was wrong, which is indirectly connected with his Copernican convictions;
(2) the idea that the telescope correctly reproduces and magnifies not only terrestrial objects but also celestial ones, an assumption that was not accepted by other scholars, as we shall see;
(3) rarely an observation in itself produces a discovery. A process of interpretation is necessary. The inferences through which Galileo came to understand the shape of the Moon surface gives support to the thesis that—at least in many circumstances—a scientist's convictions guide his observations and interpretations. In fact, Galileo was convinced that no ontological distinction existed between sublunar and superlunar objects. He pointed the telescope to the sky in order to search a confirmation of his thesis. Obviously, he was not certain a priori of the lunar aspect, but when he observed the Moon, he easily offered an interpretation of the observed phenomena since these were in line with what he expected to see.

Galileo observed that the lunar terminator (the imaginary line which separates the illuminated and the dark hemispheres of the Moon), seen at the telescope, appears to be indented and irregular. Furthermore—close to the terminator—he observed two phenomena a few days after the new Moon (Fig. 4.1 on the top): (1) obscure points in the illuminated hemisphere; (2) bright points in the dark hemisphere. With regard to the first one, let us consider the dark spots in the illuminated zone: the black part, namely the darkest one, of such spots pointed towards the Sun, whereas in the part opposite to the Sun, the spots are surrounded by bright contours. Galileo remarked that the exact same phenomenon occurs on the Earth when, at dawn, we see that the mountains opposed to the Sun are bright, while the valleys are still obscure. When, in the Moon, the amount of the illuminated area increases, the lunar spots lose their darkness. As to the second phenomenon, it is analogous to what happens on the Earth when, immediately before dawn—hence, when the Sun is still under the horizon—few Sun rays illuminate the highest summits of the mountains.

Moving from these observations, Galileo deduced that the Moon is far more similar to the Earth than the Aristotelians thought. It is anything but a perfect body composed of an ethereal substance. Its composition and conformation is similar to that of the Earth.

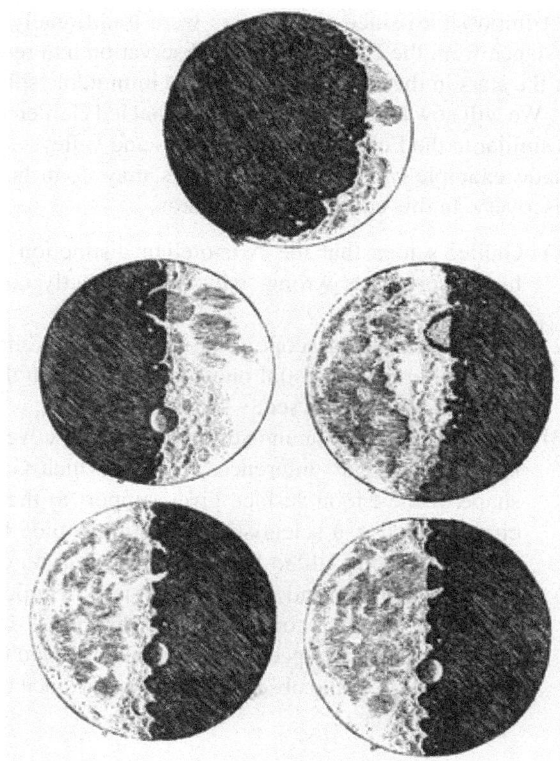

Fig. 4.1 The pictures drawn by Galileo to describe the Moon surface seen at the telescope (Galilei, 1610, EN, III/1, pp. 63, 66–67)

Galileo was also able to explain an interesting phenomenon: the cinereous light (see Fig. 4.2). Immediately before and after the new Moon, it is possible to observe that the dark part of the Moon is not black, but is illuminated by a tenuous cinereous light. Galileo explained that, since the Sun and the Moon are almost in conjunction, from the Moon you see "full Earth". The solar light reflected by the Earth hits the Moon which, if seen from the Earth, appears as enveloped in a cinereous dim light.

The *Sidereus Nuncius* can thus be regarded as a Copernican text since it undermines the most important physical-philosophical presuppositions of the Aristotelian-Ptolemaic system.

The celebrated *Dialogo* is a far more complex book than the *Sidereus Nuncius.* It is divided into four Days. Each Day tackles a specific subject. For each topic, Galileo offers a series of very interesting physical and astronomical considerations. The First Day presents a *general introduction to astronomy*. Almost at the beginning of the Day, Galileo inserts the famous "Platonic myth", where he discusses the functions of the straight and of the circular movement in the constitution of the universe. The myth is especially important for Galileo's cosmology as well as for his concept of inertia. We will discuss it in the subsection on the inertia in Galileo.

After the myth, Galileo points out the mutability of the skies that resulted from the discovery of the novae and of sunspots. He highlights the analogy between the Moon and the Earth. The Second Day deals with the *diurnal movement of the Earth*; here a series of arguments—part of which are similar to Copernicus', but most of them are Galileo's original—are presented to render the thesis of the Earth's motion more plausible than that of the motion of the skies. Galileo also conducted a far-reaching examination of Aristotle's and Ptolemy's arguments against the Earth's motion. In this context, Galileo came very close to the correct concept of the inertial system. The concept of participation in the motion of the system is clearly expressed. The tendencies of circular motion are examined too. Here, Galileo's considerations are well argued and brilliant, but, in several cases, they are not rigorously correct. Galileo also expressed his convictions about the correct methodology in science: mathematics, and, in particular, geometry is necessary. The Third Day deals with *the annual movement of the Earth around the Sun*: Galileo began his analysis by considering the nature of the two new stars appeared in 1572 and 1604; he held that

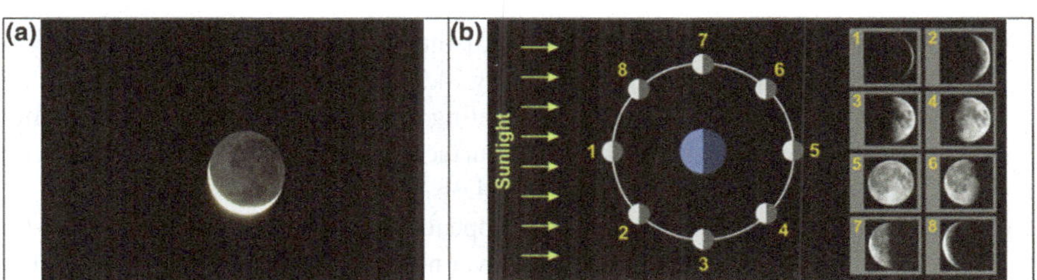

Fig. 4.2 **a** The phenomenon of the cinereous light as it appears in a modern telescope; **b** a picture which shows the geometrical-optical reason—discovered by Galileo—of such phenomenon

these phenomena took place not only beyond Moon's orbit, but beyond the solar system. In this context, we find interesting considerations on the stellar parallaxes. He then analyses the phases of Venus, the retrograde motions of the planets, the different luminosity of the planets along the year, and the importance of the discovery of Jupiter's satellites. In each case, Galileo argues that the Copernican system can explain all the phenomena in a more convincing, consistent and "economic" manner than the Ptolemaic one. The cycle of the seasons and the variable illumination of the different zones of the Earth's surface are addressed too. In the Fourth Day he discusses a *series of phenomena*, whose explanation depends, according to Galileo, on the rotation of the Earth around the Sun: the tides theory (which is wrong); the explanation of the periodical winds; the relations between the annual and the diurnal motion.

We are going to present two arguments by Galileo in support of the Copernican system. They are expounded in the *Dialogo*, but they had been developed several years before. The former concerns the sunspots[2]: it is a paradigmatic example of the subtleties and nuances of Galileo's argumentation; the latter regards the phases of Venus, which is the fundamental astronomical argument against the Ptolemaic system.

Sunspots. Sunspots are temporary phenomena in the Sun's surface, the photosphere. They appear in regions of the surface characterized by a lower temperature with respect to the neighbouring regions and, therefore, they look like little obscure zones in the Sun's surface. They were observed by telescope in 1610 independently by Thomas Harriot (1560–1621) and Johannes Fabricius, who, in 1611, also published a text in which the sunspots were described (Fabricius 1611). In the same period sunspots were observed also by Galileo and by the Jesuit astronomer Christoph Scheiner (1575–1650). At the beginning of 1612 Scheiner published—under the pseudonym of Apelles—three letters on the sunspots. He published three further letters on the same subject in September 1612: he asserted that they were due to celestial corpuscles rotating around the Sun. His aim was to defend the incorruptibility of the Sun's surface. Galileo, in *Istoria e dimostrazioni intorno alle macchie solari e loro accidenti* (1613) argued against Scheiner and guessed the real nature of the sunspots. His reasonings, furtherly clarified and specified in the *Dialogo*, run like this:

(1) Several sunspots appear in the centre of the Sun's disc and disappear before reaching the border of the disc (see Fig. 4.3a). If, as Scheiner believed, they were bodies close to the Sun and rotating around our star—something like Sun's satellites very close to the Sun's surface—they would appear to enter and exit the Sun disc at its border (see Fig. 4.3b), which is not the case.
(2) A sunspot moves at different speed: it appears to go faster when it is close to the apparent centre of the solar disc and slower near the periphery. This is an optical

[2] Galileo explained his ideas on the sunspots in 1613 in his writing *Istoria e dimostrazioni intorno alle macchie solari e loro accidenti* ("History and demonstrations about sunspots and their accidents").

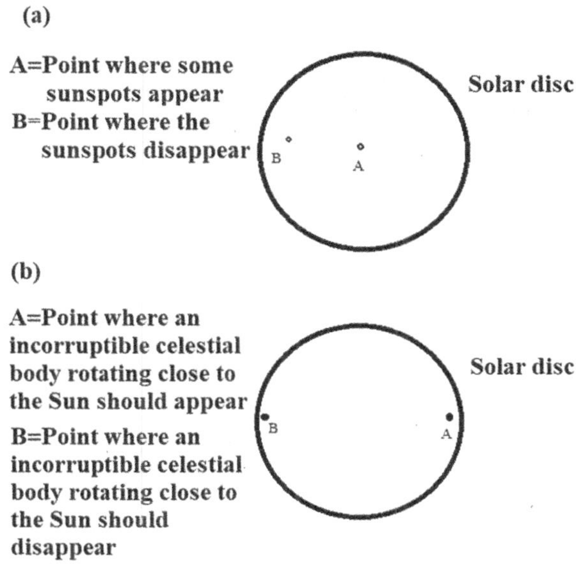

Fig. 4.3 **a** The orbit of some sunspots. **b** The orbit that an incorruptible body rotating close to the Sun should describe. The figures are ours

effect due to perspective which indicates that the sunspots belong to the Sun's surface. The arch *AB* (Fig. 4.4) is equal to the arch *CD* on the Sun's surface, but, seen from the Earth *E* the angle *AEB* looks bigger than the angle *CED*. Therefore, sunspots appear to be faster in the proximity of the central zone of the Sun's surface. This perspective effect is so evident to exclude that it is due to bodies rotating around the Sun, for those bodies should be so close to the Sun that they would be burned by our star. Furthermore, for the same optical reason due to perspective, sunspots appear to be narrower near the circumference, whereas in the centre they look wider and almost perfectly round.

(3) Finally, if sunspots were Sun's satellites, they should present phases like Venus, which is not the case.

Therefore, Galileo concluded, sunspots belong to the Sun's surface.

Galileo's reasoning represents a remarkable example of how he made use of the observations and of his acute logical skill to reach the correct conclusion or, at least, to develop an argument which worked from a logical point of view.

Venus phases. Galileo first observed the phases of Venus between the end of August and the beginning of September 1610, a few months after the publication of the *Sidereus Nuncius*. In the Ptolemaic system, Venus rotates on an epicycle along its orbit around the Earth. The deferent of Venus is set between the Earth and the Sun, which, in its turn rotates around the Earth. The centre of the epicycle of Venus rotates on the deferent in one (terrestrial) solar year. Even in this system, Venus should display some phases, but, as a matter of fact they are not those observed by Galileo. At the apogee and at the perigee, Venus should point its obscure hemisphere towards the Earth, whereas, when the elongation is maximal, Venus should present

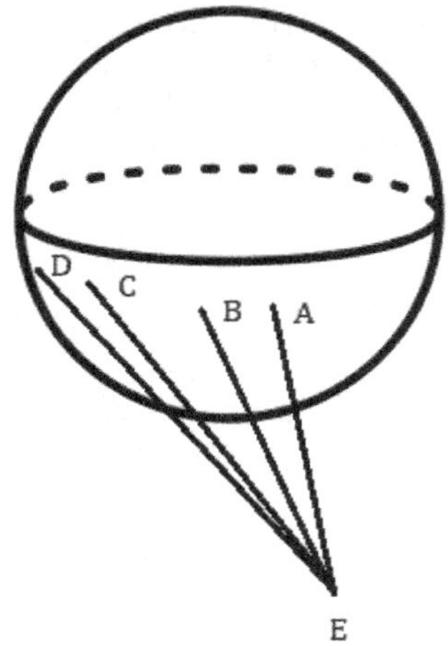

Fig. 4.4 Representation of the apparent velocity of the sunspots. The figure is ours

a little sickle, as is clear on the left side of Fig. 4.5. Venus should never be full. In contrast to this pattern, Galileo had observed that the phases of Venus are similar to those of the Moon (right side of Fig. 4.5), which is fully coherent with the rotation of both Earth and Venus around the Sun. This is a decisive argument against the Ptolemaic system, though it is not completely decisive in favour of the Copernican system because the observed phases of Venus are also explained by the Tychonic system.

Therefore, Galileo worked in two ways against the Ptolemaic system: (1) he showed a series of celestial phenomena which mined the physical and metaphysical context on which the system was based; (2) he produced the decisive optical-astronomical proof against the system. As any great scientist, Galileo tried to deduce a series of consequences from his convictions. Namely, he did not consider the Copernican system simply as a new astronomical system, but he also aimed at understanding how this new view of the universe might exert an influence on our comprehension

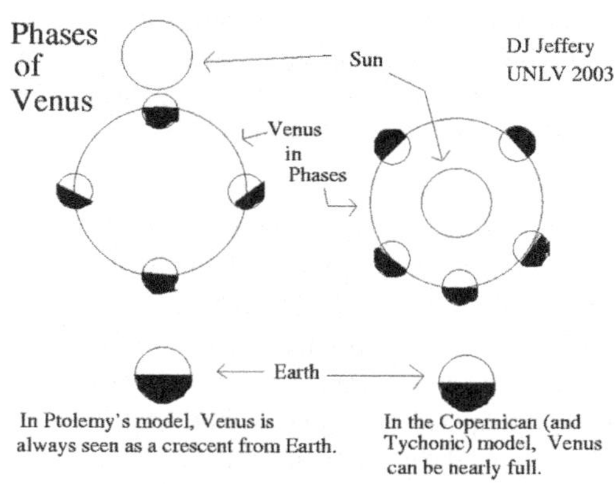

Fig. 4.5 The phases of Venus according to the Ptolemaic system (left) and to the Copernican system (right). *Source* https://www.nhn.ou.edu/~jeffery/astro/ptolemy/

and explanation of terrestrial phenomena. The tides represented a challenge for the scientists because in Galileo's epoch no shared theory existed and the phenomenon was rather mysterious. Galileo's explanation of the tides is based on the composition of the annual and the diurnal Earth's motion.

4.1.2 Galileo's Tides Theory

Most of the Fourth Day of the *Dialogo* is dedicated to the tides. In 1616 Galileo had written, though not published, the *Discorso del flusso e reflusso del mare*,[3] but the treatment offered in the *Dialogo* is incomparably more complete. Galileo started from the observation of a phenomenon that happened on the boats that carried fresh water to Venice. His reasoning can be summarized like this: as long as the motion of the boat is uniform, the surface of the water carried by the boat remains horizontal. To use modern terms which do not distort Galileo's thought: the boat and the water belong to the same inertial reference frame. However, if, for any reason, the boat decelerates, its velocity diminishes, but since water maintains its inertial velocity, it tends to accumulate towards the bow of the boat. Vice versa, if the boat accelerates, water will accumulate towards the stern. Galileo wrote:

> These [inertial] effects can be very clearly explained and made evident to the senses by means of the example of those barges which are continually arriving from Fusina filled with water for the use of this city. Let us imagine to ourselves such a barge coming along the lagoon with moderate speed, placidly carrying the water with which it is filled, when either by running aground or by striking some obstacle it becomes greatly retarded. Now the water will not thereby lose its previously received impetus equally with the barge; keeping its impetus, it will run forward toward the prow, where it will rise perceptibly, sinking at the stern. But if on the other hand the same barge noticeably increases its speed in the midst of its placid course, then the water which it contains (before getting used to this and while retaining its slowness) will stay back toward the stern, where it will consequently rise, sinking at the prow. This effect is indubitable and clear; it may be tested experimentally at any time [...].[4]

This explanation is very clear and leaves no doubt that Galileo had guessed many features of the inertial motions. He had the idea that his consideration might be useful to explain the tides by way of analogy.

---

[3] Galilei (1616, EN V, pp. 373–395).

[4] Galilei (1967, pp. 424–425). Cf. EN VII, pp. 450–451: "I quali effetti possiamo più apertamente dichiarare e manifestare al senso con l'esempio di una di queste barche le quali continuamente vengono da Lizzafusina, piene d'acqua dolce per uso della città. Figuriamoci dunque una tal barca venirsene con mediocre velocità per la Laguna, portando placidamente l'acqua della quale ella sia piena, ma che poi, o per dare in secco o per altro impedimento che le sia opposto, venga notabilmente ritardata; non perciò l'acqua contenuta perderà, al pari della barca, l'impeto già concepito, ma, conservandoselo, scorrerà avanti verso la prua, dove notabilmente si alzerà, abbassandosi dalla poppa: ma se, per l'opposito, all'istessa barca nel mezo del suo placido corso verrà con notabile agumento aggiunta nuova velocità, l'acqua contenuta, prima di abituarsene, restando nella sua lentezza, rimarrà indietro, cioè verso la poppa, dove in conseguenza si solleverà, abbassandosi dalla prua. Questo effetto è indubitato e chiaro, e puossi a tutte l'ore esperimentare [...]".

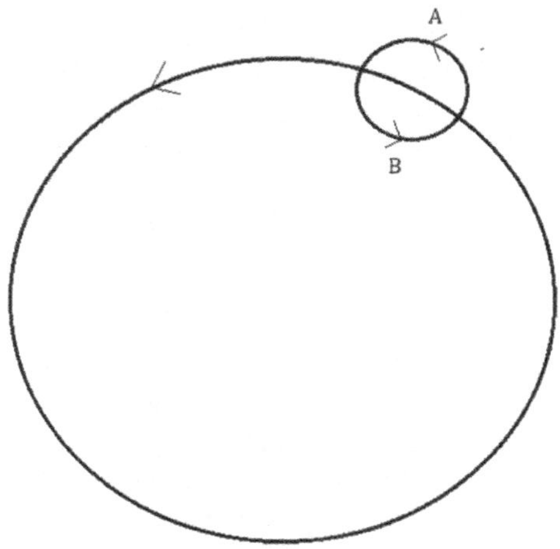

Fig. 4.6 The bigger circumference represents the annual orbit of the Earth around the Sun. The little circumference represents the Earth. Of course, the ratios between the two circumferences are far from the real ones. The figure is ours

In particular (Fig. 4.6), if the bigger circle represents the orbit of the Earth around the Sun and the smaller circle represents the Earth—being both the motions of the Earth around the Sun and the Earth's rotation around its axis counterclockwise—in point *A* the two motions have the same sense, whereas in point *B* they have opposite senses, in the points between *A* and *B* there are all the possible intermediate situations. Therefore, according to Galileo, the speed of point *A* is higher than the speed of point *B* because in *A* the annual velocity of revolution is added to the velocity of the diurnal motion, whereas in point *B* the two velocities have to be subtracted. Since the instantaneous velocity of every point belonging to the Earth's surface changes and since each points tends to maintains its inertial velocity, the consequence is that the whole mass of the sea's and ocean's water moves, so producing the tides. Once again Galileo's words are very clear:

> From this we conclude that just as it is true that the motion of the whole globe and of each of its parts would be equable and uniform if it were moved with a single motion, whether this happened to be the annual or the diurnal, so is it necessary that upon these two motions being mixed together there results in the parts of the globe this uneven motion, now accelerated and now retarded by the additions and subtractions of the diurnal rotation upon the annual revolution.
>
> Now if it is true (as is indeed proved by experience) that the acceleration and retardation of motion of a vessel makes the contained water run back and forth along its length, and rise and fall at its extremities, then who will make any trouble about granting that such an effect may — or rather, must — take place in the ocean waters? For their basins are subjected to just such alterations; especially those which extend from west to east, in which direction the movement of these basins is made. Now this is the most fundamental and effective cause of the tides, without which they would not take place.[5]

[5] Galilei (1967, pp. 427–428). Cf. EN VII, pp. 453–454: "Concludiamo per tanto, che sì come è vero che il moto di tutto il globo e di ciascuna delle sue parti sarebbe equabile ed uniforme quando elle si movessero d'un moto solo, o fusse il semplice annuo o fusse il solo diurno, così è necessario che, mescolandosi tali due moti insieme, ne risultino per le parti di esso globo movimenti

This explanation of the tides is problematic. As Aiton rightly observes, Galileo's reasoning with regard to the sum of the annual and diurnal velocities fits perfectly in a reference frame centred in the Sun. An observer on the Sun would calculate the composition of velocities exactly as Galileo stated. However, the motion of water is considered by Galileo with respect to the Earth, not with respect to the Sun; and with respect to the Earth, water receives no acceleration due to the annual motion. Furthermore, this theory is in contradiction with what Galileo had claimed more than once in his *Dialogo*, namely that every object which moves on the Earth is unaffected by the annual motion around the Sun.[6]

Commentary. Galileo believed his theory of the tides to be an important part of his cosmological view. First of all, it is founded on the two basic movements ascribed by Copernicus to the Earth and, therefore, the Copernican theory would have been sufficient to explain such a complex phenomenon as the tides, or, at least, its main features. At the same time, in a sort of reciprocal feed-back, a working theory of the tides grounded on the most elementary movements ascribed by Copernicus to the Earth (and, in general, to the planets), would have been a further confirmation of the Copernican system. As already pointed out by Kepler in the *Mysterium*, if a scientific hypothesis implies a factual truth, the validity of the hypothesis is not confirmed from a logical point of view, but its scientific value is strongly corroborated. This is the more true, the more the phenomenon explained by the hypothesis is complex, difficult and yet unexplained. Therefore, in this respect, Galileo's attempt to construct a tides theory inconceivable in the Ptolemaic or Thyconic system is perfectly understandable.

There is a further reason which highlights the importance of the tides theory for the whole of Galileo's conception: as we have seen, Galileo criticized Kepler's idea that the tides depend on the attraction exerted by the Moon on the Earth because, in

---

difformi, ora accelerati ed ora ritardati, mediante gli additamenti o suttrazioni della conversion diurna alla circolazione annua. Onde se è vero (come è verissimo, e l'esperienza ne dimostra) che l'accelerazione e ritardamento del moto del vaso faccia correre e ricorrere nella sua lunghezza, alzarsi ed abbassarsi nelle sue estremità, l'acqua da esso contenuta, chi vorrà por difficultà nel concedere che tale effetto possa, anzi pur debba di necessità, accadere nell'acque marine, contenute dentro a i vasi loro, soggetti a cotali alterazioni, e massime in quelli che per lunghezza si distendono da ponente verso levante, che è il verso per il quale si fa il movimento di essi vasi? Or questa sia la potissima e primaria causa del flusso e reflusso, senza la quale nulla seguirebbe di tal effetto". Galileo added a long series of specifications to explain many of the nuances of the tides phenomenon.

[6] Aiton (1954, pp. 44–46). A detailed work on Galileo's theory of tides is Palmieri (1998), who offers a positive re-evaluation of this theory. He argues against Aiton's interpretation that was shared, as Palmieri points out (ibid., p. 236, note 24), by Minazzi (1994, p. 254), Rossi (1989, p. 126), Morpurgo-Tagliabue (1981, p. 110), Hall (1963, p. 479). Palmieri thinks that Galileo in his theory of tides does not contradict the principle of relativity of motions, which he himself had stated. His mistake is far subtler and depends on the fact that Galileo—as it was quite natural at that time—was not skilled in the vector calculus (Palmieri 1998, pp. 227–234). Palmieri holds that Galileo's mistake is a technical-mathematical one rather than a physical–conceptual one; furthermore it is, at least in part, a remediable mistake. It seems to us that Palmieri's interpretation implies an excessive modernization of Galileo's train of thoughts. For this reason in the running text we refer to the more traditional interpretation.

Galileo's opinion, such idea belonged to the magical-astrological mentality based on mysterious and unprovable influxes among bodies, a mentality which Galileo deemed profoundly anti-scientific. Therefore, he proposed a mechanistic and kinematics theory of the tides. There is a further observation whose importance is sometimes underestimated: Galileo proposed an actual theory of the tides. Thus, we can appreciate its nuances, its strong as well as weak points. Kepler did not propose a theory; rather an outline of a theory. The outline is correct, but the theory is not developed. Galileo's criticisms are understandable: he feared a regression to a prescientific thought. Today, we might claim that the religious and philosophical presuppositions of any scientist can be an important aspect as to the genesis of his work, but that the final product of such work can be considered independently of these presuppositions. However, for Galileo the situation was completely different: since science was in its infancy, he was fighting a battle in favour of science and saw any possible trace of magical thought as a threat to his scientific programme. Anyway, no one can foresee what Galileo would had thought and written if Kepler had proposed a theory of the tides as fully developed as his own, whatever being Kepler's presuppositions. There is no doubt that Galileo was negatively impressed by those aspects of Kepler's thought he considered connected to mysterious and purely metaphysical entities and doctrines. The history behind Galileo's reception of the *Mysterium* is instructive with regard to the ways of thinking of the two scientists: on 30th May 1597 Galileo sent a letter to Jacopo Mazzoni, in which he expressed for the first time his Copernican conviction because he considered the opinion of the Pythagoreans and of Copernicus as far more likely than Aristotle's and Ptolemy's.[7] In the same year Kepler sent Galileo a copy of his *Mysterium*. Galileo, probably after having read only the initial pages of Kepler's work, where the author expresses his Copernican convictions, answered on 4 August 1597 with an enthusiastic letter, where he wrote:

> [...] I promise you that I will read your book with serenity of spirit because I am sure to find there many beautiful things. Actually, I will do that the more willingly as for many years I have accepted Copernicus's doctrine. Starting from these principles I was able to discovery the causes of many natural effects which remain inexplicable on the basis of the current theories.[8]

Galileo also claimed that he had not written anything on Copernicanism yet since he considered it dangerous to express his Copernican convictions explicitly. After this letter, a collaboration between the two scientists—or at least a discussion with regard to their results—seemed very likely. On the contrary, Galileo never sent Kepler

[7] EN II, pp. 197–202, especially p. 198.

[8] Our translation. Cf. EN X, p. 68: "[...] pollicebor me aequo animo librum tuum perlecturum esse, cum certus sim me pulcherrima in ipso esse reperturum. Id autem eo libentius faciam, quod in Copernici sententiam multis abhinc annis venerim, ac ex tali positione multorum etiam naturalium effectuum caussae sint a me adinventae, quae dubio procul per communem hypothesim inexplicabiles sunt".

a complete answer on the *Mysterium* and, even in the following years, the correspondence between the two, though cordial, was always rather poor.[9] Almost certainly, Galileo sent no other letter to Kepler on the *Mysterium* because he did not share the whole of the theory explained in that book. When Galileo read the entire text, he possibly realised that the only aspect he shared with Kepler was Copernicanism. The whole theological-metaphysical apparatus of the *Mysterium* was completely extraneous to Galileo's way of conceiving science. Galileo's reception of the *Mysterium* can be connected to his criticism of Kepler's views of tides. Galileo probably thought that Kepler's ideas on tides had been developed on metaphysical assumptions similar to those of the *Mysterium*. Therefore, he chose a mechanistic explanation of tides. It should be pointed out that with the term "mechanistic" we mean "derived only on the basis of mechanical-kinematical means, not on that of some forces whose origin is inexplicable". In this sense, Galileo's natural philosophy is mechanistic. However, this meaning of the term must not be identified with that referred to Descartes' natural philosophy (see next chapter): Galileo was an atomist, he believed in the void, and did not believe that all motions were circular. He did not feel the need to populate the skies of ethereal vortices.

Furthermore, there is an important aspect to underline: Galileo refused to resort to mysterious entities and influxes, but he did not claim explicitly that any phenomenon has to be explained mechanically in terms of collisions and movements. With regard to gravity and to the force that moves the planets around the Sun and the satellites around the planets—whatever this force be—he did not venture in mechanistic hypotheses on their origin (as Descartes did). Though these forces could not be explained with the mysterious influxes of the Renaissance tradition, Galileo seems to leave the door open to a kind of natural action which is not reducible to collisions and movements. In any case, in spite of the fact that he first offered a physical–mathematical treatment of the bodies' free falling in a gravitational field, his attitude towards the origin of gravity is surely closer to Newton's *hypotheses non fingo* than to Descartes's vorticistic theory. The following long quotation from the Second Day of the *Dialogo* clarifies Galileo's view as regards the force moving the planets and as regards gravity. We read:

> Salv. I did not say that the earth has neither an external nor an internal principle of moving circularly; I say that I do not know which of the two it has. My not knowing this does not have the power to remove it.
>
> But if this author knows by which principle other world bodies are moved in rotation, as they certainly are moved, then I say that that which makes the earth move is a thing similar to whatever moves Mars and Jupiter, and which he believes also moves the stellar sphere. If he will advise me as to the motive power of one of these movable bodies, I promise I shall be able to tell him what makes the earth move. Moreover, I shall do the same if he can teach me what it is that moves earthly things downward.
>
> Simp. The cause of this effect is well known; everybody is aware that it is gravity.

[9] Among the numerous works on the problem of the relations between Kepler and Galileo, we point out Bucciantini (2007), who is rich in details and also provides numerous indications on their correspondence. Bussotti (2011) also addresses several aspects of Kepler-Galileo epistolary.

> Salv. You are wrong, Simplicio; what you ought to say is that everyone knows that it is called "gravity." What I am asking you for is not the name of the thing, but its essence, of which essence you know not a bit more than you know about the essence of whatever moves the stars around. I except the name which has been attached to it and which has been made a familiar household word by the continual experience that we have of it daily. But we do not really understand what principle or what force it is that moves stones downward, any more than we understand what moves them upward after they leave the thrower's hand, or what moves the moon around. We have merely, as I said, assigned to the first the more specific and definite name "gravity," whereas to the second we assign the more general term "impressed force" (*virtù impressa*), and to the last-named we give "spirits" (*intelligenza),* either "assisting" (*assistente*) or "abiding" (*informante*) and as the cause of infinite other motions we give "Nature."[10]

From an epistemological and historical standpoint, this passage is important: Galileo compares our knowledge of gravity with that of the action that makes the planets move around the Sun. In both cases, we do not know the nature of these actions. Galileo seems to claim: (1) Copernicus and Kepler explained the effects of the force that causes the movements of the planets, but they did not solve the problem of its nature. (Here Galileo does not discuss the dynamical part of Kepler's theory. Didn't he know it or did he consider such a theory useless and incorrect? It is not possible to answer this question on the basis of the available documents); (2) I myself, Galileo, have explained how gravity acts, but I am not able to formulate any hypothesis on its origin. This approach is more similar to Newton's than to Kepler's and Descartes', who—though in completely different manners—tried to explain the origin of the cosmic force that moves the planets. Descartes, as we will see, also formulated a theory on the origin of gravity. Furthermore, it is interesting that Galileo compared the two forces even if he ignored the origin of both. Obviously, this does not mean that he had foreseen something as the universal gravitation, but it seems that he saw some analogies between the two forces. Newton's gravity theory might also be interpreted as an answer to Galileo. In an imaginary dialogue it is as if Newton said to Galileo: "I share your doubts on the origin of gravity and I feign no

---

[10] Galilei (1967, pp. 234–235). Cf. EN VII, p. 260 f.: "SALV. Io non ho detto che la Terra non abbia principio nè esterno nè interno al moto circolare, ma dico che non so qual de' dua ella si abbia; ed il mio non lo sapere non ha forza di levarglielo. Ma se questo autore sa da che principio sieno mossi in giro altri corpi mondani, che sicuramente si muovono, dico che quello che fa muover la Terra è una cosa simile a quella per la quale si muove Marte, Giove, e che e' crede che si muova anco la sfera stellata; e se egli mi assicurerà chi sia il movente di uno di questi mobili, io mi obbligo a sapergli dire chi fa muover la Terra. Ma più, io voglio far l'istesso s'ei mi sa insegnare chi muova le parti della Terra in giù. SIMP. La causa di quest'effetto è notissima, e ciaschedun sa che è la gravità. SALV. Voi errate, Sig. Simplicio; voi dovevi dire che ciaschedun sa ch'ella si chiama gravità. Ma io non vi domando del nome, ma dell'essenza della cosa: della quale essenza voi non sapete punto più di quello che voi sappiate dell'essenza del movente le stelle in giro, eccettuatone il nome, che a questa è stato posto e fatto familiare e domestico per la frequente esperienza che mille volte il giorno ne veggiamo; ma non è che realmente noi intendiamo più, che principio o che virtù sia quella che muove la pietra in giù, di quel che noi sappiamo chi la muova in su, separata dal proiciente, o chi muova la Luna in giro, eccettochè (come ho detto) il nome, che più singulare e proprio gli abbiamo assegnato di *gravità*, dovechè a quello con termine più generico assegnamo *virtù impressa*, a quello diamo *intelligenza*, o *assistente*, o *informante*, ed a infiniti altri moti diamo loro per cagione la *natura*". (Italics in the original).

hypothesis, but I show you that, since the force which makes the planets move around the Sun and the satellites around the planets acts exactly as gravity and since in nature the same effects are produced by the same causes, then such a force is gravity. In this way, I have reduced your two doubts on the origin of these two actions to a single doubt and, furthermore, I have continued and joined yours and Kepler's efforts".

To complete the picture of Galileo's cosmological view, we will now present the main features of his concept of inertia.

4.1.3 Galileo's Concept of Inertia

The issue of how Galileo conceived inertia is one of the most important and debated ones in the history of science. It is a prominent issue if we want to understand how Galileo regarded the universe and its laws. For this reason, we will first expound our interpretation in the light of Galileo's passages and, afterwards, we will comment on the views proposed by other historians of science.

Galileo's view on the inertial motions. Granted that we consider as "inertial" a motion that continues without the intervention of any external action, we can summarize Galileo's views as shown in the following table:

Kind of motion	Inertial/not inertial
Uniform circular motion of the planets around the Sun and of the satellites around the planets[11]	Inertial
Uniform circular rotation of a celestial object around its axis	Inertial
Uniform circular movement of an object (i.e. a ship) which moves on an ocean, a see or a lake, after receiving an initial impulse	Inertial
Local circular uniform movement, for example that of a sling	Not inertial
Uniform rectilinear motion	Inertial
Accelerated rectilinear and circular motions	Not inertial

Inertia, cosmology and the Platonic myth. Almost at the beginning of the First Day of the *Dialogo*, Salviati, in agreement with Simplicio and Sagredo, claims that the order of the world is perfect: all the positions and movements of celestial bodies are as perfect as possible. Since rectilinear motion is infinite in its nature—because an object moves continuously away from the place where it is—it cannot be the natural motion of celestial bodies:

> [...] if all integral bodies in the world are by nature movable, it is impossible that their motions should be straight, or anything else but circular; and the reason is very plain and

[11] Galileo did not believe in the ellipticity of the orbits, but was perfectly aware that the motion of the planets around the Sun is not regular. This was well known since ancient times. He, therefore, referred to the planets' mean motion.

obvious. For whatever moves straight changes place and, continuing to move, goes ever farther from its starting point and from every place through which it successively passes. If that were the motion which naturally suited it, then at the beginning it was not in its proper place. So then the parts of the world were not disposed in perfect order. But we are assuming them to be perfectly in order; and in that case, it is impossible that it should be their nature to change place, and consequently to move in a straight line. Besides, straight motion being by nature infinite (because a straight line is infinite and indeterminate), it is impossible that anything should have by nature the principle of moving in a straight line; or, in other words, toward a place where it is impossible to arrive, there being no finite end. For nature, as Aristotle well says himself, never undertakes to do that which cannot be done, nor endeavors to move whither it is impossible to arrive.[12]

Galileo, then, further clarifies the cosmological role of rectilinear motion:

But someone might say nevertheless that although a straight line (and consequently the motion along it) can be extended *in infinitum* (that is to say, is unending), still nature has, so to speak, arbitrarily assigned to it some terminus, and has given her natural bodies natural instincts to move toward that. And I shall reply that this might perhaps be fabled to have occurred in primordial chaos, where vague substances wandered confusedly in disorder, to regulate which nature would very properly have used straight motions. By means of these, just as well-arranged bodies would become disordered in moving, so those which were previously badly disposed might be arranged in order. But after their optimum distribution and arrangement it is impossible that there should remain in them natural inclinations to move any more in straight motions, from which nothing would now follow but their removal from their proper and natural places; which is to say, their disordering.

We may therefore say that straight motion serves to transport materials for the construction of a work; but this, once constructed, is to rest immovable—or, if movable, is to move only circularly.[13]

---

[12] Galilei (1967, p. 19). Cf. EN VII, p. 43: "[...] se i corpi integrali del mondo devono esser di lor natura mobili, è impossibile che i movimenti loro siano retti, o altri che circolari: e la ragione è assai facile e manifesta. Imperocchè quello che si muove di moto retto, muta luogo; e continuando di muoversi, si va più e più sempre allontanando dal termine ond'ei si partì e da tutti i luoghi per i quali successivamente ei va passando; e se tal moto naturalmente se gli conviene, adunque egli da principio non era nel luogo suo naturale, e però non erano le parti del mondo con ordine perfetto disposte: ma noi supponghiamo, quelle esser perfettamente ordinate: adunque, come tali, è impossibile che abbiano da natura di mutar luogo, ed in conseguenza di muoversi di moto retto. In oltre, essendo il moto retto di sua natura infinito, perché infinita e indeterminata è la linea retta, è impossibile che mobile alcuno abbia da natura principio di muoversi per linea retta, cioè verso dove è impossibile di arrivare, non vi essendo termine prefinito; e la natura, come ben dice Aristotile medesmo, non intraprende a fare quello che non può esser fatto, né intraprende a muovere dove è impossibile a pervenire".

[13] Galilei (1967, pp. 19–20). Cf. EN, VII, pp. 43–44: "E se pur alcuno dicesse, che se bene la linea retta, ed in conseguenza il moto per essa, è produttibile in infinito, cioè interminato, tuttavia però la natura, per così dire, arbitrariamente gli ha assegnati alcuni termini e dato naturali instinti a' suoi corpi naturali di muoversi a quelli, io risponderò che ciò per avventura si potrebbe favoleggiare che fusse avvenuto del primo caos, dove confusamente ed inordinatamente andavano indistinte materie vagando, per le quali ordinare la natura molto acconciamente si fusse servita de i movimenti retti, i quali, sì come movendo i corpi ben costituiti gli disordinano, così sono acconci a ben ordinare i pravamente disposti; ma dopo l'ottima distribuzione e collocazione è impossibile che in loro resti naturale inclinazione di più muoversi di moto retto, dal quale ora solo ne seguirebbe il rimuoversi dal proprio e natural luogo, cioè il disordinarsi. Possiamo dunque dire, il moto retto servire a condur le materie per fabbricar l'opera, ma fabbricata ch'ell'è, o restare immobile, o, se mobile, muoversi solo circolarmente.".

Rectilinear motion is not the proper condition of bodies when they are set in their natural order. It has, instead, a modifying character: by means of it a transition is possible both from order to disorder,[14] and from disorder to order. This is the case when celestial bodies, after their creation, were moved from the place in which they were created in the primitive chaos and posed in their natural position. As soon as they reached it, their uniform circular motion began.

In the quoted passages it is remarkable that Galileo has described rectilinear motion as the kind of motion able to modify a physical state, without specifying its nature: is it an accelerated or a uniform motion? This initially missing reference is not casual. Here Galileo is distinguishing the properties of rectilinear motion from those of circular motion, regardless of whether the rectilinear motion is accelerated or uniform. However, as we will see, the distinction between uniform and accelerated rectilinear motion comes to the fore as soon as Galileo specifies how celestial bodies reached their natural place and began to rotate uniformly.

Here Galileo claims that one might follow Plato's myth[15]:

> [...] these world bodies, after their creation and the establishment of the whole, were for a certain time set in straight motion by their Maker. Then later, reaching certain definite places, they were set in rotation one by one, passing from straight to circular motion, and have ever since been preserved and maintained in this. [...] Every body constituted in a state of rest but naturally capable of motion will move when set at liberty only if it has a natural tendency toward some particular place; for if it were indifferent to all places it would remain at rest, having no more cause to move one way than another. Having such a tendency, it naturally follows that in its motion it will be continually accelerating. Beginning with the slowest motion, it will never acquire any degree of speed (*velocità*) without first having passed through all the gradations of lesser speed—or should I say of greater slowness? For, leaving a state of rest, which is the infinite degree of slowness, there is no way whatever for it to enter into a definite degree of speed before having entered into a lesser, and another still less before that. It seems much more reasonable for it to pass first through those degrees nearest to that from which it set out, and from this to those farther on. But the degree from which the movable body began to move was that of most extreme slowness; that is to say, from rest.[16]

[14] Galileo offers no example of how rectilinear motion may serve to create disorder from order. We can interpret Galileo's thought by presenting a specific case: suppose that a celestial body moves in its uniform circular orbit; if it gets hit by another body, it might happen that it leaves its orbit and begins to move rectilinearly. This is an example of how disorder is produced starting from a previously ordered condition.

[15] Most scholars devoted much attention to the conceptual aspects of the Platonic myth, but little concern for ascertaining Galileo's actual reference to Plato's writings. Acerbi (2000) argues that Galileo's reference to Plato was not a generic one and that he referred to *Timaeus* 38e3-6 and 38c7-8.

[16] Galilei (1967, pp. 20–21). Cf. EN VII, p 44: "[...] i corpi mondani, dopo l'essere stati fabbricati e del tutto stabiliti, furon per alcun tempo dal suo Fattore mossi di moto retto, ma che dopo l'esser pervenuti in certi e determinati luoghi, furon rivolti a uno a uno in giro, passando dal moto retto al circolare, dove poi si son mantenuti e tuttavia si conservano. [...] Ogni corpo costituito per qualsivoglia causa in istato di quiete, ma che per sua natura sia mobile, posto in libertà si moverà, tutta volta però ch'egli abbia da natura inclinazione a qualche luogo particolare; chè quando e' fusse indifferente a tutti, resterebbe nella sua quiete, non avendo maggior ragione di muoversi a questo che a quello. Dall'aver questa inclinazione ne nasce necessariamente che egli nel suo moto si anderà continuamente accelerando; e cominciando con moto tardissimo, non acquisterà grado

God[17] used the rectilinear motion only initially, to carry the bodies towards their final natural places and to impart them their speeds. After this initial phase, He converted the rectilinear motion into a uniform circular one, which is maintained without further interventions. The rectilinear motion used by God in this operation is uniformly accelerated. After they were created, bodies were at rest in a sort of absolute space. God imparted them a rectilinear motion passing through all degrees of speed or slowness ("gradi di velocità o tardità") until each celestial object reaches its own place at a certain speed. This motion is accelerated because a body cannot reach instantaneously the speed v with which it will move circularly. It is impossible that each body acquires its speed v without passing through all the degrees of speed from 0 to v. When a planet reaches the final speed v, God converts the rectilinear accelerated motion into a circular uniform one whose speed is equal to that reached by the body in the last instant of its previous rectilinear motion. This means that the natural, inertial motion of celestial bodies is the circular uniform one.

At the end of his reasoning Galileo offers a synthesis of his whole argument:

> This assumed, let us suppose God to have created the planet Jupiter, for example, upon which He had determined to confer such-and-such a velocity, to be kept perpetually uniform forever after. We may say with Plato that at the beginning He gave it a straight and accelerated motion; and later, when it had arrived at that degree of velocity, converted its straight motion into circular motion whose speed thereafter was naturally uniform.[18]

It is clear that for Galileo the final movement of the planets, their circular uniform one, *is their real inertial motion.*[19]

The planets also rotate around their axes. In particular, Galileo addressed the diurnal rotation of the Earth in the Second Day of his *Dialogo*. In the final section of this Day, he claimed that, since both the annual and the diurnal rotations take place from West to East (counterclockwise), they were originated by the same principle.

---

alcuno di velocità, che prima e' non sia passato per tutti i gradi di velocità minori, o vogliamo dire di tardità maggiori: perchè, partendosi dallo stato della quiete (che è il grado di infinita tardità di moto), non ci è ragione nissuna per la quale e' debba entrare in un tal determinato grado di velocità, prima che entrare in un minore, ed in un altro ancor minore prima che in quello; anzi par molto ben ragionevole passar prima per i gradi più vicini a quello donde ei si parte, e da quelli a i più remoti: ma il grado di dove il mobile piglia a muoversi è quello della somma tardità, cioè della quiete".

[17] The Italian term used by Galileo in this circumstance is "Fattore", which, literally means Maker (i.e. Creator).

[18] Galileo (1967, p. 21). Cf. EN VII, pp. 44–45: "Possiamo dunque ragionevolmente dire che la natura, per conferire in un mobile, prima costituito in quiete, una determinata velocità, si serva del farlo muover, per alcun tempo e per qualche spazio, di moto retto. Stante questo discorso, figuriamoci aver Iddio creato il corpo, v. g., di Giove, al quale abbia determinato di voler conferire una tal velocità, la quale egli poi debba conservar perpetuamente uniforme: potremo con Platone dire che gli desse di muoversi da principio di moto retto ed accelerato, e che poi, giunto a quel tal grado di velocità, convertisse il suo moto retto in circolare, del quale poi la velocità naturalmente convien esser uniforme".

[19] In the Fourth Day of the *Discorsi e dimostrazoni intorno a due nuove scienze* Galileo re-proposed a synthetic version of the Platonic myth as an annotation to the Theorem III, Proposition III. The essence of the myth is the same as in the *Dialogo*.

Hence, the reason that determines the eternity and inertiality of the annual rotation is the same that determines the diurnal rotation (EN VII, p. 284).

Free movement on a portion of the Earth's surface so extended that cannot be approximated by a plane. Let us suppose that an impulse be given to a ship on an ocean or a see. How should the motion of the ship continue if the surface of the water were perfectly smooth and no continent or island were interposed? Galileo's answer is that such motion would be circular around a great circle of the Earth; it would be eternal and the crew of the ship would feel no tendency to recede along the tangent. *The motion would be horizontal and inertial.*

In *Istoria e dimostrazioni intorno alle macchie solari* Galileo wrote:

> Finally, they [the heavy bodies] are indifferent to some movements—as are the same heavy bodies to horizontal motion—to which they have neither inclination, because it is not toward the center of the Earth, nor aversion, because it is not away from the same center. And therefore, with all the external impediments removed, a heavy body on the spherical surface concentric to the Earth will be indifferent to rest and to movement toward any part of the horizon, and it will remain in the state in which it has been put; that is, if it has been put in a state of rest it will remain in it, and if it has been put in motion, toward the west, for example, it will maintain the same state. And thus a ship, for example, having received one single time some impetus, would move continuously through a quiet sea around our globe without ever stopping, and if one were to bring it gently to rest it would perpetually remain at rest, provided that in the first case all the extrinsic impediments could be removed, and in the second one that no external mobile cause came upon it.[20]

The same train of thoughts is expressed by Galileo in the Second Day of the *Dialogo*, where we read:

> Salv. Then in order for a surface to be neither downward nor upward, all its parts must be equally distant from the center. Are there any such surfaces in the world?
>
> Simp. Plenty of them; such would be the surface of our terrestrial globe if it were smooth, and not rough and mountainous as it is. But there is that of the water, when it is placid and tranquil.
>
> Salv. Then a ship, when it moves over a calm sea, is one of these movables which courses over a surface that is tilted neither up nor down, and if all external and accidental obstacles were removed, it would thus be disposed to move incessantly and uniformly from an impulse once received?
>
> Simp. It seems that it ought to be.[21]

---

[20] Galilei and Scheiner (2010, p. 125). Cf. EN V, pp. 134–135: "Finalmente, ad alcuni movimenti [i corpi pesanti] si trovano indifferenti, come pur gl'istessi gravi al movimento orizontale, al quale non hanno inclinazione, poi che ei non è verso il centro della Terra, nè repugnanza, non si allontanando dal medesimo centro: e però, rimossi tutti gl'impedimenti esterni, un grave nella superficie sferica e concentrica alla Terra sarà indifferente alla quiete e a i movimenti verso qualunque parte dell'orizonte, ed in quello stato si conserverà nel qual una volta sarà stato posto; cioè se sarà messo in stato di quiete, quello conserverà, e se sarà posto in movimento, v.g. verso occidente, nell'istesso si manterrà: e così una nave, per essempio, avendo una sol volta ricevuto qualche impeto per il mar tranquillo, si moverebbe continuamente intorno al nostro globo senza cessar mai, e postavi con quiete, perpetuamente quieterebbe, se nel primo caso si potessero rimuovere tutti gl'impedimenti estrinseci, e nel secondo qualche causa motrice esterna non gli sopraggiugnesse".

[21] Galilei (1967, p. 148). Cf. EN VII, p. 174: "SALV. Adunque una superficie che dovesse esser non declive e non acclive, bisognerebbe che in tutte le sue parti fusse egualmente distante dal centro.

This passage confirms that this kind of motion is considered inertial by Galileo.

Local circular uniform movement. So far we have considered Galileo's conception of circular motion in a cosmological context and in the case of a uniform circular motion that takes place in a vast portion of the Earth's surface. Both motions are inertial, according to Galileo. The situation is completely different with regard to the local motion. In this case Galileo recognised that a continuous action is necessary for the circular motion to be maintained. Otherwise, the tendency to recede along the tangent would carry the body along a rectilinear path with a uniform motion. Galilei clearly stated in the Second Day of the *Dialogo*:

> Circular motion has the property of casting off, scattering, and driving away from its center the parts of the moving body, whenever the motion is not sufficiently slow or the parts not too solidly attached together.[22]

The recognition that the circular motion has the property of casting off the parts of the moving body is a fundamental step towards the understanding that *the uniform circular motion is not inertial*, whereas the rectilinear uniform is indeed inertial.

In other passages of the *Dialogo*, Galileo is explicit on the tendency of the object animated by a circular motion to recede along the tangent. In the following passage of the Second Day, we read:

> Up to this point you knew all by yourself that the circular motion of the projector impresses an impetus upon the projectile to move, when they separate, along the straight line tangent to the circle of motion at the point of separation, and that continuing with this motion, it travels ever farther from the thrower. And you have said that the projectile would continue to move along that line if it were not inclined downward by its own weight from which fact the line of motion derives its curvature.[23]

This is a perfect description of what happens. The inertia principle is not explicitly formulated, but it is applied correctly, which is even more important. Circular motion is not considered inertial, while uniform rectilinear motion is. If there were no impediments, motion along the tangent would continue ad infinitum. These are the basic

---

Ma di tali superficie ve n'è egli alcuna al mondo? SIMP. Non ve ne mancano: ècci quella del nostro globo terrestre, se però ella fusse ben pulita, e non, quale ella è, scabrosa e montuosa; ma vi è quella dell'acqua, mentre è placida e tranquilla.

SALV. Adunque una nave che vadia movendosi per la bonaccia del mare, è un di quei mobili che scorrono per una di quelle superficie che non sono nè declivi nè acclivi, e però disposta, quando le fusser rimossi tutti gli ostacoli accidentarii ed esterni, a muoversi, con l'impulso concepito una volta, incessabilmente e uniformemente SIMP. Par che deva esser così"

[22] Galileo (1967, p. 132). Cf. EN VII, p. 158: "Il moto circolare ha facoltà di estrudere dissipare e scacciar dal suo centro le parti del corpo che si muove, qualunque volta o 'l moto non sia assai tardo o esse parti non sian molto saldamente attaccate insieme".

[23] Galileo (1967, pp. 193–194). Cf. EN VII, p. 220: "Sin qui avete per voi stesso saputo che il moto circolare del proiciente imprime nel proietto impeto di muoversi (quando avviene ch'e' si separino) per la retta tangente il cerchio del moto nel punto della separazione, e, continuando per essa il moto, vien sempre allontanandosi dal proiciente; ed avete detto che per tal linea retta continuerebbe il proietto di muoversi, quando dal proprio peso non gli fusse aggiunta inclinazione all'in giù, dalla quale deriva l'incurvazione della linea del moto".

properties of inertial motions. Galileo examined also other features of rectilinear uniform motion.

Inertia and rectilinear uniform motion. Galileo stated that a rotating body tends to recede along its instantaneous tangent and to continue its motion in that direction with the same speed it held in the last instant of its circular motion. This statement shows that Galileo had an intuition of rectilinear inertia. There are, moreover, passages in his works in which inertia is even more clearly expressed. At the beginning of the Fourth Day of the *Discorsi e dimostrazioni*, we read:

> Imagine any particle projected along a horizontal plane without friction; then we know, from what has been more fully explained in the preceding pages, that this particle will move along this same plane with a motion which is uniform and perpetual, provided the plane has no limits. But if the plane is limited and elevated, then the moving particle, which we imagine to be a heavy one, will on passing over the edge of the plane acquire, in addition to its previous uniform and perpetual motion, a downward propensity due to its own weight; so that the resulting motion which I call projection [*projectio*], is compounded of one which is uniform and horizontal and of another which is vertical and naturally accelerated.[24]

Two different physical situations are described in this brief but complex passage: at the beginning Galileo describes uniform rectilinear motion independent of gravity, a motion that is envisioned in an abstract physical–mathematical space. In this case the motion proceeds potentially at infinity, if no external action intervenes. Galileo defines the plane "horizontal". From our point of view, this makes no sense because a plane conceived as belonging to an abstract space does not have a position that could be defined "horizontal" unless a reference frame is established. Nonetheless, Galileo's meaning is clear: he is guessing that in an abstract space gravity is not an ontological property of bodies. The use of the term "horizontal" indicates that Galileo is not considering anymore the behaviour of bodies on Earth, but is not yet fully capable of conceiving the principles of physics as referred to an abstract physical–mathematical space. Despite these limitations, the inertia principle is clearly formulated in this sentence: "this particle will move along this same plane with a motion which is uniform and perpetual, provided the plane has no limits".

The description of the second physical situation begins with these words: "But if the plane is limited and elevated…". Here Galileo does not describe an ideal condition, rather the concrete physics on the Earth, for he adds that now the mobile is thought of as endowed with gravity. It is evident that in the previous sentences he had considered it as devoid of gravity.

Galileo, thus, guessed the main features of rectilinear inertia, though he did not develop a system of physical causality based on the inertia law. This giant step was

---

[24] Galileo (1914, p. 244). Cf. EN VIII, p. 268: "Mobile quoddam super planum horizontale proiectum mente concipio, omni secluso impedimento: iam constat, ex his quae fusius alibi dicta sunt, illius motum aequabilem et perpetuum super ipso plano futurum esse, si planum in infinitum extendatur; si vero terminatum et in sublimi positum intellegamus, mobile, quod gravitate praeditum concipio, ad plani terminum delatum, ulterius progrediens, aequabili atque indelebili priori lationi superaddet illam quam a propria gravitate habet deorsum propensionem, indeque motus quidam emerget compositum ex aequabili horizontali et ex deorsum naturaliter accelerato, quem projectionem voco".

taken only by Newton. It should also be added that Galileo considered the inertiality of rectilinear uniform motion also in cases in which motion is not horizontal, as we can see in the following passage drawn from *Della forza della percossa*. Galileo here imagines two heavy bodies in equilibrium on a pulley. If an impulsive force downwards is given to one of the two bodies, it will move *forever* rectilinearly and uniformly, if no impediment intervenes. We quote the entire passage as it is particularly significant to clarify Galileo's views on inertia:

> At all appearances, here it happens exactly what happens to a heavy and perfectly round mobile that, when located on a very clean and inclined plane, will descend naturally, always acquiring an increasing speed. In contrast to this, if from below you want to move that body upwards, it will be necessary to impart it an impetus, which progressively diminishes and finally is annihilated. However, if the plane is not inclined, but horizontal, the round body on it will do what you want. If it is posed on the plane at rest, it will remain at rest. If we give it an impetus in a direction, it will move in that direction, always maintaining the same speed received by our hand, for there is no action that might increase or decrease such a speed, as the plane is neither downward nor upward sloping. Analogously, if two bodies of the same weight are tied to the ends of a rope, they will remain in equilibrium. If an impetus downwards is imparted to one of them, the movement of that body will maintain constantly uniform. This would follow if all the external and accidental impediments were removed [...].[25]

Commentary. These passages raise a serious interpretative problem as to the uniform circular motion in Galileo. In the cosmological context—the motion of planets around the Sun and of planets around their axes—and in reference to a motion taking place on vast portions of the Earth's surface, Galileo considered the uniform circular motion to be inertial. On the other hand, he was aware of the tendency to recede along the tangent in the local circular motion. Can these views be reconciled? Obviously, from the point of view of Newtonian physics they cannot. However, Galileo did not perceive the problem of reconciling two different views, because his ideas on circular motion were not clear.[26] He did not study in depth this subject as, in contrast, he did for the free fall of heavy bodies or for the motion on inclined planes. Consequently, he did not achieve either a qualitative or a quantitative unitary view of the circular

---

[25] Our translation. Cf. EN VIII, pp. 336–337: "E qui mi pare che accada per appunto quello che accade ad un mobile grave e perfettamente rotondo, il quale, se si porrà sopra un piano pulitissimo ed alquanto inclinato, da per sé stesso naturalmente vi scenderà, acquistando sempre velocità maggiore: ma se, per l'opposito, dalla parte bassa si vorrà quello cacciare in su, ci bisognerà conferirgli impeto, il quale si anderà sempre diminuendo e finalmente annichilando; ma se il piano non sarà inclinato, ma orizontale, tal solido rotondo, postovi sopra, farà quello che piacerà a noi, cioè, se ve lo metteremo in quiete, in quiete si conserverà, e dandogli impeto verso qualche parte, verso quella si moverà, conservando sempre l'istessa velocità che dalla nostra mano averà ricevuta, non avendo azione nè di accrescerla nè di scemarla, non essendo in tal piano nè declività nè acclività: ed in simile guisa i due pesi eguali, pendenti da' due capi della corda, ponendogliene in bilancio, si quieteranno, e se ad uno si darà impeto all'in giù, quello si andrà conservando equabile sempre. E qui si dee avvertire che tutte queste cose seguirebbero quando si movessero tutti gli esterni ed accidentari impedimenti, dico di asprezza e gravità di corda, di girelle e di stropicciamenti nel volgersi intorno al suo asse, ed altri che ve ne potessero essere". For the context of this passage, we refer to Vergara Caffarelli (2007).

[26] See, e.g., Bussotti (2017, pp. 89–96).

motion. Rather, his ideas on this subject were fragmentary. He did not consider circular motion as a single unitary phenomenon. Thus, when observation or reasoning drove him to think that the uniform circular motion should be conserved without the intervention of external forces, he adhered to this view. This was the case for the cosmic uniform circular motion and for the motion of a ship on a vast ocean or sea. For these motions Galileo never spoke of the tendency to recede along the tangent, though he did with regard to local uniform circular motion. The kind of coherence we demand of physics as a full-fledged science cannot be expected from physics in its nascent state. Therefore, according to Galileo uniform circular motion is inertial in some circumstances, while is non-inertial in other circumstances. However, this view had no influence when Galileo developed concretely his mechanics in *Discorsi e dimostrazioni intorno a due nuove scienze*. In that case he dealt with local motions and, for such motions, his ideas were rather clear, at least insofar as to the problems he examined.

As for the rectilinear inertia, we think that Galileo had quite a precise concept of it, though he did not develop the consequences of such concept and though, as we will see, he made some mistakes.

Galileo's concept of inertia in secondary literature. Until the studies of Alexandre Koyré it was generally accepted that Galileo was the discoverer of the inertia principle.[27] In his *Science of mechanics*, Ernst Mach paid close attention to the theme of inertia in Galileo. He pointed out how Galileo developed his idea of rectilinear inertia, stressing that, though Galileo had the merit of discovering the inertia principle, he did not use it extensively in his physics.[28]

Koyré advanced a completely different interpretation: Galileo did not believe in rectilinear inertia. He conceived only inertia of uniform circular motion. Koyré's reading is linked to his general theses on the scientific revolution which he illustrated in his celebrated work *From the closed world to the infinite universe* (Koyré, 1957):

(1) The scientific revolution was possible because the idea of an ordered and finite cosmos was replaced by that of the infinite universe;
(2) Mathematics was systematically applied to physics;
(3) Physicists believed that the laws they were discovering were inherent to the universe;
(4) Technological and economical factors had scarce influence on the development of theoretical science;
(5) The experimental aspect was far less important than normally thought of in the birth of modern science.

Points (2)–(5) represent the Platonic approach of the scientists who lived in the sixteenth-seventeenth centuries.

As to Galileo's views on inertia, Koyré thought that he believed only in circular inertia. Galileo was impressed by the natural persistency of circular motion,

---

[27] In particular, we refer to Koyré (1939).

[28] Mach (1919, pp. 140–141).

confirmed by the daily observation of the planets which rotate around the Sun in agreement with his Copernican conviction (Koyré, 1939, p. 206). According to Koyré, the Platonic myth is the most significant text in Galileo's works that allows us to understand his concept of inertia.

Koyré also claims that, while referring to the motion along a horizontal plane on the Earth's surface, Galileo was not thinking of a geometrical plane, but of a spherical surface. Inertial motion on a vast part of the Earth's surface is, therefore, circular, not rectilinear (ibid., pp. 207–208). Galileo did not reach the concept of rectilinear inertia and remained anchored to circular inertia because, unlike Descartes, he did not accept the consequences of a complete geometrization of the universe; this means that he did not accept the infinity of the universe and the destruction of the ordered cosmos (ibid., p. 211). Koyré also adds that Galileo's conviction that every body is intrinsically heavy prevented him from reaching the concept of rectilinear inertia (ibid., pp. 239–276; in particular p. 258). He points out that inertia principle is based on the distinction between mass and weight. The weight p is a force equal to mG, where m is the mass of a body and G is the gravity acceleration. Given the mass m, the weight depends on gravity acceleration, namely it varies, while inertia, which depends only on the mass of the body, is an invariable quantity. From a physical standpoint, Koyré's observation is correct: if he had meant that no physicist before Newton had a clear concept of inertia, his remark would have been correct. However, Koyré's position is disconcerting from an historical perspective: he held that Galileo did not have a clear concept of rectilinear inertia, while other physicists, such as Descartes, had. Koyré should have noted, though, that in Descartes the distinction between mass and weight is not clearer than in Galileo. We agree with Burns' opinion that "the distinction between weight and mass emerged only at the end of the seventeenth century" (Burns, 2001, entry "Force", p. 105). Contrary to Koyré, Galileo seems to us to be the scientists who, in the period 1620–1640, had the clearest concept of rectilinear inertia, though with the limitation stressed by Koyré. No scientific notion appears all of a sudden in its perfect form, but it is developed along a genetic phase. This is particularly true for an extremely difficult concept as rectlinear inertia. During the gradual formation of this idea, Galileo was the physicist who had the most satisfying view, which was far from being perfect, but was, nonetheless, the best available.

While commenting on the numerous passages where Galileo seems to refer to rectilinear inertia, Koyré claims that he came close to conceive it, but refused to make the final step (ibid., p. 262).

Koyré's analyses of Galileo's concept of inertia are unsatisfactory from a conceptual and methodological standpoint. First of all, Koyré overestimated the importance that Galileo ascribed to circular inertia. As we have explained, we are persuaded that Galileo believed in circular inertia and that this concept is part of his cosmology. Notwithstanding this, circular inertia plays no role in the development of Galileo's arguments in the *Dialogo* and in the *Discorsi*. Therefore, though it is correct to claim that he believed in circular inertia, it is misleading to assume such conception as a milestone to interpret Galileo's physical view and work. The most misleading aspect of Koyré's interpretation is the thesis that Galileo could not reach the concept of rectilinear inertia because he did not believe in the infinity of the universe. First of all, if one analyses the Platonic myth, Galileo explicitly claims the potentially infinite character of rectilinear motion (be it indifferently uniform or accelerated motion, since he asserts the infinitary character of rectilinear motion before stating that such motion is accelerated). In this context, it is appropriate to stress a distinction that Koyré seems to have overlooked: the phrase "natural motion" is not exactly identifiable with what we call "inertial motion". The term "natural" means "appropriate" in a sense which is closer to the order of thinking of Aristotelian metaphysics than to modern physics. Uniform circular motion is natural for celestial bodies because it is appropriate for them in a metaphysical meaning. Furthermore, it is also inertial because, when it begins, it continues eternally. Rectilinear motion, be it uniform or accelerated, is not ontologically natural for celestial bodies, but this does not at all exclude that the uniform rectilinear motion is inertial in the physical meaning of this term.

Moreover, it is rather obvious that the infinity of the world has nothing to do with rectilinear inertia because inertia is a "limit-concept" in the sense that every physicist—and Galileo, too—is perfectly aware that no motion is properly inertial. At best, a motion is approximately inertial. In other terms: it can be considered inertial, granted some circumstances. The principle of inertia has a general theoretical validity, but, in fact, a motion can be inertial only locally, though the notion of "local" can be extremely extended. When Galileo claims that circular motion has the tendency to expel the rotating object along the instantaneous tangent, he is not interested in analysing what happens at infinity. Rather he is interested in stating that: (1) the sling's circular motion does not continue without an external action; (2) if the circular motion is interrupted, the rotating body moves rectilinearly and uniformly along its instantaneous tangent. What the body will do at infinity is irrelevant. The principle of inertia is a regulatory idea concerning local situations. According to Koyré's views, a physicist who nowadays believes that the universe is finite—and no one knows whether it is or not—should refuse to admit– for instance—that the potential of an electric field is null at infinity. But this phrase "at infinity" means "very far from the sources of the electric field"; it has nothing to do with the infinity of the universe. The same is true for inertia. On the other hand, Newton himself who believed in the infinity of the universe, did not connect inertia principle with the assumption of the infinity of the universe. In the bucket experiment, which we will analyse in detail, the concept of inertia plays a fundamental role, but there is no reference to the infinity of the universe.

Nonetheless, Koyré had the merit to stimulate further studies on inertia in Galileo, many of which supported or objected to his ideas. Dijksterhuis, for instance, shared in substance Koyré's view, but underlined that Galileo had the merit of guessing that in any motion there is a component that tends to maintain its kinetic state, whereas another component is produced by an external action. Furthermore, Dijksterhuis separates the problem of inertia from that of the participation of a body to the physical system to which it belongs: Galileo did not fully grasp the concept of rectilinear inertia, but—to paraphrase Dijksterhuis' expression—in "intuitively inertial conditions" a body participate to the motion of the system to which it belongs (Dijksterhuis, pp. 349–359).

Stillman Drake devoted much thought to the issue of inertia in Galileo: he claimed that Galileo was aware of the principle of rectilinear inertia, though restricted to horizontal planes. As for the Platonic myth, it was simply a concession to the Aristotelians, a sort of literary expedient to captivate the sympathies of the adversaries (Drake, 1964, p. 607). We argue that there is no reason either to overestimate or to underestimate the value of the Platonic myth. Since Galileo referred to it also in the *Discorsi*, we assume that he believed in its truth. In 1970 Drake re-examined the problem of inertia and claimed that the main question is not to decide whether Galileo believed either in the rectilinear or in the circular inertia; rather, it is to analyse his concept of neutral motion (a notion already pointed out by Coffa, 1968). This is a motion which does not begin naturally, but that, once begun, does not stop without an external action. Uniform rectilinear motion as well as the motion along a portion of Earth's surface are neutral. Drake stated that neutral motion is a fundamental step towards the idea of rectilinear inertia (Drake, 1970, pp. 248–253). However, Drake was aware that the question of inertia in Galileo is complicated and he adopted a sceptical position: we have to recognize that Galileo left the door open to both interpretations (i.e. that he conceived either rectilinear or circular inertia) and we have to avoid a final decision. Insofar as his concrete physics was concerned, Galileo applied correctly rectilinear inertia (ibid., p. 270). This sceptical position became extreme in Drake (1990) where Drake claimed that to ask whether Galileo reached a correct concept of inertia makes no sense because inertia is a dynamical concept whereas Galileo's is only kinematic. The question of inertia in Galileo's mechanics is not well posed (Drake, 1990). It seems to us that this is simply a way to circumvent a complex problem: it is true that Galileo's physics was essentially kinematic, but it makes perfectly sense to ask whether a motion can continue on its own without any external action.

Vergara Caffarelli has a clear purpose in his studies: (1) to prove, against Koyré, that the experimental aspect is fundamental in Galileo; (2) to demonstrate that Galileo clearly reached the concept of rectilinear inertia and refused that of circular inertia. Essentially, we agree with his analyses as to rectilinear inertia, while we disagree with his ideas on circular inertia. Generally speaking, it seems to us that Vergara Caffarelli offers a too modernizing view of Galileo. In particular, with regard to Galileo's idea that the motion on the Earth's surface is conserved, he ascribes to Galileo the discovery of the principle of conservation of the angular momentum (Vergara Caffarelli, 2007, p. 76). This is an excessive and unnecessary modernization of Galileo's thought. The idea of circular inertia is sufficient to explain Galileo's remarks on this topic. Boccaletti in his otherwise excellent book on Galileo also claims that the scientist did not believe in circular inertia (Boccaletti, 2016, Chap. 4).

As we have pointed out, our contention is that Galileo believed both in circular and rectilinear inertia, but that rectilinear inertia played a role in Galileo's physics, whereas circular inertia did not. However, one should avoid an unjustified modernization of Galileo's view. In a stimulating paper, Andrea Frova doubts that Galileo understood all the consequences of the principle of rectilinear inertia. Frova analyses the passage from the Second Day of the *Dialogo* where Galileo speaks of the "imberciatori d'uccelli" ("hunters of birds"). The problem is: where should the hunters point their harquebus to hit a flying bird? Galileo considers the correct answer, but rejects it and claims that the harquebus has to be pointed directly to the bird. Frova stresses that this solution is manifestly wrong because, if r indicates the length of the harquebus, R indicates the altitude of the bird, the ratio R/r is much bigger than 1, so that the tangential velocity V of the bird is far superior to the tangential velocity v of the projectile. Therefore, if one follows Galileo's rule, the projectile would not reach the bird and would not hit it. This is an evident proof that Galileo did not have a clear concept of the kinematical magnitudes connoting the uniform circular motion. Frova concludes:

> This is a simple case of application of the inertia principle and Galileo makes a trivial mistake as if he were a beginner! It's a good consolation for us common mortals and a warning to doubt even our strongest certainties.[29]

4.2 Galileo's Epistemological and Philosophical Ideas

Three themes characterize Galileo's philosophical contribution to the scientific revolution of the seventeenth century and thus to the new cosmology: (A) his view of the relation between mathematics and physics; (B) his distinction between primary and secondary properties of bodies; (C) his Biblical hermeneutics aimed at legitimizing Copernican astronomy.

A. *Mathematics and physics*

The significance and extent of mathematical knowledge are expounded in a famous passage at the end of the First Day of the *Dialogo sopra i due massimi sistemi del mondo*:

> [...] human understanding can be taken in two modes, the *intensive* or the *extensive*. *Extensively*, that is, with regard to the multitude of intelligibles, which are infinite, the human understanding is as nothing even if it understands a thousand propositions; for a thousand in relation to infinite is zero. But taking man's understanding *intensively*, in so far as this term denotes understanding some proposition perfectly, I say that the human intellect does understand some of them perfectly, and thus in these it has as much absolute certainty as Nature itself has. Of such are the mathematical sciences alone; that is, geometry and arithmetic, in which the Divine intellect indeed knows infinitely more propositions, since it knows all. But with regard to those few which the human intellect does understand, I believe that its knowledge equals the Divine in objective certainty, for here it succeeds in understanding necessity, beyond which there can be no greater sureness.[30]

---

[29] Our translation. Cf. Frova (2006, p. 44): "Si tratta di un semplice esercizio di applicazione del principio di inerzia e Galileo ci casca come un principiante! Una bella consolazione per noi comuni mortali, e un monito a dubitare sempre anche delle nostre certezze più salde".

[30] Galilei (1967, p. 115). Cf. EN VII, pp. 128–129: "[...] l'intendere si può pigliare in due modi, cioè *intensive*, o vero *extensive*: e che *extensive*, cioè quanto alla moltitudine degli intelligibili,

Galileo was aware of the audacity of his position, which equates human to divine knowledge: in the *Dialogue* Simplicio comments on this passage claiming that it is a "very bold and daring" speech, so that, to stave off any theological criticism, Salviati hastens to specify that God's mode of knowing is "more excellent than ours", because "our method proceeds with reasoning by steps from one conclusion to another while His [God's] is one of simple intuition" (ibid. p. 116). However, this qualification does not alter the main point which remains unequivocal: though both in terms of extent and mode of knowing, God's understanding of mathematics is infinitely superior to that of humans, human intellect is on a par with God's as far as the clarity of the known mathematical truths is concerned. Mathematical necessity is the ultimate truth and when human intellect reaches it, it grasps an absolute truth in a definitive manner. Thus, the strongest foundation for rational science is guaranteed.

This view would have remained without cosmological consequences had it not been linked to a second assumption: like Copernicus and Kepler, Galileo too is firmly convinced that mathematics offers the key to understanding the physical world, as he writes in another famous passage from *The Assayer* (1623):

> Philosophy is written in this grand book, the universe, which stands continually open to our gaze. But the book cannot be understood unless one first learns to comprehend the language and read the letters in which it is composed. It is written in the language of mathematics, and its characters are triangles, circles, and other geometric figures without which it is humanly impossible to understand a single word of it; without these, one wanders about in a dark labyrinth.[31]

This passage has often been read as evidence of Galileo's Platonism, but this interpretation is unnecessary.[32] Galileo does not here assert the subsistence in nature of plain geometrical structures; what he literally says is that in order to decipher the book of nature, we must learn the language of mathematics, because that is the only way to build physical knowledge. How our mathematics can be applied to nature is a question that cannot be settled a priori, but requires experiments and

---

che sono infiniti, l'intender umano è come nullo, quando bene egli intendesse mille proposizioni, perchè mille rispetto all'infinità è come un zero; ma pigliando l'intendere *intensive*, in quanto cotal termine importa intensivamente, cioè perfettamente, alcuna proposizione, dico che l'intelletto umano ne intende alcune così perfettamente, e ne ha così assoluta certezza, quanto se n'abbia l'intessa natura; e tali sono le scienze matematiche pure, cioè la geometria e l'aritmetica, delle quali l'intelletto divino ne sa bene infinite proposizioni di più, perchè le sa tutte, ma di quelle poche intese dall'intelletto umano credo che la cognizione agguagli la divina nella certezza obiettiva, poichè arriva a comprenderne la necessità, sopra la quale non par che possa esser sicurezza maggiore".

[31] Galileo (1957, p. 237f). Cf. EN VI, p. 232 "La filosofia è scritta in questo grandissimo libro che continuamente ci sta aperto innanzi agli occhi (io dico l'universo), ma non si può intendere se prima non s'impara a intender la lingua, e conoscer i caratteri ne' quali è scritto. Egli è scritto in lingua matematica, e i caratteri son triangoli, cerchi ed altre figure geometriche, senza i quali mezi è impossibile a intenderne umanamente parola; senza questi è un aggirarsi vanamente per un oscuro laberinto".

[32] The question of the link between Galileo's mathematization of physics and his supposed Platonism is widely debated. See: Burtt (1925), Bussotti (2017), Cassirer (1922) (third edition), vol I, pp. 377–418, Cassirer (1946), De Caro (1992, 2012), Dollo (2003), Feyerabend (1975), Girill (1970), Koyré (1939), Machamer (1998), Shapere (1984), Torrini (1993), Wallace (1974, 1984), Wisan (1978).

observations to be resolved. Galileo is Platonic if by this we mean that he rejects the traditional Aristotelian separation of physics from mathematics; overcoming this separation is in fact a fundamental step towards the rise of modern science. However, Galileo is certainly not an apriorist and he knows that in order fruitfully to apply mathematical calculations to material phenomena we need the ability to "deduct the material hindrances" (Galilei 1967, p. 207): this means that the mathematical basic order underlying phenomena must be extracted from the complex material conditions concealing it. To achieve this, observations and experiments are indispensable.

Let us analyse more specifically Galileo's ideas on the relation between physics and mathematics as well as Galileo's role in the development of experimental thought.

Galileo has been variously interpreted: the two most common viewpoints see him either as a Platonist or, on the contrary, as the inventor of the experimental method. One of the first and most important interpreters who regarded Galileo as a Platonist was Alexandre Koyré, who in his *Études Galiléennes* wrote: "In front of abstractive empiricism, Galileo vindicates the superior right of Platonic mathematism".[33] In his essay "Galileo and Plato" (Koyré 1943, p. 419n), Koyré adduces as a solid proof for his thesis the above-mentioned passage of *The Assayer* on the book of nature written in mathematical characters. Koyré also refers to some passages of the *Dialogo*'s Third Day, where Galilei claims that, whenever a mathematical demonstration is possible in natural philosophy, it must be preferred to any other kind of argument. Koyré states:

> Once more, we are so accustomed to mathematical science, to mathematical physics, that we no longer feel the strangeness of a mathematical approach to Being, the paradoxical daring of Galileo's utterance that the book of Nature is written in geometrical characters. For us it is a foregone conclusion. But not for Galileo's contemporaries (ibid., p. 419).

Koyré insists on the fact that the experimental aspect was secondary in Galileo's science, that the mathematical one was preponderant and that Galileo was always intent on investigating the ultimate structure of Being rather than on solving single physical problems. Mathematics is the language in which God created the laws of the universe. Koyré specifies that Galileo did not believe in the existence of an abstract world of mathematical entities subsisting independently of the laws of nature (Koyré, 1939, 1966, pp. 279–287):

> As a matter of fact, Galileo denies the premise of the discussion, which is common to the Platonists and the Aristotelians. He denies the "abstract" character of mathematical notions. He denies the ontological privilege of regular figures.[34]

It is true that Galileo often refers to Plato (and to the Pythagoreans), but in our view he did so in order to show that his view of the necessity of using mathematics in physical science was rooted in a noble and ancient philosophical tradition. Though

[33] Our translation. Cf. Koyré (1939, 1966, p. 78): "En face de l'empirisme abstractif, Galilée revendique le droit supérieur du mathématisme platonicien".

[34] Our translation. Cf. Koyré (1939, 1966, p. 282): "En effet, Galilée nie la prémisse, commune aux platoniciens et aux aristotéliciens, de la discussion. Il nie le caractère «abstrait» des notions mathématiques; et nie le privilège ontologique des figures régulières".

Galileo refused the *auctoritatis* principle, the fact that he shared his ideas with great authors of the past might have been judged favourably by his readers. As Koyré himself points out, by referring to Plato Galileo intended to emphasize the need to apply mathematics to physical science, in contraposition to the physics of Aristotle and of the Aristotelians which lacked any mathematical structure.

As to Galileo's approach to mathematics, Koyré himself recognizes that Galileo did not believe in the existence of an ideal mathematical world separate from physics; in this respect, he was simply not a Platonist. Secondly, though he was convinced that mathematics reveals the deep structure of nature, in some cases he used it in a purely instrumental way. He was not obsessed by the idea of finding the ultimate causes of phenomena. For example, with regard to the causes of accelerated motions, we read in the Third Day of *Discorsi e dimostrazioni*:

> Salv. The present does not seem to be the proper time to investigate the cause of the acceleration of natural motion concerning which various opinions have been expressed by various philosophers, some explaining the acceleration as a getting closer to the center, others as the diminishing of parts of the medium through which the body is falling, while still others attribute it to a certain push of the surrounding medium that closing itself behind the falling body presses on it and drives it from one of its positions to another. Now, all these fantasies, and others too, ought to be examined; but it is not really worth while.[35]

In other words: let us know how accelerated motions work, and what their physical laws are, but as for the causes of such laws, it is pointless to formulate metaphysical hypotheses.

To consider Galileo as a Platonist physicist, who regarded experiments only as a secondary aspect of his work, is a one-sided view. At the same time, it would be equally mistaken to say that he was an empiricist who founded science on experience and experiments from which he deduced the mathematical laws of physics. These two aspects (mathematics and experience) coexist in his intellectual activity: sometimes experiments led Galileo to think of mathematical laws, while at other times hypotheses of mathematical laws induced him to construct experiments to verify or falsify them.

An example of this variable interaction concerns the law of uniformly accelerated motion. The first proof of Galileo's famous theorem—in uniformly accelerated motion spaces are proportional to the squares of times—was elaborated in a series of manuscripts of 1604, though the final correct proof dates to 1609. (This proof was published many years later in *Discorsi e dimostrazioni* [1638].) The demonstration of 1604 derives, in part, from preceding research which was outlined in an unfinished dialogue of 1586–1587. Before arriving at his 1604 version, Galileo wrote a work on motion dating at 1590, where the themes addressed in the unfinished dialogue are

[35] Galileo (1914, p. 166), modified translation. Cf. EN VIII, p. 202: "Non mi par tempo opportuno d'entrare al presente nell'investigazione della causa dell'accelerazione del moto naturale, intorno alla quale da varii filosofi varie sentenzie sono state prodotte, riducendola alcuni all'avvicinamento al centro, altri a restar successivamente manco parti del mezo da fendersi, altri a certa estrusione del mezo ambiente, il quale, nel ricongiugnersi a tergo del mobile, lo va premendo e continuamente scacciando; le quali fantasie, con altre appresso, converrebbe andare esaminando e con poco guadagno risolvendo".

set out more clearly.[36] The general orientation of this research is not entirely clear, and Galileo changes it every time he reaches a new result. In these early writings, his most important findings concern the fall of a body through two different media. Galileo came to the same conclusions as Benedetti.[37] The main issue is to understand why some bodies fall more quickly in air than in water and why other bodies fall in air, while they remain afloat in water. In the 1590 text, Galileo asserted some propositions concerning the fall of bodies on inclined planes without friction:

> [...] the same heavy body will descend with greater force vertically than on an inclined plane, the longer the descent along the inclined plane with respect to the vertical.[38]

Galileo claims that whereas experiments with almost completely smooth inclined planes offer results that are coherent with this law, if the inclined planes are not frictionless, the data do not fit the law, but this simply means that the experiment has not been performed with appropriate tools. In Chapter 14 this interesting proposition is established:

---

[36] The text of 1604 in which Galileo formulated the law of uniformly accelerated motion, as well as all the other manuscripts on this subject were not published during Galileo's lifetime. The 1604 text was published under the title "De motu accelerato" in EN II, pp. 257–266. The unfinished dialogue and the work on motion dated around 1590 were published in EN I, pp. 243–417 under the title "De motu". In *Galileo at Work. His scientific biography*, Drake argues that the chronological order of the unfinished dialogue and of the work on motion in Galileo's National Edition is not correct (see Drake, 1978, Chapter 2, "1589–1592", pp. 18–32). On these two writings by Galileo, also known as "De motu antiquiora", see Camerota (1992), Camerota and Helbing (2000). For our purpose the chronological order of the two works of the period 1586–1590 is not significant.

[37] Giambattista Benedetti (1530–1590) in his works *Resolutio omnium Euclidis problematum aliorumque* (1553) and *Demonstratio proportionum motuum localium contra Aristotelem et omnes philosophos* (that appeared in two different editions in 1554, both of which are published in Maccagni, 1967) applied Archimedes' hydrostatic principle to the study of vertical motion in different media. He concluded that the fall speed depends on the difference between the specific weights of the falling body and of the medium through which the body falls. If two bodies are of the same material and shape (the shape is important only for the friction problem), they will fall with the same speed in the same media, provided that their specific weight is greater than that of the media. Thus, in a vacuum, all the bodies will fall along the vertical with the same speed because the hydrostatic thrust of the void is null (Benedetti in Maccagni, 1967, pp. 41–42). Furthermore, Benedetti proved that—supposing water's specific weight to be double of that of air (which is false, as Benedetti knew, but this false assumption did not invalidate his reasoning)—a body whose specific weight is double that of water's and the quadruple of air's will fall in the air with triple speed with respect to its fall speed in water, and not with double speed as Aristotelian physics predicted (ibid., p. 69). A few years after the publication of the *Demonstratio*, Simon Stevin described an experiment which proved that heavy bodies falling in a vacuum from the same altitude hit the ground at the same time (see Maccagni, 1967, Introduction, pp. XXIX–XXXI). Benedetti was also close to understanding the decomposition of motions into two components: the vertical, along which motion is uniformly accelerated, and the horizontal, which has a uniform speed. However, Benedetti interpreted his new results within the traditional *impetus* theory, as is clear from his later work *Diversarum speculationum mathematicarum et physicarum liber* (1585). On Benedetti's science see Maccagni (1967), Bordiga (1925–26), Drake (1976), Field (1985), Maccagni (1985), Roero (1997), Vailati (1897–1898). There is no doubt that Galileo knew Benedetti's works.

[38] Our translation. Cf. EN I, p. 298: "[...] et, consequenter, tanto maiori vi descendere idem grave per rectum descensum quam per inclinatum, quanto maior est inclinatus descensus quam rectus".

> [...] a mobile under the action of no external force can be moved on a plane which is not inclined either upwards or downwards by a force less than any given force.[39]

This is a particularly significant statement as it shows that already at the beginning of the seventeenth century Galileo had an incipient conception of inertia.

In 1602 Galileo discovered the fundamental theorem of isochronism, according to which a heavy body falling along the chords of a circle towards the lowest point of the circle, traverses all the chords in the same interval of time.[40] In dealing with this theorem, Galileo did not make any reference to the concept of acceleration. This means that in 1602 he did not yet know the law of uniform accelerated motion, from which the isochronism theorem is easily deducible as Galileo showed in *Discorsi e dimostrazioni* (EN VIII, pp. 221–226). The first important relation proved by Galileo with regard to uniformly accelerated motion is the so called "double speed rule",[41] according to which if a body *m* moves with uniformly accelerated motion in an interval of time *t* along a rectilinear path *ab*, then the time necessary for another body *m'* moving rectilinearly and uniformly, with a speed equal to that reached by *m* in *b*, to traverse the segment *ab* is equal to $t/2$. The "double speed rule" was proved around 1603 and finally published in *Discorsi e dimostrazioni*.

Together with this research Galileo continued to work with inclined planes and realised that, to determine a general law connecting space, time, velocity and acceleration, he needed experimental tools that enabled him to obtain precise measurements. Remarkably, Galileo had supposed, as an initial hypothesis, that in uniformly accelerated motion spaces were proportional to velocities. In late 1604 he considered this hypothesis to be valid, because in a letter to Paolo Sarpi dated 16 October 1604 we read:

> Thinking again about the problem of motion where, to prove the accidents I observed, I lacked an indubitable principle to be posited as an axiom, I came to a proposition which appears to be natural and evident. Granted this proposition, I prove the rest, namely that the spaces traversed with a natural motion are in double proportion to the times, that consequently the spaces traversed in equal times are like the odd numbers *ab unitate*, and so on. This is the principle: the natural mobile increases its velocity to the extent that it is separated from the beginning of its motion. Hence, if the body falls from the point *a* along the line *abcd*, I suppose that its speed in *c* is to its speed in *b* as the distance *ca* is at the distance *ba*.[42]

---

39 Our translation. Cf. EN I, p. 299: "[...] nempe, quodcunque mobile, nullam extrinsecam resistentiam patiens, a vi quae minor sit quacunque vi proposita, in plano quod nec sursum nec deorsum tendat, moveri posse".

40 See Drake (1978, pp. 66–73).

41 See Drake (1978, pp. 80–83).

42 Our translation. Cf. EN X, p. 115: "Ripensando circa le cose del moto, nelle quali, per dimostrare li accidenti da me osservati, mi mancava principio totalmente indubitabile da poter porlo per assioma, mi son ridotto ad una proposizione la quale ha molto del natural et dell'evidente; et questa supposta, dimostro poi il resto, cioè che gli spazzii passati dal moto naturale esser in proporzione doppia dei tempi, et per conseguenza gli spazii passati in tempi eguali esser come i numeri impari *ab unitate*, et le altre cose. Et il principio è questo: che il mobile naturale vadia crescendo di velocità con quella proportione che si discosta dal principio del suo moto; come, v.g., cadendo il grave dal termine *a* per la linea *abcd*, suppongo che il grado di velocità che ha in *c* al grado di velocità che hebbe in *b* esser come la distanza *ca* alla distanza *ba*". It is to be pointed out that, with the mathematical knowledge

Galileo dedicated the year 1604 to the construction of instruments which allowed him to obtain a good number of reliable measurements. He saw that the velocities of a free falling body increase, given a unit of time, as the odd numbers. Thus, by merely relying upon experiments, Galileo concluded that the spaces increase as the sum of the odd numbers, which means as the squares of the times. This is an experimental proposition that, in 1604, Galileo thought to be in agreement with his assumption—that he regarded as natural and evident—that the spaces are as the velocities. He also tried a mathematical deduction of the experimental proposition from his assumption, but failed since the two statements are not compatible. In the next five years he abandoned the idea that spaces and velocities are proportional because he guessed that the experimental proposition was not consistent with the proportionality between spaces and velocities. He therefore assumed it to be a primitive law that spaces are proportional to the squares of time intervals. In this case experiment guided mathematics.

Such is the intricate relationship between experiments and mathematics in Galileo's work during the eighteen years between his initial research on motion and his conclusions of 1604. Galileo acted neither as a Platonist nor as an empiricist: sometimes his mathematical assumptions led him to conceive experiments because through mathematics alone he had not been able to reach definitive results. At other times, on the contrary, Galileo guessed that the experimental results might not fit the theoretical predictions not because his theory was faulty but because experimental conditions were not sufficiently fine-tuned. The problem of friction is typical in this respect.

There is no doubt that mathematical assumptions guided Galileo to conceive experiments which might verify or falsify them, but it is also true that experimental data led him to reflect on a possible mathematical law coherent with them. Galileo sometimes believed in the coherence of experimental results with a mathematical law, which however turned out not to be consistent with the experiments. It is rather strange that an allegedly Platonic natural philosopher changes his understanding of the a priori eidetic universe because of results obtained through experiments. Therefore, Galileo was not a Platonist, at least if by that term we refer to the Platonic theory of ideas. Galileo wasn't an apriori theorist either, because his actual procedures were far from the Platonic undervaluation of empirical knowledge and methods. Galileo favoured mathematics as he understood that it was the only conceptual instrument that permitted the discovery of physical laws. This view went together with the belief that God had created nature according to a mathematical structure, and that there

of differential equations achieved at the end of the seventeenth century, it thereby follows that a motion in which spaces are proportional to velocities cannot even begin. Of course, Galileo did not know differential equations, but he changed his mind on the proportionality of spaces and velocities so that on the Third Day of *Discorsi e Dimostrazioni* he even offered a demonstration to show that, in fact, such a law cannot work (EN, VIII, 202–204). Galileo's proof is not straightforward and has been widely debated in the literature.

was an harmonic correspondence between our mind and the world.[43] However, what matters for the history of science is that he was aware that the rational order underlying phenomena could not be revealed merely by means of a priori mathematical reflection (by associating, for instance, the Platonic solids to the structure of the five elements, as Plato had suggested in the *Timaeus*); to understand the mathematical order of nature, the study of bodies and of their behaviour through observation and experiments was necessary.

A viewpoint opposite to that of Koyré was adopted by Geymonat:

> [Galileo] considers mathematics to be not so much a superior reality to that of experience, but a very precious aid to apply to it.[44]

Geymonat sees in Galileo the scientist who, through a synthesis of experience and mathematics, invented the modern scientific method. Geymonat summarised Galileo's method in two features: (1) the rejection of hypotheses which cannot be verified; (2) the moving towards the verification and confirmation of a theory by successive steps.

Geymonat reads Galileo as if he were a twentieth-century physicist embracing the rules of the neo-positivist philosophy of science. This is not the method followed by Galileo, as Feyerabend rightly argued.[45] For his astronomical observations Galileo used an instrument, the telescope, whose optical theory was not known at that time, to validate the truth of an astronomical system that was not supported by sufficient observations (in particular, because of the lack of any detectable stellar parallax).[46] In *Against Method,* Chapter 10, Feyerabend holds that Galileo focused on those telescopic observations which were consistent with the Copernican system and assumed, on this basis, that the telescope offered a faithful image of the sky. However, Galileo disregarded other observations (such as the lack of parallax of the stars) which might contradict the Copernican system. Feyerabend's considerations can be opposed to Geymonat's theory that Galileo followed a precise scientific method. Had Galileo followed the methodological canons ascribed to him by Geymonat, he should have suspended his judgement on the validity of the Copernican system, which Galileo certainly did not do.

Descartes' assessment of Galileo's most systematic work, *Discorsi e dimostrazioni*, is indicative in this respect. Descartes wrote in a detailed letter to Mersenne dated 11 October 1638:

> Generally speaking, I find he philosophizes much more ably than is usual, in that, so far as he can, he abandons the errors of the Schools and tries to use mathematical methods in the

---

[43] See Galileo, *Diversi Fragmenti attenenti al Trattato delle cose che stanno su l'acqua*, EN, vol. IV 1894, p. 52: "Posuit Deus omnia in numero, pondere et mensura". Galileo quotes from the *Bible, Liber Sapientiae* 11, 21. On this topic see Dijksterhuis (1961, p. 404).

[44] Our translation. Cf. Geymonat (1981, p. 119): "[Galileo] vede nella matematica non tanto una realtà superiore all'esperienza, quanto un ausilio preziosissimo per operare sull'esperienza".

[45] Feyerabend (1975, Chaps. 8–10), where Galileo's use of the telescope in connection with his arguments in favour of the Copernican system is analysed.

[46] It should be noted that no astronomical system was free from objections in the first four decades of the seventeenth century.

> investigation of physical questions. On that score, I am completely at one with him, for I hold that there is no other way to discover the truth. But he continually digresses, and he does not take time to explain matters fully. This, in my view, is a mistake: it shows that he has not investigated matters in an orderly way, and has merely sought explanations for some particular effects, without going into the primary causes in nature; hence his building lacks a foundation.[47]

Descartes' assessment of Galileo's scientific procedure can be summarized in these terms:

(1) Galileo made use of mathematics as the necessary instrument to develop a consistent physics;
(2) Galileo's approach to research and, as a consequence, Galileo's exposition are not systematic and methodical, or, at least, they did not correspond to Descartes' idea of a systematic approach to research;
(3) Galileo does not look for the first causes of the effects he studied.

A third position between the interpretations of Koyré and Geymonat is that held by Ernst Cassirer who thought that Galileo is a Platonist in a weak meaning of this term. Cassirer claims that Galileo entertained a new conception of truth, based on the idea that a profound interconnection exists between mathematics and nature and that mathematics represents, so to say, the immanent essence of the world. However, the idea of the mathematical character of physical truth is in Galileo a subjective preconception that cannot be referred to any philosophical school:

> We have to emphasize from the beginning that Galileo, despite having developed and applied a new method of knowledge, is not a systematic thinker either in philosophy or in the theory of knowledge. The general agreement between mathematics and nature, the harmony between thought and reality is for him a subjective conviction anterior to any philosophical reflection. We will see that this conviction permeates all parts of his doctrine and keeps them intimately unified. But, in spite of the fact that this conviction is developed and determined ever more precisely, the problem of its foundation and justification is never posed. What connotes Galileo's research is precisely the self-assurance with which scientific thought realizes its own nature without leaving its sphere.[48]

---

[47] Descartes (1984–1991, vol. III, p. 124). Cf. AT II, p. 380: "Ie trouve en general qu'il philosophe beaucoup mieux que le vulgaire, en ce qu'il quitte le plus qu'il peut les erreurs de l'Eschole, et tasche a examiner les matieres physiques par des raisons mathematiques. En cela ie m'accorde entierement avec luy et ie tiens qu'il n'y a point d'autre moien pour trouver la verité. Mais il me semble qu'il manque beaucoup en ce qu'il fait continuellement des digressions et ne s'areste point a expliquier tout a fait une matiere; ce qui monstre qu'il ne les a point examinées par ordre, et que, sans avoir consideré les premieres causes de la nature, il a seulement cherché les raisons de quelques effets particuliers, et ainsy qu'il a basti sans fondement".

[48] Our translation. Cf. Cassirer (1922, p. 383): "Eins freilich müssen wir uns hier von Anfang an gegenwärtig halten: daß Galilei, so sehr er eine neue Methode der Erkenntnis handhabt und zur Anwendung bringt, selbst kein Systematiker der Philosophie und Erkenntnistheorie ist. Die durchgängige Übereinstimmung zwischen Mathematik und Natur, die Harmonie zwischen dem Gedanken und der Wirklichkeit steht ihm, vor aller philosophischen Reflexion, als subjektive Überzeugung fest. Wir werden sehen, wie diese Grundüberzeugung alle Teile seiner Lehre gleichmäßig durchdringt und innerlich zusammenhält. Aber wenngleich sie sich immer prägnanter ausprägt und fortschreitend entfaltet: nach ihrem Grund und ihrer Rechtfertigung wird nicht gefragt.

Cassirer grasps two fundamental aspects of Galileo's ontology and epistemology: (a) that Galileo believed in the inherence of mathematical laws in the physical world, but not in a transcendent realm of mathematical objects *à la* Plato. There is an agreement between mathematics and nature in the sense that nature is immanently mathematical and does not receive mathematical forms from a superior ideal world; (b) that, on the other hand, Galileo did not write anything explicit to examine and clarify the conceptual relations between mathematics and physics.

Cassirer's thesis is debatable since it is very difficult to ascertain whether Galileo's assertion of the inherence of mathematics to nature was a subjective preconception or a belief developed during his studies on motion when Galileo realised that, without quantifying the magnitudes, a satisfying solution to the problems of mechanics was impossible.

To summarize, our view of Galileo's scientific procedures and philosophical ideas is the following:

(1) Galileo reached the conviction that mathematics is indispensable to physics because it is the only way we have of measuring exactly any magnitude. Therefore, the first and essential use of mathematics is an operative use.
(2) He very likely believed that physical laws mathematically expressed exist in the universe. In this respect he was a "weak Platonist". In any case, Galileo's achievements obtained by means of the use of mathematics in his physical investigations do not depend on a philosophical view about the nature of mathematical objects or about the relationship between mathematics and physics.
(3) He did not believe either in anything like "mathematical archetypes inscribed in the universe" *à la* Kepler or in a transcendent world of mathematical entities separate from physical reality and its functioning laws. He was not, therefore, a Platonist in the classical sense of the term.
(4) He had a clear idea of the importance of experiments for physical investigations. However, he did not adopt a scientific method in which experiments and mathematics are coordinated in a precise and systematic way. Of great significance in this regard is his *Capitolo contro il portar la toga*,[49] where Galileo claimed that no privileged method exists in scientific research and that if the usual procedures are not suitable for solving a problem, you have to work with art and invention because thousands of other ways can help you to discover the truth, if you have the patience to search for them. Galileo devised experiments to prove hypotheses that he had formed by pure intuition, not through a deductive process. On the other hand, sometimes, appropriately carried out experiments led Galileo to the discovery of physical laws. We can conclude that the link between the intuitive, the deductive and the experimental character of science is anything but methodical in Galileo's work.

Gerade in der Selbstgewißheit, mit der das wissenschaftliche Denken sich hier erfaßt und bei sich selbst verweilt, liegt das Auszeichnende der Galileis Forschung".

[49] EN IX, pp. 213–223. For an English translation, see Galileo 2009. The *Capitolo* was written around 1590, though the composition date cannot be established with an absolute precision.

(5) Several a priori assumptions contributed to the development of Galileo's science. From a cosmological perspective, more significant than Galileo's belief in the existence of a mathematical order in nature were his convictions of the correctness of the Copernican system and of the telescope's reliability, though a correct and accepted theory of refraction did not exist. Feyerabend rightly identified these assumptions, but he overestimated their aprioristic character. While admitting that the Copernican theory was not fully confirmed, the arguments against the Ptolemaic one—and against the whole system of physical and philosophical ideas connected to Ptolemaic cosmology—were stronger. Furthermore, since the telescope was reliable on Earth, it was reasonable to think that it would have been so also when directed to the skies, though, admittedly, this was not fully certain.

B. *Primary and secondary properties of bodies*

Galileo highlights the distinction between quantitative and qualitative aspects of experience, which dated back to Democritus and which, after Galileo, would be developed by Descartes, Boyle and Locke. By means of this distinction the qualitative features of experience are interpreted as the effect of the interaction between the external bodies and the senses of the human observer, whereas the quantitative features are regarded as pertaining to the objective structure of the bodies. In *The Assayer* we read:

> Now I say that whenever I conceive any material or corporeal substance, I immediately feel the need to think of it as bounded, and as having this or that shape; as being large or small in relation to other things, and in some specific place at any given time; as being in motion or at rest; as touching or not touching some other body; and as being one in number, or few, or many. From these conditions I cannot separate such a substance by any stretch of my imagination. But that it must be white or red, bitter or sweet, noisy or silent, and of sweet or foul odor, my mind does not feel compelled to bring in as necessary accompaniments. Without the senses as our guides, reason or imagination unaided would probably never arrive at qualities like these. Hence, I think that tastes, odors, colors, and so on are no more than mere names so far as the object in which we place them is concerned, and that they reside only in the consciousness. Hence if the living creature were removed, all these qualities would be wiped away and annihilated.[50]

[50] Galilei (1957, p. 274). Cf. EN VI, pp. 347–348: "Per tanto io dico che ben sento tirarmi dalla necessità, subito che concepisco una materia o sostanza corporea, a concepire insieme ch'ella è terminata e figurata di questa o di quella figura, ch'ella in relazione ad altre è grande o piccola, ch'ella è in questo o quel luogo, in questo o quel tempo, ch'ella si muove o sta ferma, ch'ella tocca o non tocca un altro corpo, ch'ella è una, poche o molte, nè per veruna imaginazione posso separarla da queste condizioni; ma ch'ella debba essere bianca o rossa, amara o dolce, sonora o muta, di grato o ingrato odore, non sento farmi forza alla mente di doverla apprendere da cotali condizioni necessariamente accompagnata; anzi, se i sensi non ci fussero scorta, forse il discorso o l'immaginazione per sé stessa non v'arriverebbe già mai. Per lo che vo io pensando, che questi sapori, odori, colori etc., per la parte del suggetto nel quale ci par che riseggano, non sieno altro che puri nomi, ma tengano solamente lor residenza nel corpo sensitivo, sì che rimosso l'animale, sieno levate ed annichilate tutte queste qualità".

This distinction goes together with a corpuscularian view of the structure of matter which Galileo discussed both in *Il saggiatore* and in *Discorsi e dimostrazioni matematiche intorno a due nuove scienze*. Investigating the nature of "those materials which produce heat in us and make us feel warmth, which are known by the general name of 'fire'", Galileo asserts that they are composed of "a multitude of minute particles having certain shapes and moving with certain velocities". The fire-corpuscles are extremely subtle, swift and penetrating, and therefore can dissolve the structure of other substances. Heat, considered as a quality of fire, is entirely resolved in the quantitative properties of the fire-corpuscles, their "shape, number, motion, penetration, and touch". Apart from these, the quality of heat "belongs so intimately to us that when the live body is taken away, heat becomes no more than a simple name" (Galilei, 1957, pp. 276–78).

In the First Day of the *Discorsi e dimostrazioni matematiche intorno a due nuove scienze* Galileo advanced his conjectures on the composition of matter as a series of an infinite number of atoms and empty spaces:

> But if we consider the line resolved into an infinite number of indivisible parts that are not quantities, we shall be able to conceive the line extended indefinitely by the interposition, not of a finite, but of an infinite number of indivisible empty spaces. Now this which has been said concerning simple lines must be understood to hold also in the case of surfaces and solid bodies, it being assumed that they are made up of an infinite, not a finite, number of atoms that are no quantities.[51]

This attempt at conceiving the continuum by overlapping mathematical and physical atomism is undoubtedly problematic, but it is further evidence that in Galileo's view the 'new science' was born thanks to the mathematical grounding of physical investigations.

C. *Defence of Copernicanism*

In three private letters which he left unpublished, the so called "Copernican Letters", Galileo defended Copernican cosmology against the attacks of the Dominicans, who upheld that Copernicanism was incompatible with the teachings of the Bible. These letters were addressed to Benedetto Castelli,[52] a pupil of Galileo and teacher of mathematics at the University of Pisa; to Piero Dini, a Florentine academician and a priest, who lived in Rome and was the nephew of an eminent Cardinal; and to the duchess Christine of Lorraine, the wife of the Grand Duke of Tuscany. Galileo wrote

---

[51] Galilei (1914, p. 25). The English translation has been modified. The original version translated erroneously Galileo's "parti non quante" as "infinitely small parts", rather than as "parts that are not quantities" (which should be the correct translation). Cf. EN VIII, p. 72: "[...] ma imaginandola risoluta in parti non quante, cioè ne' suoi infiniti indivisibili, la possiamo concepire distratta in immenso senza l'interposizione di spazii quanti vacui, ma sì bene d'infiniti indivisibili vacui. E questo, che si dice delle semplici linee, s'intenderà detto delle superficie e de' corpi solidi, considerandogli composti di infiniti atomi non quanti". On this question see Knobloch (1999).

[52] The famous letter by Galileo to Benedetto Castelli written on the 21st of December 1613 has an amazing story: it was copied several times and in slightly different versions. However, Galileo's original version was missing and it has been found only recently in the Library of the Royal Society. On this discovery see Camerota et al. (2019).

these letters with the purpose of asserting the physical truth of the Copernican system as well as the autonomy of science from religion. The way to render Copernicanism compatible with the conflicting literal sentences of the Bible was found by Galileo in the traditional idea that the Bible should be interpreted beyond its literal formulation. Two more premisses supported Galileo's argument: (a) the Bible is not a book of science, but contains only a revelation of moral and religious truths, or, to put it incisively, according to the epigram that Galileo attributed to Cardinal Baronio: "the intention of the Holy Ghost is to teach us how one goes to heaven, not how heaven goes"; (b) if there is an apparent contradiction between the Bible's literal expression and scientific truth, the Bible should be interpreted in such a way as to accommodate its meaning to the scientific truth because the latter is achieved through experiment and mathematical deduction which ascertain natural necessity to a degree of certainty that cannot be cast into doubt by the opposite literal meaning of the Bible. This second premiss is a direct consequence of Galileo's epistemological views about mathematics and physics (see A. above). A clear statement of Galileo's position is found in this passage from the Letter to the Grand Duchess Christine of Lorraine (1615):

> I think that in discussions of physical problems we ought to begin not from the authority of scriptural passages, but from sense-experiences and necessary demonstrations; for the holy Bible and the phenomena of nature proceed alike from the divine Word, the former as the dictate of the Holy Ghost and the latter as the observant executrix of God's commands. It is necessary for the Bible, in order to be accommodated to the understanding of every man, to speak many things which appear to differ from the absolute truth so far as the bare meaning of the words is concerned. But Nature, on the other hand, is inexorable and immutable; she never transgresses the laws imposed upon her, or cares a whit whether her abstruse reasons and methods of operation are understandable to men. For that reason it appears that nothing physical which sense-experience sets before our eyes, or which necessary demonstrations prove to us, ought to be called in question (much less condemned) upon the testimony of biblical passages which may have some different meaning beneath their words. For the Bible is not chained in every expression to conditions as strict as those which govern all physical effects; nor is God any less excellently revealed in Nature's actions than in the sacred statements of the Bible.[53]

Galileo's attempt was unsuccessful as he made the mistake of encroaching on the prerogative of the clergy, that claimed exclusively for itself the task of interpreting the

[53] Cited from Drake (1957, pp. 182–183). Cf. EN V, pp. 316–317: "Stante, dunque, ciò, mi par che nelle dispute di problemi naturali non si dovrebbe cominciare dalle autorità di luoghi delle Scritture, ma dalle sensate esperienze e dalle dimostrazioni necessarie: perché, procedendo di pari dal Verbo divino la Scrittura Sacra e la natura, quella come dettatura dello Spirito Santo, e questa come osservantissima essecutrice de gli ordini di Dio; ed essendo, di più, convenuto nelle Scritture, per accommodarsi all'intendimento dell'universale, dir molte cose diverse, in aspetto e quanto al nudo significato delle parole, dal vero assoluto; ma, all'incontro, essendo la natura inesorabile ed immutabile, e mai non trascendente i termini delle leggi impostegli, come quella che nulla cura che le sue recondite ragioni e modi d'operare sieno o non sieno esposti alla capacità degli uomini; pare che quello degli effetti naturali che o la sensata esperienza ci pone dinanzi a gli occhi o le necessarie dimostrazioni ci concludono, non debba in conto alcuno esser revocato in dubbio, non che condennato, per luoghi della Scrittura che avessero nelle parole diverso sembiante; poi che non ogni detto della Scrittura è legato a obblighi così severi com'ogni effetto di natura, nè meno eccellentemente ci si scuopre Iddio negli effetti di natura che ne' sacri detti delle Scritture.".

Bible. Besides, the most influential theologian at the papal court, Cardinal Roberto Bellarmino, held a position which was precisely the opposite of Galileo's. He believed "that the word of Scripture, which speaks the truth of God, outweighs the hypothetical conclusions of science, not just in questions of theology, but also in questions of science" (Westfall, 1989, p. 13). Christian theologians today commonly accept Galileo's arguments as a way of ensuring the compatibility of faith and science through the distinction of their respective spheres. However, this is too easy an escape from the intellectual predicament triggered by the Copernican revolution in cosmology.

The issue of the relationship between Copernican cosmology and Christian theology has philosophical implications which were and are often left undiscussed and which go well beyond the historical circumstances of Galileo's trial and the gradual (though slow and hard-fought) acceptance of Copernicanism during the 16th and 17th centuries: an acceptance which occurred earlier in Protestant countries and later in the Catholic ones (only in 1835 were the works of Copernicus, Galileo and Kepler removed from the Index of forbidden books). The real issue is how much of the Christian doctrine, including the Biblical narration of the Old and New Testaments, is tightly linked to the ancient cosmology or at least to an anthropocentric view of a unique universe. To focus on this problem we must look beyond Copernicus' astronomical system to the developments of Copernican cosmology which became apparent at the end of the 16th century (in Bruno's and Gilbert's infinitism, for instance) and even more in the 17th and 18th centuries. Already in 1612 a poet and cleric of the Church of England like John Donne (who was the author of at least 160 extant sermons) voiced in a poem (*An Anatomy of the World*) his anguished reaction to the world-view of the new science:

> And new philosophy calls all in doubt, / the element of fire is quite put out, / the sun is lost, and th'earth, and no man's wit / can well direct him where to look for it. / And freely men confess that this world's spent, / when in the planets and the firmament / they seek so many new; they see that this / is crumbled out again to his atomies. / 'Tis all in pieces, all coherence gone, / all just supply, and all relation... (Donne, 1996, p. 163 f.).

The question that was (and still is) left undiscussed concerns the meaning of the incarnation of Christ in a universe in which Earth has lost any cosmological significance and in which other 'inhabitants' could exist in planetary systems surrounding the innumerable stars. How can Christianity as a universal truth be reconciled with the idea of a plurality of worlds if Christianity is the religion of the incarnation of God in Jesus Christ in this world? In the famous *Entretiens sur la pluralité des mondes* (*Conversations on the Plurality of Worlds*) (1686) by Bernard de Fontenelle we find an instance of the theological embarrassment which accompanied the new view of the universe triggered by the Copernican revolution. The author was well aware of the theological problems which his thesis might raise:

> When I say to you that the moon is inhabited, you picture to yourself men made like us, and if you're a bit of a theologian, you're instantly full of qualms. The descendants of Adam have not spread to the moon, nor sent colonies there. Therefore, the men in the moon are not

sons of Adam. Well, it would be embarrassing to theology if there were men anywhere not descended from him, it's not necessary to say any more about it [...].[54]

Fontenelle evaded the difficulty simply asserting that when he talks of inhabitants of other planets he does not refer to 'other men': "I put no men there at all: I put inhabitants there who are not like men in any way. What are they, then? I've never seen them". What mattered to Fontenelle were not the theological concerns, which he wanted simply to dismiss, but the idea of "the infinite diversity that Nature has placed in her works. This idea governs the whole book, and cannot be contested by any philosopher".

Hans Blumenberg wrote that all "the basic constituents of Christian dogma—the creation, the unity of mankind in guilt, God's incarnation and redemption of man—[...] presuppose the uniqueness of a world to which God's saving care relates and in which the 'salvation story' takes place" (Blumenberg, 2000, p. 372). Therefore, it is not surprising that, historically, the cosmological consequences of the Copernican revolution (the theses of the infinity of the universe and of the plurality of the words) paved the way to the demise of the Christian God along an itinerary that, from the end of the 16th to the end of the 18th centuries, comprises pantheism, rational Deism and, finally, materialistic atheism.

References

Acerbi, F. (2000). Le fonti del mito platonico di Galileo. *Physis, 37*, 359–392.

Aiton, E. J. (1954). Galileo's theory of the tides. *Annals of Science, 10*(1), 44–57.

Blumenberg, H. (2000). *The Genesis of the Copernican World* (R. M. Wallace, Trans.). London (England): The MIT Press.

Boccaletti, D. (2016). *Galileo and the equations of motion.* Cham-Heidelberg-New York-Dordrecht-London: Springer.

Bordiga, G. (1925–26). Giovanni Battista Benedetti, filosofo e matematico veneziano del secolo XVI. *Atti del Reale Istituto Veneto di Scienze Lettere ed Arti, 85*, 585–754.

Burns, W. E. (2001). *The scientific revolution. An encyclopedia.* Santa Barbara: ABC-Clio.

Burtt, E. A. (1925). *The metaphysical foundations of modern physical science: A historical and critical essay.* London: Kegan Paul, Trench, Trubner & Co.

Bussotti, P. (2011). The circulation of Kepler's cosmological ideas in Italy during Kepler's lifetime. In É. Mehl (Ed.), *Kepler: La Physique Celeste. Autour de l'*Astronomia Nova *(1609)* (pp. 209–229). Paris: Les Belles Lettres.

Bussotti, P. (2017). Galileo e il suo concetto di inerzia. In A. Neri (Ed.), *Là dove il pensiero incontra l'esperienza* (Vol. I, pp. 51–110). ETS.

Camerota, M. (1992). *Gli scritti* De Motu Antiquiora *di Galileo Galilei. Il Ms. Gal 71: un'analisi storico-critica.* Cagliari: CUEC.

[54] Fontenelle (1990, p. 6). Cf. Fontenelle (1852, p. 36): "Quand on vous dit que la Lune est habitée, vous vous y représentez aussitôt des hommes faits comme nous; et puis, si vous êtes un peu théologien, vous voilà plein de difficultés. La postérité d'Adam n'a pu s'étendre jusque dans la Lune, ni envoyer des colonies en ce pays-là. Or, il serait embarrassant, dans la théologie, qu'il y eût des hommes qui ne descendissent pas de lui. Il n'est pas besoin d'en dire davantage [...]".

Camerota, M., & Helbing, M. (2000). Galileo and Pisan Aristotelianism: Galileo's *De Motu Antiquiora* and the *Questiones de Motu Elementorum* of the Pisan Professors. *Early Science and Medicine, 5*(4), 319–365.

Camerota, M., Giudice, F., & Ricciardo, S. (2019). *Galileo ritrovato. La lettera a Castelli del 21 dicembre 1613*. Brescia: Morcelliana.

Cassirer, E. (1922). *Das Erkenntnisproblem in der Philosophie und Wissenschaft der neueren Zeit* (Vol. I). B. Berlin: Cassirer.

Cassirer, E. (1946). Galileo's Platonism. In M. Ashley-Montagu (Ed.), *Studies and Essays in the History of Science and Learning Offered in Homage to George Sarton in the Occasion of His Sixtieth Birthday* (pp. 277–297). New York: Schuman.

Coffa, J. A. (1968). Galileo's concept of inertia. *Physis, X*(4), 261–281.

De Caro, M. (1992). Galileo's mathematical Platonism. In G. Czermak (Ed.), *Philosophy of mathematics* (pp. 1–9). Wien: Hoelder-Pichler-Tempsky.

De Caro, M. (2012). Galileo e il platonismo fisico-matematico. In R. Chiaradonna (Ed.), *Il platonismo e le scienze* (pp. 123–142). Roma: Carocci.

Descartes, R. (1897–1913). *Oeuvres de Descartes* (12 vols) publiées par C. Adam & P. Tannery. Vrin. Abbreviated as AT.

Descartes, R. (1984–1991). *The philosophical writings of Descartes* (J. Cottingham, R. Stoothoff, & D. Murdoch, Trans.) (in 3 vols.). Cambridge-London-New York-New Rochelle-Melbourne-Sydney: Cambridge University Press.

Dijksterhuis, E. J. (1961). *The mechanization of the world picture*. Clarendon Press (Original Dutch ed., 1950).

Dollo, C. (2003). *Galileo Galilei e la cultura della tradizione*. Soveria Mannelli: Rubbettino.

Donne, J. (1996). *Selected poetry*, ed. with an Introduction and Notes by J. Carey. Oxford: Oxford University Press.

Drake, S. (Ed.). (1957). *Discoveries and opinions of Galileo* (pp. 173–216). New York: Anchor-Doubleday.

Drake, S. (1964). Galileo and the law of inertia. *American Journal of Physics, 32*, 601–608.

Drake, S. (1970). *Galileo studies*. Ann Arbor: University of Michigan Press.

Drake, S. (1976). A further reappraisal of impetus theory: Buridan, Benedetti, and Galileo. *Studies in History and Philosophy of Science, 7*(4), 319–336.

Drake, S. (1978). *Galileo at work. His scientific biography*. Chicago and London: The University of Chicago Press.

Drake, S. (1990). *Galileo: Pioneer scientist*. Toronto: University of Toronto Press.

Feyerabend, P. K. (1975). *Against method: Outline of an anarchistic theory of knowledge*. London: Verso.

Field, J. V. (1985). Giovanni Battista Benedetti on the mathematics of linear perspective. *Journal of the Warburg and Courtauld Institute, 48*, 71–99.

Fontenelle, Bernard le Bovier de (1852). *Entretiens sur la pluralité des mondes*. In *Oeuvres de Fontenelle* (pp. 33–149). Paris: Eugène Didier (Original 1686).

Fontenelle, Bernard le Bovier de (1990). *Conversations on the plurality of worlds*. Translated by H. A. Hargreaves. Introduced by N. R. Gelbart. Berkeley: University of California Press.

Frova, A. (2006). Anche Galileo può errare. In *Atti di «Pianeta Galileo 2006»* (pp. 35–47). Firenze: Regione Toscana.

Galilei, G. (1610). *Sidereus Nuncius* (pp. 55–96). EN, III, Parte Prima.

Galilei, G. (1616). *Discorso del flusso e reflusso del mare* (pp. 373–395). EN, V.

Galilei, G. (1890–1909). *Edizione Nazionale delle Opere*. Direttore A. Favaro, (20 vols). Firenze: Barbèra. Abbreviated as EN.

Galilei, G. (1914). *Dialogues concerning two new sciences* (H. Crew & A. De Salvio, Trans.). New York: Macmillan.

Galilei, G. (1957). *The Assayer*. Selections. Translated by S. Drake, *Discoveries and opinions of Galileo* (pp. 231–280). New York: Doubleday & Co.

Galilei, G. (1967). *Dialogue concerning the two chief world systems: Ptolemaic and Copernican* (2nd ed.). Translated and with notes by S. Drake. Foreword by A. Einstein. Berkeley and Los Angeles: University of California Press.

Galilei, G. (2009). *Capitolo Contro il portar la toga—Against the Donning of the Gown*. Italian text plus English translation by G. Bignami. Edited by L. Tongiorgi Tomasi. With essays by G. Bignami, V. Finucci, M. Rippa Bonati, F. Tognoni, L. Tongiorgi Tomasi, R. Vergara Caffarelli. ETS. EN IX: 213–223.

Galilei, G., & Scheiner, C. (2010). *On Sunspots*. Translated and introduced by E. Reeves & A. Van Helden. Chicago and London: Chicago University Press.

Geymonat, L. (1981). *Per Galileo*. Verona: Bertani.

Girill, T. R. (1970). Galileo and Platonistic methodology. *Journal of the History of Ideas, 31*(1970), 501–520.

Hall, A. R. (1963). *From Galileo to Newton*. New York: Harper and Row.

Knobloch, E. (1999). Galileo and Leibniz—Different approaches to infinity. *Archive for History of Exact Sciences, 54*, 87–99.

Koyré, A. (1939). *Études Galiléennes*. Paris; Hermann.

Koyré, A. (1957). *From the closed world to the infinite universe*. Baltimore: John Hopkins Press.

Maccagni, C. (1967). *Le speculazioni giovanili "de motu" di Giovan Battista Benedetti*. Pisa: Domus Galilaeana.

Maccagni, C. (1985). *Le due edizioni della* Demonstratio proportionum motuum loculium contra Aristotelem et omnes philosophos *di Giovanni Battista Benedetti*. Istituto Veneto di Scienze, Lettere ed Arti.

Mach, E. (1919). *The science of mechanics. A critical and historical account of its development*. Chicago-London: The Open Court Publishing Co. English translation of the fourth edition of Mach's *Die Mechanik in ihrer Entwickelung historisch-kritisch dargestellt* (1st ed., 1883).

Machamer, P. (1998). Galileo's machines, his mathematics, and his experiments. In P. Machamer (Ed.), *The Cambridge companion to Galileo* (pp. 53–79). Cambridge: Cambridge University Press.

Marcacci, F. (2018). *Cieli in contraddizione. Giovanni Battista Riccioli e il terzo sistema del mondo*. Perugia: Aguaplano.

Minazzi, F. (1994). *Il Flauto di Popper*. Milano, Franco Angeli.

Morpurgo-Tagliabue, G. (1981). *I processi di Galileo e l'epistemologia*. Roma: Armando.

Palmieri, P. (1998). Re-examining Galileo's theory of the tides. *Archive for History of Exact Sciences, 53*, 223–375.

Roero, C. S. (1997). Giovanni Battista Benedetti and the scientific environment of Turin in the 16th century. *Centaurus, 39*(1), 37–66.

Rossi, P. (1989). *La Scienza e la Filosofia dei Moderni*. Torino: Bollati Boringhieri.

Schofield, C. J. (1965). The geoheliocentric mathematical hypothesis in sixteenth-century planetary theory. *The British Journal for the History of Science, 2*(8), 291–296.

Schofield, C. J. (1981). *Tychonic and semi-Tychonic world systems*. New York: Ayer Co Pub.

Torrini, M. (1993). Galileo, Platone e la filosofia. In P. Prini (a cura di), *Il neoplatonismo nel Rinascimento* (pp. 233–243). Roma: Istituto dell'Enciclopedia Italiana.

Vailati, G. (1897–1898). Le speculazione di Giovanni Battista Benedetti sul moto dei gravi. *Atti dell' Accademia delle Scienze di Torino, 33*, 559–576. Reprinted in G. Vailati, *Scritti di G. Vailati* (pp. 161–178). Leipzig-Firenze: Barth-Seeber.

Vergara Caffarelli, R. (2007). Il principio d'inerzia negli ultimi scritti di Galilei. *Cronos, 10*, 63–88.

Wallace, W. A. (1974). Galileo and the reasoning *ex suppositione*. The methodology of the Two New Sciences. In S. Cohen (Ed.), *Proceedings of the* 1974 *biennial meeting of the Philosophy of Science Association* (pp. 79–104). Reidel.

Wallace, W. A. (1984). *Galileo and his sources: The heritage of the Collegio Romano in Galileo's science*. Princeton: Princeton University Press.

Westfall, R. S. (1989). The trial of Galileo: Bellarmino, Galileo and the clash of two worlds. *Journal for the History of Astronomy, 20*(1), 1–23.

Wisan, W. L. (1978). Galileo's scientific method: A reexamination. In R. E. Butts & J. C. Pitt (Eds.), *New perspectives on Galileo* (pp. 1–58). Dordrecht: Reidel.

Chapter 5
Descartes and the New Mechanistic Paradigm

Descartes' philosophy is one of the most studied and debated subject in the whole history of philosophy because of the innovative character of his metaphysics and theory of knowledge. What matters for our purpose is to be aware that Descartes' intention of reforming the traditional learning on a solid foundation beyond any doubt—on the intellectual evidence instantiated and granted by the elementary reflexive act of the *cogito*—aimed at the construction of a new science which could transform human life and mankind's relation to nature. Whatever the significance of Descartes' metaphysics—to which hundreds of scholars addressed their attention, we should not forget what Descartes wrote in the sixth and final part of his *Discours de la méthode* (1637):

> But as soon as I had acquired some general notions in physics and had noticed, as I began to test them in various particular problems, where they could lead and how much they differ from the principles used up to now, I believed that I could not keep them secret without sinning gravely against the law which obliges us to do all in our power to secure the general welfare of mankind. For they opened my eyes to the possibility of gaining knowledge which would be very useful in life, and of discovering a practical philosophy which might replace the speculative philosophy taught in the schools. Through this philosophy we could know the power and action of fire, water, air, the stars, the heavens and all the other bodies in our environment, as distinctly as we know the various crafts of our artisans; and we could use this knowledge – as the artisans use theirs – for all the purposes for which it is appropriate, and thus make ourselves, as it were, the lords and masters of nature.[1]

[1] Descartes (1984–1991, I, pp. 142–143). Cf. AT VI, pp. 61–62: "Mais, sitost que i'ay eu acquis quelques notions generales touchant la Physique, & que, commençant a les esprouver en diverses difficultez particulieres, i'ay remarqué iusques où elles peuvent conduire, & combien elles different des principes dont on s'est servi iusques a present, i'ay creu que ie ne pouvois les tenir cachées, sans pecher grandement contre la loy qui nous oblige a procurer, autant qu'il est en nous, le bien general de tous les hommes. Car elles m'ont fait voir qu'il est possible de parvenir a des connoissances qui soient fort utiles a la vie, & qu'au lieu de cete Philosophie speculative, qu'on enseigne dans les escholes, on en peut trouver une pratique, par laquelle connoissant la force & les actions du feu, de l'eau, de l'air, des astres, des cieux, & de tous les autres cors qui nous environnent, aussy distinctement que nous connoissons les divers mestiers de nos artisans, nous les pourrions employer

P. Bussotti and B. Lotti, *Cosmology in the Early Modern Age: A Web of Ideas*,
Logic, Epistemology, and the Unity of Science 56,
https://doi.org/10.1007/978-3-031-12195-1_5

This 'Baconian' intention proves how much Descartes was interested in the knowledge of nature which he believed he could achieve by virtue of his 'mathematical' method. In a letter to Mersenne written on 13 November 1629, Descartes hinted at his work, *Le Monde* (that remained unpublished for prudential reasons) as at a little treatise where he intended to "explain all the natural phenomena, that is the whole physics". He added that this project "makes me happy more than any other I have ever had" (Descartes, 2009, p. 87). There is no doubt that Descartes did not devote the greater part of his time and intellectual energy to elaborate a well refined and complete system of metaphysics in the traditional manner; rather, he wanted to develop a system of the world based on the assumption that all phenomena are explicable by mechanisms, depending exclusively upon motion and collisions of particles. Because of this, his theory, commonly referred to as mechanistic science and philosophy, represented a turning point in early-modern philosophy of nature, since it provided a new general paradigm that was radically alternative to the traditional views. The merit of Descartes as a scientist and as a philosopher is outstanding: he tried to remove from natural philosophy those mysterious attractions, forces, and influxes which characterised the Medieval and Renaissance views of nature; in their place he outlined the view of a unified material universe ruled by three simple laws of motion. His limits are twofold: (1) he did not place his mechanistic cosmology within a mathematical structure, so that he did not subject his physics to a more rigorous analysis that would have shown that several of his presuppositions and reasonings could not work; (2) his physics remained, in substance, aprioristic; namely it is a physics in which the role of observations and experiments is secondary.

His system of the world is kinematical and structural, not dynamical. This was a deliberate choice, for Descartes was convinced that any reference to physical concepts which were not reducible to motions and impacts were unavoidably connected to mysterious entities that should not be admitted in natural philosophy. He did not think that the original causes of motions could be explained within physics, and he referred them directly to God. In other terms: he never understood that dynamics could and should be a part of natural philosophy.

5.1 Descartes: A Mechanical Model of Universe Devised Without Mathematics

The works in which Descartes explained his cosmology are *Le Monde* and *Principia Philosophiae* (see Descartes, 1644). Several of his letters also deal with this topic. It is a well-known fact that *Le Monde* was not printed during Descartes's lifetime since he was so worried about Galileo's condemnation that he decided not to publish his manuscript. Descartes's cosmology is thoroughly explained in the *Principia Philosophiae*, which is a broader and more complete treatise than *Le Monde*.

---

en mesme façon a tous les usages ausquels ils sont propres, & ainsi nous rendre comme maistres & possesseurs de la Nature".

Despite the differences between these two books, for the purpose of our work it will suffice to refer to the *Principia*. This text is divided into four parts: the first concerns the principles of human knowledge and outlines a general metaphysical frame that summarizes the contents of his *Meditationes de Prima Philosophia* (1641); the second deals with the principles of material things (the principles of physics); the third regards the visible world, and here Descartes develops his cosmology; the fourth is dedicated to terrestrial phenomena.

5.1.1 Fundamental Principles of Descartes' Physics

Descartes outlines a real cosmogony, a history of the universe, which is based on some physical presuppositions:

1. The movement of a body can be determined only by referring it to the reciprocal positions and changes of position of neighbouring bodies. It makes no sense to think of a space existing independently of the bodies' mutual positions. Descartes claimed (*Principia* II, 28):

> In my definition I specified that the transfer occurs from the vicinity of contiguous bodies to the vicinity of other bodies; I did not say that there was a transfer from one place to another. This is because, as explained above, the term 'place' has various meanings, depending on how we think of it.[2]

It follows that even the concept of rest is relative. This means that the idea of a body, or several bodies, at rest is one of convenience, created to answer various scientific needs; strictly speaking, one might think of a body at rest only with respect to another body in movement. Descartes wrote (*Principia* II, 29):

> I further specified that the transfer occurs from the vicinity not of any contiguous bodies but from the vicinity of those which 'are regarded as being at rest'. For transfer is in itself a reciprocal process: we cannot understand that a body AB is transferred from the vicinity of a body CD without simultaneously understanding that CD is transferred from the vicinity of AB. Exactly the same force and action is needed on both sides. So if we wished to characterize motion strictly in terms of its own nature, without reference to anything else, then in the case of two contiguous bodies being transferred in opposite directions, and thus separated, we should say that there was just as much motion in the one body as in the other.[3]

---

[2] Descartes (1984–1991, I, p. 234). Cf. AT VIII, p. 55: "Addidi praeterea, traslationem fieri *ex vicinia corporum contiguorum in viciniam aliorum*, non autem ex uno loco in alium: quia, ut supra explicui, loci acceptio varia est, ac pendet a nostra cogitatione" (italics in the text). The English translation of the *Principia* is based on the Latin text. Since Picot's French translation (1647, AT IX), approved by Descartes himself, presents considerable differences with respect to the Latin original, the English translators inserted in diamond brackets < > the parts of Picot's version at variance with the Latin text (Descartes 1984–1991, I, p. 178).

[3] Descartes (1984–1991, I, p. 235). Cf. AT VIII, pp. 55–56: "Addidi denique, translationem illam fieri ex vicinia, non quorumlibet corporum contiguorum, sed *eorum* duntaxat, *quae tanquam quiescentia spectantur*. Ipsa enim translatio est reciproca, nec potest intelligi corpus AB transferri ex vicinia corporis CD, quin simul etiam intelligatur corpus CD transferri ex vicinia corporis AB:

2. The universe is full of matter. No void exists. According to Descartes, this implies that every motion takes places in closed paths because as soon as a particle leaves its position, it is necessarily replaced by another particle. He commented thus (*Principia*, II, 33):

> I noted above [*Principia* II, 18 and 19] that every place is full of bodies, and that the same portion of matter always takes up the same amount of space, <so that it is impossible for it to fill a greater or lesser space, or for any other body to occupy its place while it remains there>. It follows from this that each body can move only in a <complete> circle <of matter, or ring of bodies which all move together at the same time>: a body entering a given place expels another, and the expelled body moves on and expels another, and so on, until the body at the end of the sequence enters the place left by the first body at the precise moment when the first body is leaving it.[4]

3. At the beginning of the history of the universe, God created the whole of matter imparting a series of movements to it. We will not enter into the question of whether God created matter as a continuum or divided into particles. To grasp the model of Descartes' universe it is important to understand that, at least starting from a certain moment, matter is divided into particles. God created some of them in motion, others at rest (*Principia*, II, 36; cfr. *Le Monde*, chap. VI). One might hypothesize that, for Descartes, it is only in this initial condition that it makes sense to speak of absolute movement and rest. From that moment onwards, all movements are determined by the collisions of the particles set in motion at the beginning of the universe. The movements and collisions do not take place casually, but according to precise natural laws imparted by God. There is no further intervention by God. Thus, the quantity of motion imparted at the beginning is conserved. Descartes erroneously considered this magnitude to be a scalar quantity.

5.1.2 The Structure of Descartes' Universe

Given these principles, how is the universe structured? First of all, Descartes highlights the similarity between the Sun and the fixed stars: these celestial bodies emit light, whereas the planets and the satellites do not. The stars appear to us as being different from the Sun only because of their distance from us. In fact, each star is a "sun". This means that the space around any star is structured exactly like the space

---

ac plane eadem vis et actio requiritur ex una parte atque ex altera. Quapropter si omnino propriam, et non ad aliud relatam, naturam motui tribuere vellemus, cum duo corpora contigua unum in unam, aliud in aliam partem transferuntur, sicque a se mutuo separantur, tantundem motus in uno quam in altero esse diceremus. (Italics in the text)".

4 Descartes (1984–1991, I, pp. 237–238). Cf. AT VIII, p. 58: "Ex hoc autem quod supra fuerit animadversum, loca omnia corporibus plena esse, semperque easdem materiae partes aequalibus locis coaequari, sequitur nullum corpus moveri posse nisi per circulum, ita scilicet ut aliud aliquod corpus ex loco quem ingreditur expellat, hocque rursus aliud, & aliud, usque ad ultimum, quod in locum a primo derelictum, eodem temporis momento quo derelictus est, ingrediatur".

around the Sun. No star is close to the Sun. Therefore, analogically, stars are not close to each other, but each of them is in the centre of an empty space ("empty", obviously, in a relative sense). As a consequence, no sphere of the fixed stars exists (*Principia*, III, 21–23). Having outlined this general picture, Descartes analyses the material composition of the skies: he had already claimed that the Sun and the stars are made up of a fluid matter (*Principia* III, 21). In fact, the skies are composed of a fluid matter, too, which, as Descartes will clarify later on, is different from that of the stars. This fluid is continuous in the sense that no void exists within its structure (principle 2), but it is separated in rotating regions of different densities and velocities. Each region constitutes a fluid rotating vortex. In particular, each star has a rotating vortex of which it is the centre (not necessarily the geometrical centre, but the centre around which the rotation occurs) (Fig. 5.1).

The velocity of any rotating vortex decreases from the star outwards until it reaches a minimum at the surface of a sphere of which the star is the centre. From this spherical surface to the external limit of the vortex, velocity increases. This explains why the

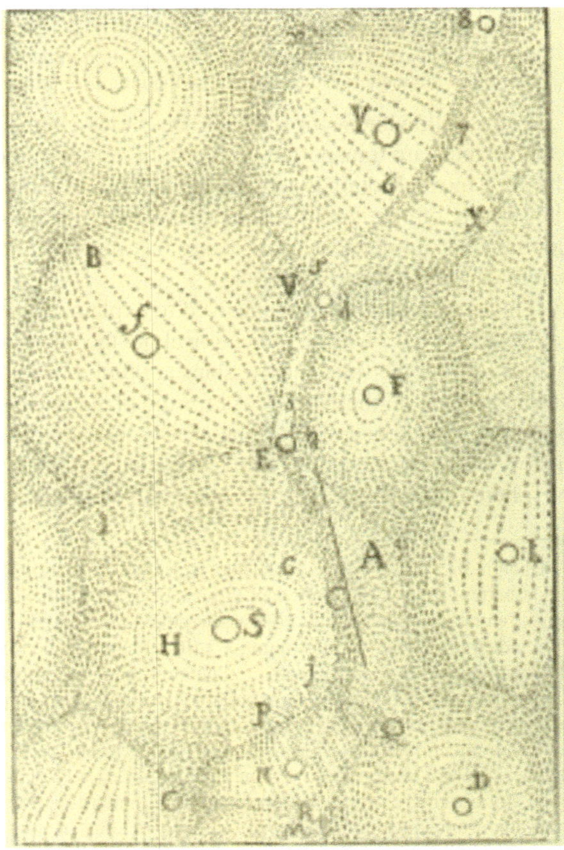

Fig. 5.1 The image used by Descartes to illustrate the vortices surrounding every star. The letters associated to a circle as Y, *f*, *F*, *L*, *S*, *D*, … indicate a star (*Principia* III, 23, AT VIII, p. 88)

velocity of the planets decreases with their distance from the Sun. As a matter of fact, all the planets are confined within the region where velocity decreases. Thus, all the planets are transported around the Sun by the vortex of our star (*Principia*, III, 24–30). What Descartes claims in *Principia* III, 26–27 is worth noting. On the one hand, he highlights the complete analogy between the Earth and the other planets. On the other hand, he introduces his famous assertion: the Earth, as well as the other planets, is transported by a vortex around the Sun, and this is coherent with the idea that the Earth does not have a proper movement around the Sun. The Earth only participates to the movement of the solar vortex. This is a clear verbal ploy to avoid a possible condemnation of his theory by the Catholic Church:

> We see that the earth isn't supported by columns or held up by cables, but is completely immersed in a very fluid heaven. Let us assume that the earth is at rest, having no innate tendency to motion (because we don't see any such propensity). But let's not think that this prevents the earth from being carried along by the current of that heaven, following the motion of the heaven without itself moving. Compare this with an unanchored ship that isn't driven by the wind or by oars, floating motionlessly in the middle of the ocean, though it may be imperceptibly carried along by the ebb and flow of this great mass of water. And just as the other planets resemble the earth in being opaque and reflecting the rays of the sun, there's reason to believe that they also resemble it in remaining at rest, each in its own part of the heaven, and that the observed variation in their positions results solely from the motion of the matter of the heaven in which they are immersed.[5]

In the following section III, 28 Descartes claimed that since the Earth and the other planets are transported by the solar vortex in their orbits, it is incorrect to say that they move. Motion is actually defined by Descartes as the transfer of a body from the vicinity of contiguous bodies considered at rest towards other bodies. Descartes continued claiming that each planet has its own "little" vortex, around which its satellites rotate (*Principia* III, 33).

This vortex theory was conceived to explain the macro-structure of the skies. So far Descartes has given no indication about the micro-structure of the skies and of the bodies: if both the stars and the skies are liquid, what is their microphysical composition, in what do they differ? As for the planets: what is the nature of their matter? Specifying his conception of matter, Descartes explains that at the beginning of the history of the universe, the particles created by God could not be round because a set of spheres cannot tessellate the space and would leave some empty gaps (*Principia*

---

[5] Principia III, 26–27. We have used here Bennett's translations because this section is not translated in Descartes (1984–1991, I). Bennet's translation (2008, amended 2012) is available on the Internet: https://www.earlymoderntexts.com/assets/pdfs/descartes1644.pdf. Cf. AT, VIII, pp. 89–90: " [...] cum videamus Terram nullis columnis suffultam, nullisque funibus appensam, sed circumquaque fluidissimo tantum coelo cinctam effe, putemus quidem illam quiescere, ac nullam habere propensionem ad motum, quandoquidem nullam advertimus. Sed ne putemus hoc obstare, quominus ab isto coelo deferatur, & ejus motibus immota obsequatur ut navis, nullis ventis nec remis impulsa, nullisque anchoris alligata, in medio mari quiescit, etsi forte aquae ingens moles, occulto cursu delabens, ipsam secum ferat. Et quemadmodum caeteri Planetae in hoc cum Terra conveniunt, quod sint opaci & radios Solis reflectant, non immerito arbitrabimur illos etiam in hoc ei similes esse, quod unusquisque quiescat in ea coeli regione in qua versatur; quodque omnis variatio situs quae in illis observatur, ex eo tantum procedat, quod omnis materia coeli, quae illos continet, moveatur".

III, 48). However, since all motions are circular, the particles shaped by these motions have progressively assumed a round form, or better, a form without any angle. To avoid the problem of tessellation, Descartes assumed that not all the particles were round. This implies that some fragments of original matter did not have a circular form. The idea is that the particles did not initially have a regular form. Colliding with each other through circular motions, the particles became spherical. During this process, they lost a part of their matter, which composed little particles that were not spherical. These fragments are so small that they can fit into any possible gap among the larger round particles (*Principia* III, 49 and 50). This particular composition of the matter of the universe is connected to the plentitude principle as Descartes clearly states:

> Because there can't be any empty space in the universe, and because spherical particles can't unite so closely as to leave no spaces around them, these spaces must be filled by scrapings of matter that are extremely tiny and able to change their shapes at any moment so as to fit into the spaces they enter. The scheme of things I am putting forward easily provides for these needed tiny and malleable chips of matter, because this is the story it tells: Matter that is knocked off the corners of the particles of matter that are becoming spherical is gradually ground down to form particles that are so tiny and so fast-moving that through the sheer force of their motion they come to have no exact shape and size and can easily fill spaces that other parts of matter can't enter.[6]

There are, therefore, two kinds of matter: one composed of round particles above a certain dimension (Descartes called them the second element); another one represented by particles which are very little, very quick and adaptable, whose form is irregular (Descartes called them the first element). This physical situation allowed the formation of structures larger than the particles. In particular: the little and irregular particles of the first element composed the Sun and the fixed stars; the round particles of the second element composed the skies. According to Descartes, this is consistent with the fact that the stars are luminous and the skies are transparent. To explain the form of the celestial bodies, it is necessary to point out that any circular motion has, according to Descartes, a centrifugal tendency (*Principia*, III, 48). Therefore, the natural motions of the original universe also had such a tendency. Descartes imagines a situation like this: a region of particles of the second element is animated by a circular motion. They have a centrifugal tendency. However, other contiguous regions are also animated by circular motion. Therefore, this centrifugal tendency cannot become a real centrifugal motion since each region of particles moving circularly exerts a certain pressure on the contiguous rotating regions and thus each region, pressed by the contiguous ones, reaches a sort of equilibrium. This

---

[6] *Principia* III, 49 (Bennett's translation). Cf. AT, VIII, p. 104: "Cum autem nullibi spatia omni corpore vacua esse possint, cumque rotundae illae materiae particulae, simul junctae, perexigua quaedam intervalla circa se relinquant,: necesse est ista intervalla quibusdam aliis materiae ramentis minutissimis, figuras ad ipsa implenda aptas habentibus, easque pro ratione loci occupandi perpetuo mutantibus, impleri. Nempe, dum earum materiae particularum, quae fiunt rotundae, anguli paulatim atteruntur, id quod ex ipsis eraditur adeo est minutum, & tantam celeritatem acquirit, ut sola vi sui motus in ramenta innumerabilia dividatur; sicque impleat omnes angulos, quos aliae materiae particulae subingredi non possunt".

centrifugal tendency has nonetheless an effect within a single region because the particles tend to abandon the centre of the region and to leave a spherical hole in the centre. As Descartes explains, this centrifugal pressure might potentially generate what we call the light, and the centre of any region (he calls each region a 'vortex') might be a star. Paradoxically, due to such a mechanism, the centre of any vortex would emit light, even if it were void (*Principia* III, 58–64). As we said, the skies are composed of rotating vortices made up of particles belonging to the second element. The particles of the first element are very little in comparison to those of the second. The centrifugal tendency of each vortex is sufficient to expel those little particles of the first elements whose motion out of the vortex is not impeded by the contiguous vortices because these tiny particles enter among the interstices of the particles of the second element. Thence, several particles of the first element moving from a contiguous vortex reach the centre of each vortex. The problem is that, when these particles reach the centre of the vortex, there is no apparent mechanism which would prevent them from escaping toward another vortex. To avoid this effect Descartes invented a complicated device, the basis of which is that the particles of the first element entering a vortex from the poles of the vortex are subject to a pressure exerted by the particles belonging to the second element. This pressure is directed towards the centre of the vortex, so that the particles of the first element remain in the centre of the vortex. Conversely, the particles of the first element, which do not enter the vortex along the pole-line, but from a different direction, are not subject to such a pressure and, thus, escape towards another vortex. Therefore, no hole can exist: the stars are composed of particles belonging to the first element (*Principia* III, 69–72).

The emission of light is due to the pressure exerted by these small and quickly rotating particles of the first element, set in the centre of a vortex, on the particles of the second element composing the vortex. The issue concerning the origin and transmission of light is consequent upon Descartes' idea that the centrifugal tendency of the rotating vortices plays a fundamental role. This tendency is a unifying element of Descartes' cosmology and it is worth considering how it is applied to light theory. Furthermore, light theory—as we will point out in the commentaries to this section (see Sect. 5.1.3)—might help us to guess the problematic role of the inertia principle within Descartes' cosmology. The transmission-mechanism of light is analysed by Descartes in *Principia* III, 64 and had been discussed in details in the 13th chapter of *Le Monde* (Fig. 5.2). To summarize, the situation is as follows.[7]

Descartes considers the Sun S surrounded by its rotating vortex. E is a point of the vortex. Because of the vortex's rotation, a particle in E is pushed towards M by the particles on the straight line SEM. In fact, the situation is more complicated because Descartes claims that such a particle is not pushed only towards M, but feels a pressure toward the whole zone LEN that is exerted by the particles of the zone $ADFG$. However, this aspect is not fundamental in the context we are analysing. Let

[7] The explanation of the origin and transmission of light is drawn, with some slight modifications, from Bussotti and Lotti (2018, pp. 105–107). On Descartes' theory of light in the context of the distinction between emission theory and continuum theory of light see Giudice (2010, pp. 45–48).

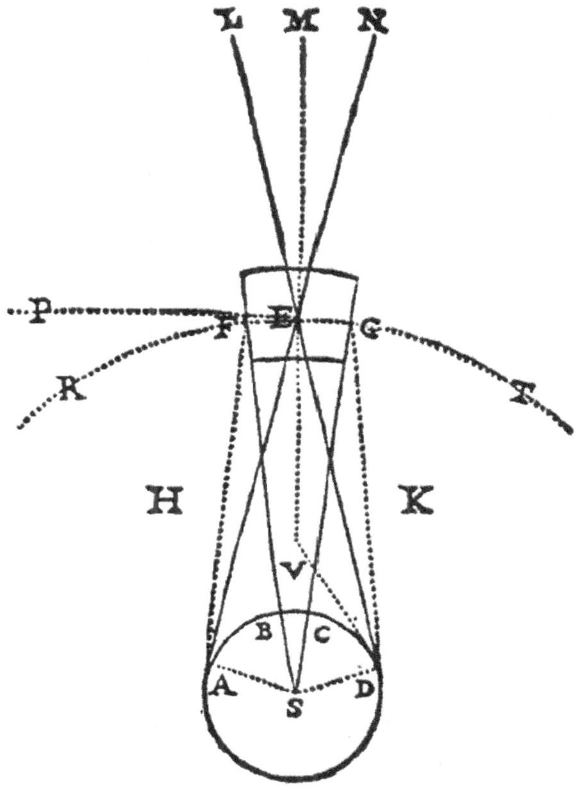

Fig. 5.2 The mechanism of light transmission expounded in *Le Monde* (AT IX, p. 47)

us now suppose that the space is void of the particles of the first element and that only particles of the second element exist. To understand how a body is pushed by other bodies, the best method is to imagine what happens if the place which will be occupied by the body is considered void. Descartes imagines, thence, a completely void region around *E*. After a long series of tortuous and unconvincing arguments, he concludes that all the particles in the zone *ADFG* proceed *in act* towards the free zone around *E*. The cause of this movement is the centrifugal inclination from *S* outwards. Thus, the centrifugal tendency produces a real motion, and not merely a tendency to motion, if it is not impeded by empirical circumstances.

Having explained the mechanism of light transmission, Descartes analyses how comets and planets are born: the particles of the first element have different forms and, when they are pressed together within a star, can mutually join and make up progressively bigger particles. These bigger particles compose the third element. When they have reached a sufficient dimension, they begin to form the sunspots (*Principia*, III, 93–110). It can happen that the sunspots or stars-spots become progressively bigger. If they become big enough, they get slow and obscure so that they cannot exert pressure on the particles of the second element. At this point, the vortex becomes unstable. It cannot bear the pressure of other vortices and collapses. The star of a collapsed vortex—that has become obscure—is pushed within another vortex where it assumes the form of a planet or of a comet, according to a series of different circumstances (*Principia*, III, 112–119). One might say, therefore, that the skies are a first generation, the stars a second generation, and the planets and comets a third generation objects. Comets are particularly interesting as they give us indication of the vastity of the universe: Descartes guesses that the orbits of the comets extend far away from the Sun and go beyond Saturn. Therefore, a vast space between the Sun and other stars is needed to account for their long and various excursions (*Principia*, III, 41).

Descartes tries to explain the celestial phenomena mechanically: the formation of the stars, of the planets, of the comets, the transmission of light and so on, are explained through the movements of the different kinds of particles. In the fourth part of his work, he gives a mechanical account of several terrestrial phenomena such as gravity and magnetism. Each explanation resorts to circular motions and collisions.

5.1.3 Commentaries

(1) Some commentaries concerning the numerous problematic aspects of Descartes' principles are necessary: the idea to consider the kinetic state of a body as a physical condition relative to the kinetic state of other bodies was not new: we have seen that Copernicus and Galileo too expressed this conviction. However, Descartes' formulation is more general, as is clear from his statement of the first two laws of motion:

> *The first law of nature: each and every thing, in so far as it can, always continues in the same state; and thus what is once in motion always continues to move.* From God's immutability we can also know certain rules or laws of nature, which are the secondary and particular causes of the various motions we see in particular bodies. The first of these laws is that each thing, in so far as it is simple and undivided, always remains in the same state, as far as it can, and never changes except as a result of external causes.

And:

> *The second law of nature: all motion is in itself rectilinear; and hence any body moving in a circle always tends to move away from the centre of the circle which it describes.* The second law is that every piece of matter, considered in itself, always tends to continue moving, not in any oblique path but only in a straight line. This is true despite the fact that many particles are often forcibly deflected by the impact of other bodies; and, as I have said above, in any motion the result of all the matter moving simultaneously is a kind of circle.[8]

[8] *Principia* II, 37, Descartes (1984–1991, I, pp. 240–241), Cf. AT VIII, p. 62: "*Prima lex naturae: quod unaquaeque res, quantum in se est, semper in eodem statu perseveret: sicque quod semel movetur, semper moveri pergat.* Atque ex hac eadem immutabilitate Dei, regulae quaedam sive leges naturae cognosci possunt, quae sunt causae secundariae ac particulares diversorum motuum, quos in singulis corporibus advertimus. Harum prima est, unamquamque rem, quatenus est simplex & indivisa, manere, quantum in se est, in eodem semper statu, nec unquam mutari nisi a causis externis".

One might expect a treatment of natural philosophy based on the relativity of any motion (*Principia* II, 28) and on the inertia principle asserted in the first and second law of motion. However, these rules play an ambiguous role, and the explanation of curved motions—which is a fundamental issue in a cosmological perspective since the motions of celestial bodies are curved—is extremely problematic.[9] The main difficulty lies in Descartes' plentitude principle because: (1) in a full universe, it is very difficult to guess what the meaning of the inertia principle should be. At best, it might be a sort of ideal law which would hold in an abstract and absolute space devoid of any matter, a sort of limit law. However, this solution is not accepted by Descartes even in principle, since he denies the distinction between space and matter, as will be clarified in Sect. 5.2; thence, the postulation of a space separated—also ideally—from matter is insubstantial for Descartes. To be more precise: in *Principia* II, 10 Descartes specifies that the distinction between space and bodies is only a subjective conception and that the idea of a space separated from matter is an abstraction devoid of physical meaning:

> There is no real distinction between space, or internal place, and the corporeal substance contained in it; the only difference lies in the way in which we are accustomed to conceive of them. For in reality the extension in length, breadth and depth which constitutes a space is exactly the same as that which constitutes a body. The difference arises as follows: in the case of a body, we regard the extension as something particular, and thus think of it as changing whenever there is a new body; but in the case of a space, we attribute to the extension only a generic unity, so that when a new body comes to occupy the space, the extension of the space is reckoned not to change but to remain one and the same, so long as it retains the same size and shape and keeps the same position relative to certain external bodies which we use to determine the space in question.[10]

In defence of Descartes one might claim that his principle is valid in an abstract geometrical space, but the problem is precisely that, according to him, space is not separable from matter. However, this is contradictory because no physical object

---

(AT VIII, p. 62). *Principia* II, 39, Descartes (1984–1991, I, pp. 241–242). Cf. AT VIII, p. 63. Second law: "*Altera lex naturae: quod omnis motus ex se ipso sit rectus: et ideo quea circulariter moventur tendere semper ut recedant a centro circuli quem describunt.* Altera lex naturae est: unamquamque partem materiae, seorsim spectatam, non tendere unquam ut secundum ullas lineas obliquas pergat moveri, sed tantummodo secundum rectas; etsi multae saepe cogantur deflectere propter occursum aliarum, atque, ut paulo ante dictum est, in quolibet motu fiat quodammodo circulus, ex omni materia simul mota". (The sentences in italics are the titles of the original sections).

[9] For the problem of circular or, to be more precise, curved motion in Descartes see Bussotti and Lotti (2018). On this subject, see also: Shea (1991), in particular Chap. 10; Garber (1992, pp. 220–225), Garau (2014), Gaukroger (2002, pp. 114–121), Schuster (2013, pp. 171–172), Shapiro (1974), Slowik (1999).

[10] Descartes (1984–1991, I, p. 227). AT VIII, p. 45: "Non etiam in re differunt spatium, sive locus internus, & substantia corporea in eo contenta, sed tantum in modo, quo a nobis concipi solent. Revera enim extensio in longum, latum & profundum, quae spatium constituit, eadem plane est cum illa quae constituit corpus. Sed in hoc differentia est, quod ipsam in corpore ut singularem consideremus, & putemus semper mutari quoties mutatur corpus: in spatio vero unitatem tantum genericam ipsi tribuamus, adeo ut, mutato corpore quod spatium implet, non tamen extensio spatii mutari censeatur, sed remanere una & eadem, quamdiu manet ejusdem magnitudinis ac figurae, servatque eundem situm inter externa quaedam corpora, per quae illud spatium determinamus".

can exist in an abstract geometrical space. Descartes argues "that the extension that constitutes the nature of a body is exactly the same as the extension that constitutes the nature of a space" (*Principia* II 11). Descartes seems unaware that inertia has nothing to do with the extension of a body, but rather with its mass, which is an intrinsic property of the body and cannot be in any way reduced to space and geometry.

(2) There is a further specific issue to be considered: we will see that Newton referred the inertia principle to absolute space. In the celebrated scholium to the *Definitions* in the *Principia* (see Sect. 7.1). Newton states that the forces of receding from the axis of circular motion (centrifugal forces) distinguish absolute from relative motion. This implies that absolute motions have to be referred to absolute space, so that centrifugal forces also have to be referred to absolute space. Since these forces are due exclusively to the bodies' inertia, inertia is to be referred to the absolute space too. The problem of a physical entity to which to refer inertia has been fundamental in the history of physics. Without such a reference, it is problematic to separate the inertial reference frames (which, as is well known, are fundamental in Newton's physics and in any physical theory where an inertia principle exists) from the non-inertial ones. Ernst Mach, who was highly critical of Newton's absolute space, did not think that the problem of inertia was a mere metaphysical issue. In his *The Science of Mechanics*, he explained the thesis that inertia has to be referred to all the masses of the universe. Einstein himself clearly argued that without a radical change of paradigm in physics, the reference to an absolute space is difficult to avoid. As he wrote in 1930:

> The physical space was so completed in a four-dimensional space, which also entails the temporal dimension. The four dimensional space of special relativity is as rigid and absolute as Newton's space.[11]

This is correlated to the fact that the special relativity extended coherently the concept of inertial system to the electromagnetic phenomena as well, but such a concept remains fundamental. Einstein wrote:

> Now the special relativity theory came with its knowledge of the physical equivalence of all the inertial systems.[12]

It is only in the general relativity that the treatment of inertial and non-inertial reference frames was unified.

---

[11] Our translation. Cf. Einstein (1988, 1930) p. 144): "Der physikalische Raum wurde so zu einem vierdimensionalen Raum ergänzt, der auch die zeitliche Dimension enthält. Der vierdimensionale Raum der speziellen Relativitätstheorie ist ebenso starr und absolut wie der Raum Newtons". Originally this paper appeared in 1930 on the journal *Forum Philosophicum* 1, pp. 173–180, and was republished in *Mein Weltbild* (1934). Our reference is to the 1988 edition of *Mein Weltbild*, available on the Internet: https://gedankenfrei.files.wordpress.com/2009/01/mein-weltbild-albert-einstein.pdf.

[12] Our translation. Cf. Einstein (1988, 1930, p. 144): "Nun kam die spezielle Relativitätstheorie mit ihrer Erkenntnis der physikalischen Gleichwertigkeit aller Inertialsysteme".

Now, if, as Descartes thought, no absolute space and no absolute motions exist, it is simply impossible to refer inertia to a physical space (using Einstein's words). Inertia remains an abstract concept that cannot be coherently framed in a system of physical causality. The idea that any motion is relative becomes an impediment to develop a consistent physics if it is not inserted in a general frame compatible with such a conception. This happened only with Einstein's general relativity or, at least, with the Machian vision of classical physics, not with Descartes's and neither with Leibniz's physics. The plentitude principle and relativity principle converged to make a consistent use of inertia at best problematic within Descartes's physical universe.

In an almost void universe such as the Newtonian one, the principle of inertia can make sense from a physical point of view, even without believing in the existence of absolute space, as Newton himself did. As a matter of fact, the inertia principle holds approximately. For example, a body moving in space and distant from big masses has an "almost" uniform rectilinear motion with respect to the barycentre of those masses: clearly the notion of inertia remains problematic from a theoretical standpoint (this topic will be addressed in the section on Newton), but in most physical cases this principle has a clear operative sense. However, even an operational solution seems problematic in Descartes's case since his universe is completely full, so full that the motions are determined by the instantaneous impact of a particle with other particles and that no rectilinear motion exists, not even approximately. Descartes is convinced that any motion is curvilinear and that when we perceive a motion which appears rectilinear, this is, in fact, only a rectilinear part of a trajectory which, in its totality, is curvilinear. Descartes repeatedly stated that the curvilinear motions depended on the plentitude principle. For example, in a letter to Reneri dated 2 July 1634 he wrote:

> [...] first of all, I would like to claim the necessity to consider that no void exists in nature. As a consequence, when a body moves, it necessarily enters the place of some other body, from which the latter body is expelled. In the same instant, the displaced body enters the place of a third body, and this one the place of a further body and so on, until the last body occupies the place left by the first one, so that any movement of the universe is, in a sense, circular.[13]

On several other occasions he tried to deduce from the principle of plentitude the idea that all motions take place in closed paths. As we have pointed out, the plentitude principle does not imply as a logical consequence the necessity that every motion is circular.[14]

Thus, in Descartes' physics, the only possible interpretation of inertia is the following: it is a principle which holds for any motion. If the vacuum existed, inertial motions would also exist. However, since no void is given, no inertial motion actually

[13] Our translation. Cf. AT, I, p. 301: "[...] ie vous diray que premierement il faut considerer qu'il n'y a point de vuide en la nature, et que par consequent lors qu'un cors se meut, il doit necessairement entrer en la place de quelque autre, de laquelle celuy qui en est chassé, doit au mesme instant occuper celle d'un autre, et celuy-cy derechef celle d'un autre, et ainsy de suite, iusques a ce que le dernier occupe la place qui est laissee par le premier, de façon que tous les mouvemens qui se font au monde sont en quelque façon circulaires".

[14] Bussotti and Lotti (2018, § 2.2, pp. 83–85).

exists. Inertia is a tendency which is never actualized, but which, in some phenomena, such as light and the movement of the vortices, exerts an important influence on the actual events. Inertia is a principle that regulates all motions, but it manifests itself only as a tendency that can never be actualized, due to the fact that our universe is full. It should be highlighted that the status of inertia principle in Descartes is rather complex and not clear in all its facets.

Another crucial problem should be considered: Descartes repeatedly argued that no motion can take place in an instant. For example, in the 12th Rule, we read:

> It is in this way that shape is conjoined with extension, motion with duration or time, etc., because we cannot conceive of a shape which is completely lacking in extension, or a motion wholly lacking in duration.[15]

We have proved that this (correct) idea is in contradiction with the other (incorrect) idea that a particle replaces instantaneously another particle and that this contradiction cannot be avoided, whatever interpretative hypotheses you add.[16]

Thus, if you analyse the fine tuning of Descartes's basic principles and assumptions, they do not form a coherent and unitary logical-argumentative structure. In great part, this depends on the fact that Descartes did not give a mathematical formulation of his principles and their consequences. Presumably he thought that, in the case of cosmology, his duty was to offer just a general picture within which all phenomena might be framed. Some physicists would later develop the necessary mathematical specifications. Had he transcribed into mathematical terms his principles, probably he would have realized that they were not mutually consistent. It is not a coincidence that Cartesian scientists had great difficulties in adapting mathematics to such principles. The greatest of these scientists, Huygens, as we will see, had to modify many of them. Thus, Huygens's theories are very different from those proposed originally by Descartes, though the general frame in which Huygens worked was a Cartesian one.

(3) The mechanism of light transmission is a further evidence of the ambiguity of Descartes's inertia principle. Since he thought that light is due to a centrifugal tendency, what better occasion to point out the identity of the centrifugal tendency with the inertia principle? What better occasion to recall the second law of motion? But Descartes did not follow this line of reasoning. He thought that the tendency to move along the tangent is due to the tendency of a body to move along a circumference plus a radial centrifugal tendency, an assumption which involves his theory in something very problematic and wrong from a physical point of view.[17]

---

[15] Descartes (1984–1991, I, pp. 45–46). Cf. AT X, p. 421: "[...] hoc pacto figura extensioni conjuncta est, motus durationi, sive tempori, &c., quia nec figuram omni extensione carentem, nec motum omni duratione, concipere licet".

[16] Bussotti and Lotti (2018, § 2.3.1, pp. 86–87).

[17] We have offered a complete and detailed analysis of the way in which Descartes interpreted the tendency along the tangent in a curvilinear motion in Bussotti and Lotti (2018, pp. 92–97 and 104–112).

It is clear that Descartes intuited the importance of the inertia principle, since he considered it one of the fundamental laws of nature. This was, undoubtedly, a great merit of Descartes. A further merit was the idea to state clearly a series of laws of nature which should represent the initial bases of physical deductions. But, since the entire context of his physics was not coherent with the inertia principle, he was not able to exploit the potential of such a productive idea. Obviously, the lack of a transcription in mathematical terms of the physical magnitudes considered by Descartes gave him a certain amount of conjectural freedom in creating concepts whose inconsistency might have been exposed by a deeper mathematical and experimental analysis.[18]

(4) Descartes tried to outline a mechanistic physics which excluded mysterious influxes and forces. Every phenomenon must be explained by clear and detectable mechanisms. This is a great step forward in comparison to most of his predecessors and contemporaries. However, the impression of clarity vanishes when one examines the details of Descartes's physics because, as we have pointed out, he did not try to offer a precise mathematical treatment of the phenomena he studied. He harboured the grandiose vision of explaining mechanically all physical phenomena, but his natural philosophy and cosmology consist of qualitative descriptions with no predictive power. In a sense, the idea of explaining every phenomenon by simple kinematical means, like motions and collisions, is the dream of every physicist, but Descartes's extremely simplified picture was bound to reveal its weakness within a few years. A comparison with Kepler may be helpful to understand the strengths and weaknesses of Descartes's view. In some aspects of his thought, Kepler was still tied to vague speculative conceptions and to theological assumptions that conditioned his idea of the universe and that Descartes avoided. However, Kepler realized that a cause or a series of causes had to be found for celestial movements. In other words: it seems that Kepler understood that, to explain the motions of the heavens, it is not enough to enrich the number of objects that make up the universe, as Descartes did by introducing a micro-structure of three kinds of particles. According to Kepler, it was also necessary to postulate some interactions among the objects different from motions and impacts. These interactions, which are more basic, should explain motions and impacts.

[18] Our analysis of Descartes's inertia principle concerns its physical meaning. Slowik (2005, 2017), in particular sections 3–6, shows that the origin and meaning of the principle of inertia in Descartes is not only tied to physical questions, but also to metaphysical ones.

This means the introduction of a new ontological level which is tied neither to mysterious influences, nor to the actions of souls or minds,[19] nor to the material composition of the objects, but to their relations.

(5) We have seen that the most unsatisfactory aspect of Descartes's natural philosophy is the lack of a mathematical treatment. This does not mean that Descartes was not able to offer a mathematical analysis and solution to some physical problems: the second discourse of the *Dioptrique* offers an interesting treatment of refraction, though based on some wrong assumptions. In the same work the discourses concerning the mechanism of vision (the 5th and the 6th ones) are developed through rigorous reasonings and acute empirical observations. The 8th discourse concerning the dependence of refraction on the form of the refracting surfaces is a little masterpiece where the use of mathematics is decisive. Descartes' fundamental contribution to optics is his rainbow theory explained in the 8th discourse of *Les Météores* where Descartes used experience, experiments and mathematics as a great scientist.[20] In the *Principia* Descartes proposed a too ambitious project based on the conviction of the simplicity of nature without realizing that, in natural philosophy, it is relatively easy to offer a general picture, but it is difficult to frame a correct explanation and prevision of phenomena within this picture. Mathematics and experiments, which are the instruments which allow us to study the relations among events and to understand if a general theory is coherent with the phenomenal world, are the two missing aspects of Descartes' *Principia.* From a scientific and epistemological standpoint, this work can be interpreted as a sort of research program, that inaugurated a new era of mechanical investigations, rather than a complete physical system.

5.2 Philosophical Issues of Descartes' Cosmology

Of the authors here examined Descartes and Leibniz are by far the most significant in the history of philosophy. Descartes was a mathematician, a natural philosopher (with a predominant interest in physics and medicine) and a metaphysician who profoundly influenced early modern thought, reshaping traditional metaphysics with a new subjectivist approach (see his *Meditationes de prima philosophia* 1641). Scholars debate the relationship between metaphysics and science in Descartes' philosophy:

---

[19] We do not claim that Kepler did not believe in the existence of souls and minds of the celestial bodies. He certainly did. But while he thought—at least starting from the *Astronomia Nova*—that the rotation of each body around its axis was explicable only in connection to the existence of each body's soul, the rotation of the planets around the sun has to be explained by a purely physical force. On this problem, see Pisano and Bussotti (2018, § 16.5.2), where vast references to the literature are also present.

[20] The literature on Descartes' optics is abundant. Without any claim to be exhaustive, we mention: Armogathe (2000), Bussotti and Pisano (2013), Maitte (2006), Malet (1990), Ronchi (1956), Sabra (1967), Schuster (2000), Tiemersma (1988).

when a reader opens the *Principles of Philosophy*, that is the general exposition of the Cartesian philosophy, he or she will find hard to escape the impression that the first part is only a metaphysical preamble to those contents that really interested Descartes. These are minutely examined in the remaining three parts, dedicated to physics, which are disproportionally larger compared to the metaphysical introduction. It is indisputable that Descartes asserted that metaphysics, grounded on the unshakable evidence of the cogito, offered a solid theoretical foundation for natural science but, in fact, it is to science that he dedicated most of his research. Descartes' well-known dualism between thinking and extended substance influenced the development of philosophy until Kant—the Kantian distinction between phenomenon and noumenon being only the last outcome of the original Cartesian antithesis. For our purpose it is sufficient to note that Cartesian dualism broke entirely with the philosophical tradition, in so far as it conferred the status of a substance to matter itself and made it autonomous from any spiritual substance: neither God nor the soul nor other spiritual entities are present or active in the material mechanisms of nature. A further consequence of his dualism was the problematic condition of man in the universe: through his body, conceived as an automaton, man was an integral part of a mechanistic universe, but, because of his mind, he was entirely independent of it. The antithesis between phenomenal necessity and noumenal freedom in Kant's philosophy is a further expression of this dualism, which still persists in contemporary philosophical debates on the so-called mind–body problem and on the issue of determinism versus freedom in human agency. The question of Cartesian dualism is often a source of misunderstanding. It is true that in the Cartesian universe there are only two types of substances with opposite attributes, but man is a compound substance that constitutes the unique connection point between the two substances. Descartes was perfectly aware of the tight causal interaction between mind and body, as he amply demonstrated in his last work, *Passions of the Soul* (1649). In a letter of 21 May 1643 to the princess Elisabeth of Bohemia, Descartes stated that the union of soul and body is a primitive notion, to be posited on the same level of intellectual evidence of the notions we have of the thinking substance and of the extended substance.[21] Therefore, the union of the two substances has for Descartes the same degree of evidence of their distinction. For this reason some interpreters think that "trialism" instead of dualism is a more appropriate definition of Descartes' ontology.[22]

In narrowing our focus on the philosophy of nature and cosmology, we must deal with several topics to give an outline of Descartes' views. The integral mechanization of nature, and in particular the theory of the evolution from the initial chaotic state of matter to an ordered cosmos – a transition determined only by the laws of motion – entailed a significant reduction of God's role in relation to nature. Descartes' mechanistic theism still assigned the cause of motion and of its laws to God's creative power but expelled any presence of divine providence from natural phenomena, in

[21] See Descartes (1984–1991 III, p. 218): "Primitive notions [...] are, as it were, the patterns on the basis of which we form all our other conceptions" (cf. AT III, p. 665).

[22] Descartes (1984–1991 III, pp. 218–230).

stark contrast with the Aristotelian and Platonic traditions. As Pascal noted in one of his thoughts: "I cannot forgive Descartes. In all his philosophy he would have been quite willing to dispense with God. But he had to make Him give a fillip to set the world in motion; beyond this, he has no further need of God" (Pascal, 1958, p. 23). Pascal's hostile remark was shared by many other Christian philosophers who feared that Descartes' mechanicism was the first step towards a fully-fledged atheistic materialism. Descartes held that by the sheer operation of the laws of nature "matter must successively assume all the forms of which it is capable; and if we consider these forms in order we'll eventually arrive at the form that characterizes the universe in its present state".[23] This idea triggered a negative reaction since it appeared to be too close to a Lucretian cosmogony (see *De rerum natura* 5, 55 ff. 516 ff.) and led concerned and diffident readers to regard the Cartesian system as atheism in disguise. Leibniz shared this concern and accused Descartes of supporting, like Hobbes and Spinoza, a system of material necessity that eliminated the very distinction between possible and actual existence:

> I do not think that a more dangerous proposition than that [see *Principia* III 47] can be formed. For if matter receives all possible forms successively it follows that nothing can be imagined sufficiently absurd nor bizarre, nor contrary to what we call justice which has not either already happened or will happen. These are exactly the sentiments which Spinoza has explained more clearly, namely that justice, beauty, order are only things which relate to us, but that the perfection of God consists in the plenitude of His action such that there is nothing possible or conceivable which is not actually produced. This is also the opinion of Mr Hobbes who maintains that all that is possible is past or present or future… Mr Descartes took very good care not to speak too clearly, but that did not prevent him from revealing his opinions in passing with such cunning that they would only be understood by those who

[23] Descartes (1984–1991 I, p. 258) (*Principia* III 47). See also *Le Monde* chap. VI: "Let us suppose that God really divides it [sc. matter] into many such parts, some larger and some smaller, some of one shape and some of another, however we care to imagine them. […] From the first instant of their creation, he causes some to start moving in one direction and others in another, some faster and others slower (or even, if you wish, not at all); and he causes them to continue moving thereafter in accordance with the ordinary laws of nature. For God has established these laws in such a marvellous way that even if we suppose he creates nothing beyond what I have mentioned, and sets up no order or proportion within it but composes from it a chaos as confused and muddled as any the poets could describe, the laws of nature are sufficient to cause the parts of this chaos to disentangle themselves and arrange themselves in such good order that they will have the form of a quite perfect world – a world in which we shall be able to see not only light but also all the other things, general as well as particular, which appear in the real world" (Descartes, 1984–1991, vol. I, p. 91). See AT XI, pp. 34–35: "Et supposons de plus, que Dieu la divise veritablement en plusieurs telles parties, les unes plus grosses, les autres plus petites; les unes d'une figure, les autres d'une autre, telles qu'il nous plaira de les feindre. […] dés le premier instant qu'elles sont creées, les unes commencent à se mouvoir d'un costé, les autres d'un autre; les unes plus viste, les autres plus lentement (ou mesme, si vous voulez, point du tout), & quelles continuent par aprés leur mouvement suivant les loix ordinaires de la Nature. Car Dieu a si merveilleusement estably ces Loix, qu'encore que nous supposions qu'il ne crée rien de plus que ce que j'ay dit, & mesme qu'il ne mette en cecy aucun ordre ny proportion, mais qu'il en compose un Cahos, le plus confus & le plus embroüillé que les Poëtes puissent décrire: elles sont suffisantes pour faire que les parties de ce Cahos se démélent d'elles-mesmes, & se disposent en si bon ordre, qu'elles auront la forme d'un Monde tres-parfait, & dans lequel on pourra voir non feulement de la Lumiere, mais aussi toutes les autres choses, tant generales que particulieres, qui paroissent dans ce vray Monde".

> make a profound study of this kind of thing. It is, in my opinion, the πρωτον ψευδος and the foundation of the Atheistical Philosophy, which nevertheless appears to say some beautiful things about God.[24]

The anti-Cartesians reacted by maintaining or reviving traditional finalism (see Leibniz for instance) and by incrementing in every possible way the so called 'argument from design', which was typical of natural theology particularly in British thought, in which it remained popular until Darwin, notwithstanding Hume's demolishing critique.[25] It is noteworthy that, due to the combination of subjectivist metaphysics and mechanistic physics, Descartes abandoned all the traditional proofs of the existence of God drawn from the structure and order of the world.

Descartes had explicitly rejected any consideration of final causes as useless in science and inappropriate in metaphysics. To explain particular natural phenomena final causes must be left aside in favour of efficient causes, the only ones required in a mechanistic view of nature. Moreover, we cannot presume to know God's goals in creating the universe and it would be an act of foolish arrogance to believe that God created the world only for the benefit of men. Descartes insisted on the infinite inscrutable omnipotence of God to reject finalism and anthropocentrism. The actions of the Cartesian God have to be measured in the cosmic realm, which is the natural manifestation of his creative power, and not with reference to man; the physical world has value in itself as a work of God's creative will and it is not merely the background of the relation between God and man.

In *Principia* I 28, Descartes refused to base natural philosophy on the research of final causes:

> When dealing with natural things we will, then, never derive any explanations from the purposes which God or nature may have had in view when creating them <and we shall entirely banish from our philosophy the search for final causes>. For we should not be so arrogant as to suppose that we can share in God's plans. We should, instead, consider him as the efficient cause of all things; and starting from the divine attributes which by God's will we have some knowledge of, we shall see, with the aid of our God-given natural light, what

[24] Our translation. Letter to Philipp, January 1680: *Philosophischen Schriften*, ed. Gerhardt, vol. 4, p. 283f.: "Je ne crois pas qu'on puisse former une proposition plus perilleuse que celle là. Car si la matiere reçoit toutes les formes possibles successivement, il s'ensuit qu'on ne se puisse rien imaginer d'assez absurd ny d'assés bizarre et contraire à ce que nous appelons justice, qui ne soit arrivé et qui n'arrive un jour. Ce sont justement les sentimens que Spinosa a expliqués plus clairement, sçavoir que justice, beauté, ordre ne sont que des choses qui se rapportent a nous, mais qui la perfection de Dieu consiste dans cette amplitude de son operation en sorte que rien ne soit possible ou concevable qu'il ne produise actuellement. Ce sont aussi les sentimens de M. Hobbes qui soutient que tout qui est possible, est passé ou present ou futur, et il n'y aura pas lieu de se rien promettre de la providence, si Dieu produit tout et ne fait point de choix parmy les estres possibles. Mons. des Cartes s'est bien donné de garde de parler si nettement, mais il n'a pù s'empecher de decouvrir ses sentimens en passant avec une telle addresse qu'il ne sera entendu que de ceux qui examinent profondement ces sortes des choses. C'est à mon avis le πρωτον ψευδος et le fondement de la Philosophie Athée, qui ne laisse pas de dire de Dieu des belles choses en apparence".

[25] For the most outstanding instances of English natural theology based upon the argument from design see Ray (1691), Bentley (1692–93), Derham (1713, 1715), Paley (1802). Hume moved a devastating criticism to all the commonplaces of natural theology in his *Dialogues concerning natural religion* (1779). For an overview of this topic see Emerton (1989).

> conclusions should be drawn concerning those effects which are apparent to our senses. At the same time we should remember, as noted earlier, that the natural light is to be trusted only to the extent that it is compatible with divine revelation.[26]

On the same line of thought, as to the relation between God and man, it is significant what Descartes claimed in *Principia* III 1–3, where (specifically in III, 3) we read:

> In ethics it may admittedly be an act of piety to assert that God made everything for our benefit, since this may impel us all the more to give him thanks and burn with love for him. Indeed the assertion is in a sense true, since we can make some use of all things, if only by exercising our minds in contemplating them, and thus adoring God for his marvellous works. But nevertheless it is wholly improbable that all things were in fact made for our benefit, in the sense that they have no other use. And in the study of physics such a supposition would be utterly ridiculous and inept, since there is no doubt that many things exist, or once existed, though they are now here no longer, which have never been seen or thought of by any man, and have never been of any use to anyone.[27]

Overcoming Descartes' theology, Spinoza was to bring to completion the discarding of finalism and anthropocentrism in the relationship between man and the infinity of nature (see Spinoza, Appendix to *Ethica* Part I).[28] The demise of finalism in Descartes' world-view represents a significant historical change that cannot be underrated. In his seminal work, Edwin Burtt highlighted it in terms that are still valid:

> The vortex theory was [...] a most significant achievement historically. It was the first comprehensive attempt to picture the whole external world in a way fundamentally different from the Platonic-Aristotelian-Christian view which, centrally a teleological and spiritual conception of the processes of nature, had controlled men's thinking for a millenium and a half. God had created the world of physical existence, for the purpose that in man, the highest natural end, the whole process might find its way back to God. Now God is relegated to the position of first cause of motion, the happenings of the universe then continuing *in*

---

[26] Descartes (1984–1991, I, pp. 202–203). Cf. AT VIII, pp. 15–16: "Ita denique nullas unquam rationes, circa res naturales, a fine quem Deus aut natura in iis faciendis sibi proposuit, desumemus: quia non tantum nobis debemus arrogare, ut ejus consiliorum participes esse putemus. Sed ipsum ut causam efficientem rerum omnium considerantes, videbimus quidnam ex iis ejus attributis, quorum nos nonnullam notitiam voluit habere, circa illos ejus effectus qui sensibus nostris apparent, lumen naturale, quod nobis indidit, concludendum esse ostendat; memores tamen, ut jam dictum est, huic lumini naturali tamdiu tantum esse credendum, quandiu nihil contrarium a Deo ipso revelatur".

[27] Descartes (1984–1991, I, pp. 248–249). Cf. AT VIII, 6, p. 81: "Quamvis enim in Ethicis sit pium dicere omnia a Deo propter nos facta esse, ut nempe tanto magis ad agendas ei gratias impellamur, ejusque amore incendamur; ac quamvis etiam suo sensu sit verum, quatenus scilicet rebus omnibus uti possumus aliquo modo, saltem ad ingenium nostrum in iis considerandis exercendum, Deumque ob admiranda ejus opera suspiciendum: nequaquam tamen est verisimile, sic omnia propter nos facta esse, ut nullus alius sit eorum usus; essetque plane ridiculum & ineptum id in Physica consideratione supponere; quia non dubitamus, quin multa existant, vel olim extiterint, jamque esse desierint, quae nunquam ab ullo homine visa sunt aut intellecta, nunquamque ullum usum ulli praebuerunt." See also Descartes' letter to Chanut, The Hague, 6 June 1647: "I do not believe that we are compelled to believe that man is the aim of Creation" (cfr. Descartes, 2009, p. 2470: AT V 53).

[28] Dessì (2012, p. 61f).

> *aeternum* as incidents in the regular revolutions of a great mathematical machine. Galileo's daring conception is carried out in fuller detail. The world is pictured concretely as material rather than spiritual, as mechanical rather than teleological (Burtt, 1925, p. 105).

It is true that the supremacy of the Christian God is preserved in Descartes' metaphysics and that the Cartesian view of the relationship between God and the world should not be confused with the deistic image of the remote architect of the universe.[29] God is not only the first cause, but also the only true infinite being endowed with all perfections. In the Third Meditation Descartes emphasized the radical contingency of every created being that is preserved in its existence instant after instant by the will of God, so that the conservation of beings is like their continuous creation (AT VIII p. 48 f.). The very existence of the external world is demonstrated at the end of the itinerary of the *Metaphysical Meditations* thanks to the previous proof of the existence of a benevolent God. Finally, Descartes upheld that the logical and mathematical truths are not intrinsic to the divine essence, but are freely established by the imperscrutable will of God (the so-called doctrine of the "creation of the eternal truths"). However, all these metaphysical tenets belong to a frame that doesn't affect the entirely material and mechanical view of the physical universe, in which everything is determined by the transmission of motion according to the immutable laws of nature as second causes. The voluntarism of Descartes' theology is neutralized by Descartes himself when he states that, once God has fixed the laws of nature, his will remains unchanged and the divine immutability grants the invariance of the laws of nature and the conservation of the quantity of motion, that "constitues the supreme natural law", as Dijksterhuis remarked.[30] As for the principle of causality, it is undoubtedly the backbone both of Descartes' physics and metaphysics, and it must be considered from an ontological and epistemological perspective. Ontologically, the idea of "efficient causality" is metaphysically founded on God who is the first cause. However, epistemologically, Descartes thinks that we are allowed to apply the principle of causality to everything, including God itself:

> Nothing exists concerning which one can't ask 'Because of what cause does it exist?' This question can even be asked about God, not because he needs any cause in order to exist, but because the immensity of his nature is the cause or reason why he doesn't need any cause in order to exist.[31]

Thomas Aquinas had asserted that "nothing is the cause of itself; if it were, it should be prior to itself, which is impossible".[32] Descartes is so fond of the principle

---

[29] Williams (1978 [2005], , p. 134).

[30] Dijksterhuis (1961), p. 410. See Descartes, *Principia Philosophiae* II, §§ 36–37: AT VIII, pp. 61–63.

[31] "Objections to the Meditations and Descartes's Replies", transl. by J. Bennett, p. 38: http://www.earlymoderntexts.com/assets/pdfs/descartes1642_1.pdf. This assertion is drawn from the remarkable exposition in geometric fashion of his system of philosophy given by Descartes in his Replies to the Second Objections. Cf. AT VII pp. 164f: "Nulla res existit de qua non possit quaeri quaenam sit causa cur existat. Hoc enim de ipso Deo quaeri potest, non quod indigeat ulla causa ut existat, sed quia ipsa ejus naturae immensitas est causa sive ratio, propter quam nulla causa indiget ad existendum".

[32] Aquinas (1918, I, c. 18, n. 4, p. 49).

of efficient causality that he does not hesitate to apply it even to our understanding of God. This is hardly surprising when we think that Descartes' mechanicism depended entirely on the adoption of this sole explanatory principle restricted to local motion. The other three types of Aristotelian causes were put aside, every kind of change was interpreted as an effect of local motion, following the rules of impacts, and the final outcome of this approach was the dissolution of the Aristotelian qualitative philosophy of nature. The prominence of the principle of causality in Descartes' metaphysics seems to be the generalization of a principle whose actual explanatory efficacy was evident above all in the realm of the new physics.

To give a general assessment of the relation between metaphysics and philosophy of nature in Descartes' cosmology, we agree with Gaukroger when he writes that "Descartes' system of natural philosophy is not generated by inference from first principles, but has been established quite independently of these first principles. In structuring his system around first principles, what Descartes hopes to achieve is a particular kind of legitimation of the whole project" (Gaukroger, 2002, p. 2). Descartes' foundational metaphysics is a legitimatory device that Descartes tried to harmonize as more as possible with the features of his physics. This does not imply in any way that the latter depended on the former for its conceptual structure: "there was nothing internal to Descartes' project in natural philosophy that required metaphysical foundations, and there was nothing crucial to his natural philosophy that could only be generated from such metaphysical foundations" (ibid., p. 3). The main links between Descartes' metaphysics and physics concern the origin of motion, the preservation of the quantity of motion and the institution of the natural laws ruling the transmission of motion. With regard to quantity of motion, Descartes argued that it depended on the immutability of God's nature:

> Thus, God imparted various motions to the parts of matter when he first created them, and he now preserves all this matter in the same way, and by the same process by which he originally created it; and it follows from what we have said that this fact alone makes it most reasonable to think that God likewise always preserves the same quantity of motion in matter.[33]

As for the laws of nature, he asserted that there could not be "a more firm and solid foundation" of their truth than "the very firmness and immutability which is in God".[34] It is obvious that these assumptions are neither metaphysically unquestionable nor physically required. One could hold either that a metaphysic foundation

[33] Descartes (1984–1991, I, p. 240); *Principia* II 36: AT VIII, p. 62: "[...] quod Deus diversimode moverit partes materiae, cum primum illas creavit, jamque totam istam materiam conservet eodem plane modo eademque ratione qua prius creavit, eum etiam tantundem motus in ipsa semper conservare". Hatfield held that only the presupposition of a *continued* action of God in nature could justify Descartes' thesis that the conservation of the quantity of motion depended on God's immutability: "For if God put a certain quantity of motion into the universe at the beginning and were no longer actively involved in the production of that movement, so that the passage of motion from one particle to another took place independently of God's operation, there seems to be no reason for *God's* immutability to imply conservation. In such a case Descartes would have to base his argument for conservation on the immutability of nature itself, not on that of God". (Hatfield, 1979, p. 122).

[34] Ibid., I, p. 96 (*Le Monde*, chap. VII: AT XI, p. 43).

is unnecessary or that another kind of metaphysics—in which motion, its quantity and regular transmission are inherent features of nature—could replace Descartes' theistic assumptions to accomodate his mechanist natural philosophy. Finally, though a connection between theology and science entailed inevitably ascribing to God the origin of motion and the institutions of the laws of nature, it did not require necessarily the grounding of the laws of nature and of the preservation of the quantity of motion on the immutability of God's nature. Newton is a clear example of a very different approach to the issue of the relation of God to nature.[35]

To complete an overview of those aspects of Descartes' philosophy of nature which are especially significant for his cosmology, we address here three more topics: the identity of matter and space, the indefinite extension of universe and the central significance of light. Descartes' ontology of matter is characterized by a corpuscularism that asserted the potentially infinite divisibility of matter and denied the existence of the void. This view derives from the identification of matter and space: these two concepts can be distinguished only because we talk of 'matter' when we consider a particular portion of space that is in motion, while we talk of 'space' when we consider the same portion as immobile. The identification of space and matter is one of the fundamental tenets of the entire Descartes' natural philosophy—as we have explained in Sect. 5.1—and is clearly stated by Descartes in *Principia* II, 10 where we read:

> There is no real distinction between space, or internal place, and the corporeal substance contained in it; the only difference lies in the way in which we are accustomed to conceive of them. For in reality the extension in length, breadth and depth which constitutes a space is exactly the same as that which constitutes a body. The difference arises as follows: in the case of a body, we regard the extension as something particular, and thus think of it as changing whenever there is a new body; but in the case of a space, we attribute to the extension only a generic unity, so that when a new body comes to occupy the space, the extension of the space is reckoned not to change but to remain one and the same, so long as it retains the same size and shape and keeps the same position relative to certain external bodies which we use to determine the space in question.[36]

The essence of material substance is defined exclusively by the attribute of extension in the three dimensions of space and its property of being infinitely divisible is a consequence of this essential attribute. We may ask what prompted Descartes to identify matter and space in the first place. His mathematical achievements (the

---

[35] Garber (1992, p. 230) remarked that Descartes was close to Newton on his view of the principle of inertia ("both see the persistence of rectilinear motion as one of the basic properties bodies have"), but that "Descartes' proofs of his principles from the immutability of God are far from anything Newton would have used".

[36] Descartes (1984–1991, I, p. 227). Cf. AT VIII, p. 45: "Non etiam in re differunt spatium, sive locus internus, & substantia corporea in eo contenta, sed tantum in modo, quo a nobis concipi solent. Revera enim extensio in longum, latum & profundum, quae spatium constituit, eadem plane est cum illa quae constituit corpus. Sed in hoc differentia est, quod ipsam in corpore ut singularem consideremus, & putemus semper mutari quoties mutatur corpus: in spatio vero unitatem tantum genericam ipsi tribuamus, adeo ut, mutato corpore quod spatium implet, non tamen extensio spatii mutari censeatur, sed remanere una & eadem, quamdiu manet ejusdem magnitudinis ac figurae, servatque eundem situm inter externa quaedam corpora, per quae illud spatium determinamus".

invention of analytic geometry) certainly played a role in so far as assimilating matter to space guaranteed in principle a rigorous science of figures and motions of bodies. Analytic geometry favours the identification between space and figures which are immersed in space because both points of space and points of figures are identified through a triple of real numbers (or a couple, if our mathematical space is the plane). To be clearer, let us consider an instructive limit-case: suppose that only a cube and the rest of space exist. You might consider the cube and the rest of space as two figures, one finite, the other infinite, both of which identified through the planes which represent the faces of the cube and their intersections. To each plane an equation is associated. Rather than considering the cube and the rest of space, one is allowed to conceive two figures, one internal, the other external, delimited by six planes. Thus, we may understand how in analytic geometry the difference between space and figures becomes more tenuous than in the classical Euclidean geometry. Descartes extended this conception from the geometrical to the physical world and thought that such a difference was so tenuous as to become inexistent, so that he identified space and matter. Obviously, this interpretation is not intrinsic to analytic geometry, but analytic geometry lends itself to be interpreted in this way. Moreover, a metaphysical motivation was also relevant, because the infinite divisibility of extension (of space and, consequently of matter identified with space), opposed to the indivisibility of the thinking substance, enabled Descartes to distinguish the two substances in the sharpest manner.

Finally, a rigorous application of the distinction between primary and secondary properties of bodies induced Descartes to conclude that the only immutable property of matter was its extension in length, width and depth, while solidity and hardness varied according to perceptive conditions (*Principia* II 4; II 11). As for the denial of any void, this was motivated by the metaphysical truism that since void is nothing, then it cannot exist; but, even more, it was an inevitable consequence of the identity of space and matter, as clearly emerges from *Principia* II 16 where Descartes claimed:

> The impossibility of a vacuum, in the philosophical sense of that in which there is no substance whatsoever, is clear from the fact that there is no difference between the extension of a space, or internal place, and the extension of a body. For a body's being extended in length, breadth and depth in itself warrants the conclusion that it is a substance, since it is a complete contradiction that a particular extension should belong to nothing; and the same conclusion must be drawn with respect to a space that is supposed to be a vacuum, namely that since there is extension in it, there must necessarily be substance in it as well.[37]

An aspect of Descartes' cosmology that needs to be mentioned is the issue of the infinity of the universe. This was a dangerous topic, involving theological questions, and so Descartes treated it very cautiously. The identity of matter and space should have induced Descartes to conceive the extension of the universe as infinite (*Principia*

---

[37] Descartes (1984–1991, I, pp. 229–230). Cf. AT VIII, p. 49 "Vacuum autem philosophico more sumptum, hoc est, in quo nulla plane sit substantia, dari non posse manifestum est, ex eo quod extensio spatii, vel loci interni, non differat ab extensione corporis. Nam cum ex hoc solo quod corpus sit extensum in longum, latum & profundum, recte concludamus illud esse substantiam, quia omnino repugnat ut nihili sit aliqua extensio, idem etiam de spatio, quod vacuum supponitur, est concludendum: quod nempe, cum in eo sit extensio, necessario etiam in ipso sit substantia".

II 21), but the theological concern to avoid the admission of a material infinity besides the infinity of God pushed him in the opposite direction. His solution to this dilemma was to state that the universe was neither finite nor infinite, but 'indefinite', meaning by this that while the infinity of God is actual and positive, the infinity of the universe is only negative: in the first case we know that there are no limits, while in the second we don't know of any limit (*Principia* I 26–27). The 26th proposition is particularly interesting because Descartes stated that no mathematical or scientific treatment of the actual infinity is possible and that, in contrast to this, the concept of "potential" is positively applicable both to potentially infinite quantities and to potentially infinitesimal quantities. For, we read:

> Thus we will never be involved in tiresome arguments about the infinite. For since we are finite, it would be absurd for us to determine anything concerning the infinite; for this would be to attempt to limit it and grasp it. So we shall not bother to reply to those who ask if half an infinite line would itself be infinite, or whether an infinite number is odd or even, and so on. It seems that nobody has any business to think about such matters unless he regards his own mind as infinite. For our part, in the case of anything in which, from some point of view, we are unable to discover a limit, we shall avoid asserting that it is infinite, and instead regard it as indefinite. There is, for example, no imaginable extension which is so great that we cannot understand the possibility of an even greater one; and so we shall describe the size of possible things as indefinite. Again, however many parts a body is divided into, each of the parts can still be understood to be divisible and so we shall hold that quantity is indefinitely divisible. Or again, no matter how great we imagine the number of stars to be, we still think that God could have created even more; and so we will suppose the number of stars to be indefinite. And the same will apply in other cases.[38]

This issue was discussed by Henry More in his exchange of letters with Descartes in 1648. The Cambridge Platonist tried to force Descartes out of what appeared to him an ambiguous position, since the world in itself must be either finite or infinite; Descartes replied that his notion of 'indefinite' is the most appropriate to the epistemological limitation of the human mind that is not capable of deciding which horn of the dilemma is true. For Descartes, the correct statement on this issue is to say that we do not have any reason to prove that the universe is limited, while at the same time we cannot prove and do not know for certain that the universe is infinite. Indeed, there could be limits of the universe which are known to God and unknown

[38] Descartes (1984–1991, I, p. 201–202). Cf. AT VIII, pp. 14–15: "Ita nullis unquam fatigabimur disputationibus de infinito. Nam sane, cum simus finiti, absurdum esset nos aliquid de ipso determinare, atque sic illud quasi finire ac comprehendere conari. Non igitur respondere curabimus iis, qui quaerunt an, si daretur linea infinita, ejus media pars esset etiam infinita; vel an numerus infinitus sit par anve impar, & talia: quia de iis nulli videntur debere cogitare, nisi qui mentem suam infinitam esse arbitrantur. Nos autem illa omnia, in quibus sub aliqua consideratione nullum finem poterimus invenire, non quidem affirmabimus esse infinita, sed ut indefinita spectabimus. Ita, quia non possumus imaginari extensionem tam magnam, quin intelligamus adhuc majorem esse posse, dicemus magnitudinem rerum possibilium esse indefinitam. Et quia non potest dividi aliquod corpus in tot partes, quin singulae adhuc ex his partibus divisibiles intelligantur, putabimus quantitatem esse indefinite divisibilem. Et quia non potest fingi tantus stellarum numerus, quin plures adhuc a Deo creari potuisse credamus, illarum etiam numerum indefinitum supponemus; atque ita de reliquis".

to us.[39] More insisted on his dilemma: even if we cannot easily establish whether the world is finite or 'non-finite' (a predicate that More regarded as equivalent to 'infinite'), only one of the two options must be true in reality. He added that the fact that Descartes' vortices did not dissolve was a clear evidence of the infinity of the world, which is an erroneous remark since Descartes admitted the dissolution of single vortices. Descartes replied that he could not conceive how to attribute a limit to the world and for this reason he termed it 'indefinite' or 'indeterminate'; however, he did not dare to call it 'infinite' since this property pertains to God who is greater than the world ("Deum esse mundo majorem") by virtue of his perfections and not by his extension, since God does not have any extension.[40] The exchange with More on this delicate subject shows how Descartes tried to keep together epistemological caution and theological prudence.

Another significant aspect of Descartes' cosmology is the relevance given to the light in his view of the different regions of the universe, which is manifested by the title itself of his first work on physics, *Le Monde ou le Traité de la Lumière* (written between 1629 and 1633). Koyré noted, with reference to Kepler and Henry More, that in the seventeenth century light was still regarded as a perfect example of something immaterial and incorporeal, and yet extended throughout the space.[41] Descartes's equation of extension and matter broke with this Platonic tradition and entailed a material conception of light whose nature consisted in a tendency to motion due to the pressure of particles. Furhtermore, light marked the distinction among the three material elements. The significance of light in the new Cartesian physics is made clear by this passage from *Le Discours de la Méthode* in which Descartes summarizes the conception of light he had outlined in *Le Monde*:

> I undertook [...] to reveal fully my conception of light; thereafter, I took the opportunity of adding something about the sun and the fixed stars, because light proceeds almost wholly from them; something about the heavens, because they transmit it; something about the planets, the comets, and the earth, because they reflect it; and in particular something about terrestrial bodies, because they are either coloured, or transparent, or luminous; and finally something about man, because he is the spectator of all this.[42]

The whole of physics as well as man's place in the universe as knowing subject are referred to light, which appears to be the key to understand the nature of heavenly and terrestrial body as well as the medium enabling us to contemplate the universe and

---

[39] More to Descartes 11 December 1648, Descartes (2009, pp. 2598–2600) (AT V 242f); Descartes to More, 5 February 1649, ibid., pp. 2620–2622 (AT V 274f.). Cfr. Descartes to Chanut, 6 June 1647, ibid., p. 2468 (AT V 51f.).

[40] More to Descartes 5 March 1649, Descartes (2009, p. 2646) (AT V 304); Descartes to More, 15 April 1649, ibid., p. 2685 (AT V 344). Cfr. Koyré (1957, pp. 110–124).

[41] Koyré (1957, p. 130f).

[42] Descartes (2006, p. 36). Cf. AT VI, p. 42: "[...] i'entrepris seulement d'y exposer bien amplement ce que ie concevois de la Lumiere; puis, a son occasion, d'y adiouster quelque chose du Soleil & des Estoiles fixes, a cause qu'elle en procede presque toute; des Cieux, a cause qu'ils la transmettent; des Planetes, des Cometes & de la Terre, a cause qu'elles la font refleschir; & en particulier de tous les Cors qui sont sur la terre, a cause qu'ils sont ou colorez, ou transparens, ou lumineux; & enfin de l'Homme, a cause qu'il en est le spectateur".

build our knowledge of it. We might say that light continues to play the traditional role of connecting macrocosm and microcosm, to use a phraseology that was typical of the Renaissance philosophy, but in Descartes' physics both light, macrocosm and microcosm are unified in the form of a material mechanicism.

Given the identity of matter and space, Descartes affirmed that the principles of physics coincide with those of geometry (*Principia* II 64). In this respect, at least in his statements, Descartes carried on and even pushed to the extreme the mathematization of physics that Galileo and Kepler had initiated. However, as we have seen, he did not implement this methodological programme in actual physical research. Apart from the law of refraction in *La Dioptrique*, from which he derived his explanation of the rainbow in the eight discourse of *Les Météores*, Descartes never managed to explain phenomena through mathematical treatment; given his merely descriptive and conjectural exposition of physics in the *Principia*, it's not surprising that a scientist of the stature of Christian Huyghens—though being a Cartesian—referred to "the fine romance of the Cartesian physics".[43] Descartes was not unaware that his hypothetical reconstructions might have been at variance with reality and that his method showed how all things in nature could have arisen, without ensuring the absolute certainty that they in fact were made in the way he had described them to be. After all, "there are countless configurations that God might have chosen, and experience will have to tell us which ones he actually chose" (*Principia* 46). Thus, at least in principle, Descartes preserved the role of experience and the joining of mathematics to experience in physical science: "we're free to make any assumption we like about this, provided that its consequences agree with our experience" (ibid.). However, his confidence that he had provided a coherent account of phenomena, deriving them in an unbroken chain from the absolute principles of human knowledge, led him to conclude the *Principia* with the bold assertion that at least the most general features of the world and the Earth hardly seemed to be understandable but in the way he had explained them (*Principia* IV 204–206).

The legacy of Descartes was extremely significant not only for philosophy in general, but also specifically for cosmology. We give here two instances (D'Alembert and Kant) of how philosophers and scientists in mid-eighteenth century still reflected on Descartes' views and how they came to terms with his ideas. D'Alembert's article on 'Attraction' in the *Encyclopédie* is an example of how one of the most acute scientists of his age regarded the Cartesian theory of vortices. With a turn of phrase revealing his methodological caution, D'Alembert wrote that "the system of the world rightly allows us to suppose that perhaps the impulse is not the only cause of the motions of bodies".[44] The development of the Newtonian science has rendered the idea of cosmic vortices untenable, "however nice their spectacle was".[45] Nonetheless, even if the existence of a force of attraction among the heavenly bodies is admitted by every philosopher, the attempts at explaining it by means of a fluid, in continuity with the Cartesian theory, were justified by the epistemic precept that "one must not

---

[43] Rossi (1971, p. 221). See here below, Section 6.2.

[44] D'Alembert (1751, p. 855).

[45] Ibid., p. 854.

multiply the principles". The hypothesis to explain gravity resorting only to impulse was an intelligent approach to the issue, but the Cartesian theory failed because it did not fit adequately particular phenomena: "It is not sufficient for an hypothesis to correspond to phenomena roughly, as it were, and vaguely: the details are its touchstone and the details have ruined the Cartesian system".[46] With great acumen d'Alembert grasps the conceptual strength of the theory of vortices (that explained its prolonged fortune after Descartes), while at the same time pinpointing the fatal defect of the original Cartesian physics.

As is known, in the second book of the *Principia* Newton argued that the Cartesian theory of vortices is not compatible with Kepler's laws and with the law of universal gravitation. Newton's critique did not put an immediate end to the Cartesian theory. In the eighteenth century attempts were made to offer a vortex theory coherent with Kepler's laws. The most important cosmologist who worked in this direction was Philippe Villemot. In his *Nouveau système ou nouvelle explication du movement des planets* (Villemot, 1707), he tried not only to reconcile the vortex theory with Kepler's laws, but also to explain through the use of vortices specific phenomena such as the diurnal Earth's rotation around its axis. He also gave a mathematical treatment of this topic.[47] Starting from the late 30s–40s of the eighteenth century the vortex theory declined definitively in the French cultural milieu too.

It has been argued that Newton's proof of the instability of the vortices was not conclusive, since it was based on assumptions of which the most important are: (1) to be stable, the vortices would require densities and velocities of rotations that appeared to Newton extremely implausible from a physical point of view; (2) Newton treated the vortices as if their fluid layers were composed of solid matter, which is not the case (Aiton, 1972, pp. 110–113). It is true that Newton's argument is not conclusive because it does not exclude categorically that a vortex might satisfy in theory the physical conditions for its stability. However, as a matter of fact, Newton's proof makes the vortices' stability extremely unlikely by subjecting it to strong constraints on their density and kinetic conditions which are highly implausible from a physical standpoint. Though the vortex theory was abandoned from the 40s of the eighteenth century, the cosmogenesis devised by Descartes survived and represented a significant progress along the way of a scientific explanation of the origin of the universe. When Immanuel Kant wrote the *Universal Natural History and Theory of the Heavens* (1755) to illustrate his nebular hypothesis on the origin of the solar system, he cited Descartes as his predecessor in this daring speculation that superseded the tale of the *Genesis* to which Newton's cosmology had remained faithful. Kant invokes Descartes' example while defending his theory from the predictable attacks of theologians: "One will not at any rate berob me of the right which Cartesius, when he dared to explain the formation of celestial bodies from purely mechanical laws, enjoyed on the part of equitable judges of those times".[48] The evolutionary

---

[46] Ibidem.

[47] Aiton (1972), Chapter Seven, pp. 152–193 offers an exhaustive treatment of Villemot's theory as well as of the contributions of other scientists to the vortex theory in the period 1700–1729.

[48] Kant (1981, p. 86).

theory of the formation of the universe is not incompatible with theism, Kant argued by pointing out to Descartes' example, since the material development obeys the laws of nature that testify to the existence of an ordering rational First Cause of the universe. The rough Cartesian sketch of a mechanical cosmogony and the more sophisticated hypothesis by Kant will finally reach in the hands of Laplace the status of a scientific theory that did not need to be concerned anymore about theological justification.

References

Aiton, E. J. (1972). *The vortex theory of planetary motions*. New York: American Elsevier Publishing Company.

Aquinas, T. (1918). *Summa contra Gentiles*. In *Opera Omnia* (editio Leonina, Vol. 13). Rome: Typis Riccardi Garroni.

Armogathe, J. R. (2000). The rainbow: A privileged epistemological model. In S. Gaukroger, J. Schuster, & J. Sutton (Eds.), *Descartes' natural philosophy* (pp. 249–257). London: Routledge.

Bentley, R. (1692–93). *A confutation of atheism from the origin and frame of the world*. London: H. Mortlock.

Burtt, E. A. (1925). *The metaphysical foundations of modern physical science: A historical and critical essay*. London: Kegan Paul, Trench, Trubner & Co.

Bussotti, P., & Pisano, R. (2013). On the conceptual and civilization frames in René Descartes' physical works. *Advances in Historical Studies, 2*(3), 106–125.

Bussotti, P., & Lotti, B. (2018). The problem of circular motion in René Descartes. In *Giornale Critico della Filosofia Italiana* (settima serie, Vol. XIV, anno XCVII (XCIX), fasc. I, pp. 76–114). Gennaio-Aprile 2018.

D'Alembert, J. L. R. (1751). Attraction. In *Encyclopédie ou Dictionnaire Raisonné des Sciences, des Arts, et des Métiers* (Vol. I, pp. 846–855). Briasson, David, Le Breton, Durand.

Derham, W. (1713). *Physico-theology, or, a demonstration of the being and attributes of God from his works of creation being the substance of XVI sermons preached in St. Mary le Bow-Church, London, at the Hon'ble Mr. Boyle's lectures in the years 1711 and 1712*. London; W. Innys.

Derham, W. (1715). *Astro-theology, or, a demonstration of the being and attributes of God: From a survey of the heavens*. London: W. Innys.

Descartes, R. (1644). *Principia Philosophiae*, Latin original, AT VIII-I (pp. 1–348). French translation AT IX (pp. 1–352).

Descartes, R. (1897–1913). *Oeuvres de Descartes* (12 vols.) publiées par C. Adam & P. Tannery. Paris: Vrin. Abbreviated as AT.

Descartes, R. (1984–1991). *The philosophical writings of Descartes* (J. Cottingham, R. Stoothoff, & D. Murdoch, Trans.) (in 3 vols.). Cambridge-London-New York-New Rochelle-Melbourne-Sydney: Cambridge University Press.

Descartes, R. (2006). *A discourse on the method of correctly conducting one's reason and seeking truth in the sciences*. Translated with and introduction and notes by I. Mclean. Oxford: Oxford University Press.

Descartes, R. (2009). *Tutte le Lettere 1619–1650*, a cura di G. Belgioioso. Testo francese, latino e olandese. Milano: Bompiani.

Dessì, P. (2012). *Causa/Effetto*. Bologna: Il Mulino.

Einstein, A. (1988, 1930). Das Raum-, Äther- und Feld-Problem der Physik. In A. Einstein & C. Seelig (Ed.), *Mein Weltbild* (pp. 139–147). Ullstein.

Emerton, N. (1989). The argument from design in early modern natural theology. *Science and Christian Belief, 1*(2), 129–147.

Garau, R. (2014). Late-scholastic and Cartesian *conatus*. *Intellectual History Review, XXIV*, 479–494.
Garber, D. (1992). *Descartes' metaphysical physics*. Chicago and London: The University of Chicago Press.
Gaukroger, S. (2002). *Descartes' system of natural philosophy*. Cambridge: Cambridge University Press.
Giudice, F. (2010). The visible world: Aspects of optics in the seventeenth century. In E. Agazzi, E. Giannetto, & F. Giudice (Eds.), *Representing light across arts and sciences: Theories and practices* (pp. 41–55). V&R Unipress.
Hatfield, G. (1979). Force (God) in Descartes' physics. *Studies in History and Philosophy of Science, X*, 113–140.
Kant, I. (1981). *Universal natural history and theory of the heavens*. Translation with introduction and notes by S. L. Jaky. Scottish Academic Press.
Koyré, A. (1957). *From the closed world to the infinite universe*. Baltimore: John Hopkins Press.
Maitte, B. (2006). *Histoire de l'arc–en–ciel*. Paris: Seuil.
Malet, A. (1990). Gregoire, Descartes, Kepler and the law of refraction. *Archives Internationales D'histoire Des Sciences, 40*, 278–304.
Paley, W. (1802). *Natural theology: Or evidences of the existence and attributes of the Deity collected from the appearances of nature*. London: R. Faulder.
Pascal, B. (1958). *Pascal's Pensées*. Introduction by T. S. Eliot. Translated by W. F. Trotter. New York: E. P. Dutton & Co.
Ray, J. (1691). *The wisdom of God manifested in the works of the creation being the substance of some common places delivered in the Chappel of Trinity-College, in Cambridge*. London: Samuel Smith.
Rossi, P. (1971). *Aspetti della rivoluzione scientifica*. Napoli: Morano.
Sabra, A. I. (1967). *Theories of light from Descartes to Newton*. London: Oldbourne.
Schuster, J. (2000). Descartes opticien: The construction of the law of refraction and the manufacture of its physical rationales, 1618–1629. In S. Gaukroger, J. Schuster, & J. Sutton (Eds.), *Descartes' natural philosophy* (pp. 258–312). London: Routledge.
Schuster, J. (2013). *Descartes-agonistes: Physico-mathematics, method & corpuscular-mechanism 1618–33*. New York-London: Springer.
Shapiro, A. (1974). Light, pressure, and rectilinear propagation: Descartes' celestial optics and Newton's hydrostatic. *Studies in History and Philosophy of Science, V*, 239–296.
Shea, W. R. (1991). *The magic of numbers and motion. The scientific career of René Descartes*. Science History Publications.
Slowik, E. (1999). Descartes, spacetime and relational motion. *Philosophy of Science, LXVI*, 117–139.
Slowik, E. (2005, 2017). Descartes' physics. In E. N. Zalta (Ed.), *Stanford Encyclopedia of Philosophy*. https://plato.stanford.edu/entries/descartes-physics/2005, rev. 22 August 2017.
Tiemersma, D. (1988). Methodological and theoretical aspects of Descartes' Treatise on the Rainbow. *Studies in History and Philosophy of Science, 19*, 347–364.
Villemot, P. (1707). *Nouveau système, ou nouvelle explication du mouvement des planetes*. Lyon: Louis Declaustre.
Williams, B. (2005 [1978]). *Descartes: The project of pure enquiry*. London: Routledge.

Chapter 6
Huygens: The Greatest Cartesian Scientist

In our interdisciplinary research concerning the relations between science and philosophy in the context of early modern cosmology, seven scientists are considered: Copernicus, Kepler, Galileo, Descartes, Huygens, Newton and Leibniz. Among them, Christiaan Huygens (1629–1695) is surely the one who is less involved in philosophical considerations: he was not particularly troubled by the idea of reconciling science and theology, he wrote nothing significant with regard to the role of God in the structure and laws of universe,[1] it seems that the influence of the classical philosophers Plato and Aristotle was negligible in his thought. Huygens gave fundamental contributions to science: he studied the quadrature of several curves; in *De Motu Corporum ex Percussione* (1656, published posthumously 1703), he offered a correct theory of the collisions, a subject which had evaded the efforts of many scientists; in *De Vi Centrifuga* (1659, published posthumously 1703), he offered the correct laws of the circular motion and some important elements of the pendulum theory; in the *Horologium Oscillatorium* (1673), one of his two most famous masterpieces, Huygens presented a complete theory of the pendulums (simple, compound and conic pendulums); in 1690 he published his second masterpiece, the celebrated *Traité de la lumière*, where the undulatory theory of light is advanced; in the same year he published the *Discours de la cause de la pesanteur*, a very interesting text in which Huygens, in the light of Newton's *Principia*, reshaped his theory of gravity presented in 1669 in a series of discussions with other scientists at the *Académie Royale des Sciences* in Paris. Huygens also dealt with astronomy and cosmology: his most important works are *Systema Saturnium* (1659), where he published his observations of Saturn's ring and of Saturn's satellites, and *Cosmotheoros* (1698, posthumous) where he proposed his ideas on the structure of the universe and on

[1] On this topic see Huygens' short and not particularly significant notes published in his complete works, volume XXI, pp. 339–344 under the title "Que penser de Dieu"? Some references also in the *Cosmotheoros*. The complete works by Huygens (*Oeuvres completes de Christiaan Huygens* abbreviated as OH) were published in 22 volumes by the Société Hollandaise des Sciences between 1888 and 1950. Unless otherwise indicated, all translations are ours.

P. Bussotti and B. Lotti, *Cosmology in the Early Modern Age: A Web of Ideas*,
Logic, Epistemology, and the Unity of Science 56,
https://doi.org/10.1007/978-3-031-12195-1_6

possible inhabitants of other worlds. The *Cosmotheoros* is probably the only work by Huygens that has a philosophical flavour and, in fact, we will deal with this treatise in Sect. 6.2.3. However, with the exception of *Cosmotheoros*, it might seem that the other works by Huygens are more significant for the history of science than for the relations between science and philosophy. The problem is that "philosophy" is a polysemous term. Therefore, if by "philosophy" of a scientist in the early modern centuries one means the connections of his physical theories with theology and metaphysics, then Huygens is not a philosopher. However, if by "philosophy" one refers to the a-priori convictions (i.e. convictions that precede and frame empirical data) of a scientist with regard to interactions among bodies and to the structure of the universe, then Huygens is a true natural philosopher. His philosophy was mechanicism: the idea that all the movements arise by contact and that no action at a distance exists. His philosophical mentor was Descartes. Huygens was by far superior to Descartes as a scientist. Though remaining within the context of mechanistic natural philosophy, he offered a mathematical treatment of phenomena such as collisions, gravity, circular motion, and so on, with regard to which Descartes had proposed vague and often erroneous speculations. In this process, he also changed many of those wrong ideas and presuppositions offered by Descartes. Nonetheless, he remained a Cartesian. As Barbour wrote:

> The key to both the strengths and weakness of Huygens' approach to dynamics are to be found in his mechanical philosophy of nature. He read Descartes' *Principles of Philosophy* at the impressionable age of 15 or 16. [...] Huygens was carried away by the idea of a mechanical explanation of the phenomena of nature along the lines proposed by Descartes. In accordance with this programme, all phenomena must be given a microscopic interpretation in terms of invisible particles of matter, the only elements allowed in this explicatory enterprise being, as we have seen, the size, shape and motion of the particles. (Barbour, 2001, p. 452)

These mechanistic convictions led Huygens to formulate a precise theory of centrifugal forces, which in its turn, was crucial for his gravitational theory as well as for his ideas concerning the structure of the world; this was fundamental for his cosmology. This aspect reveals the profound connections between Huygens's science and natural philosophy. Therefore, in the next section we will deal with three topics: (1) centrifugal forces; (2) gravity; (3) nature of motion. Centrifugal forces play a crucial role in Huygens' natural philosophy and their complete theory is expounded in *De Vi Centrifuga*. As for gravity, our main sources will be the discussion on gravity at the *Académie* and the *Discours de la cause de la pesanteur*. The nature of motion concerns the problem whether every motion is relative or absolute. The fundamental source on this issue are the manuscripts belonging to the *Codex Hugeniorum 7A*.[2]

In Sect. 6.2 we will analyse the philosophical implications of Huygens' natural philosophy as well as the structure of his *cosmos*. The two most important sources we will examine are the *Cosmotheoros* and the *Traité de la lumière*.

---

[2] Mormino (1993) edited Huygens' manuscripts on motion belonging to *Codex Hugeniorum 7A*: some manuscripts are published by Mormino for the first time while others were already present in OH, vols. XVI, XVIII and XXI. We will refer to Mormino's edition and, when existing, to OH as well. For a complete history of these manuscripts see Mormino (1993, pp. 105–115).

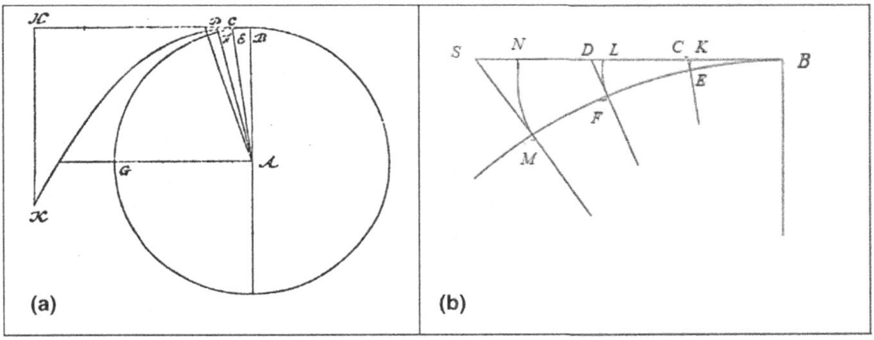

Fig. 6.1 **a** On the left and **b** on the right. Two of the figures used by Huygens to explain the nature of centrifugal forces, see OH XVI, pp. 259 and 265 respectively

6.1 Huygens' New Physics: Centrifugal Forces, Gravity, Nature of Motion

In Huygens' cosmology centrifugal forces are used to explain the mechanism on which gravity depends. Thus, to describe Huygens's road to gravity by means of centrifugal forces, it should be recalled that this idea dates back to Descartes: according to the latter's view an ethereal matter rotating around the Earth pushes the body downwards. We have already clarified that the general lines of his conception were clear, but the details were rather obscure and any mathematical treatment was missing. Therefore, first of all, Huygens (*De Vi Centrifuga*) provided all the mathematical details concerning centrifugal forces and, afterwards, applied these results to the problem of gravity.

6.1.1 De Vi Centrifuga

Huygens considered a rotating wheel parallel to the horizon (Fig. 6.1a). Suppose that a ball, tied to a rope, is rotating counterclockwise on the circumference of the wheel.[3] According to the inertia principle, when the ball is in *B*, it has the tendency to escape along the tangent *BH* at the wheel in *B*. Huygens clearly stated that the ball would continue its movement along *BH* if it were separated from the rope unless it were pulled downwards by gravity or by the collision with other bodies (OH XVI, p. 259). The problem posed by Huygens is the following: how much is the rope *AB* stretched by the revolution of the ball around *A*?

[3] On the theory of centrifugal forces in Huygens, see: Barbour (2001, pp. 451–497); Bertoloni Meli (2005); Bertoloni Meli (2006, pp. 319–322); Chareix (2004); Stinner (2001). Huygens coined the expression *vis centrifuga*.

He proposed the following ingenious reasoning: suppose that a man stands on the circumference in *B* and that he is so well attached at the wheel that he cannot fly away. The man holds in his hand a string. At the other end of the string, a lead shot is attached. The situation is exactly the same as if the string were attached in *B* at the wheel. This means that the centrifugal force will stretch the string exactly with the same intensity experienced by the chord *AB* in the revolution of the ball around *A*. Suppose that the man leaves the string with the lead shot. Let us now analyse the situation after the man has left the shot lead. Be τ the infinitesimal time during which the man completes the arch of circumference *BE* (Fig. 6.1b). The arch *BF*, which is the double of *BE*, will be completed in the infinitesimal time 2τ and so on. The lead shot will complete, on the tangent at *B*, in the same infinitesimal times $\tau, 2\tau, \ldots$ the segments *BK*, *BL*, and so on, respectively equal to the arches of circumference *BE*, *BF*, ... This motion depends on the inertia principle.[4] Huygens claimed that the infinitesimal curvilinear segments[5] *EK*, *FL*, *MN*, ..., which represent the trajectory of the shot lead as seen from the rotating man at the infinitesimal times $\tau, 2\tau, 3\tau, \ldots$ indicate the action of the centrifugal force because they represent the deviation of the lead shot from its instantaneous tangent in *B*. The problem is thus how to calculate the length of these curvilinear segments. Huygens argued like this:

(1) *First step*: He begins his proof by calculating the length of the segments *BC*, *BD*, *BN*, .., in the case in which their lengths are infinitesimal. He uses Euclid's theorem of tangent and secant, (*Elements*, III, 36) according to which if from a point *C* or (*D*) of the plane, a tangent *BC*, (or *BD*) and a secant *CEE'* (or *DFF'*) are drawn (Fig. 6.2) it holds that the tangent segment *CB* (or *BD*) is the mean proportional between the segment of secant included between the tangent and the circumference *CE* (or *DF*) and the whole secant *CE'* (or *DF'*).

This means that it holds

$$EC : CB = CB : CE' \rightarrow CB^2 = EC \times CE'$$

$$DF : DB = DB : DF' \rightarrow DB^2 = DF \times DF'$$

Now the fact that all segments—excluded the diameter *EE'* or *FF'* of the circumference—are infinitesimal becomes important. In the ratio $\frac{DB^2}{CB^2} = \frac{DF \times DF'}{EC \times CE'}$ the finite segments *CE'* and *DF'* can be considered equal because *CE* and *DF* are infinitesimal. Thence, this ratio can be written as $\frac{DB^2}{CB^2} = \frac{DF}{EC}$. This means, as Huygens wrote, that the spaces *EC*, *DF*, ... increase as the squares of the spaces *CB*, *DB*, etc. calculated

[4] With regard to the motion of the lead on the instantaneous tangent at *B* and to the fact that the arches *BE*, *EF*, ... are as long as the segment lines *BC*, *CD*, ... Huygens stated: "Whence we say correctly that this tendency is in the lead". Translation by Mahoney, see OH XVI, p. 261.

[5] Since Huygens is dealing with infinitesimal elements, the curvilinear segments *EK*, *FL*, *MN*, ..., can be considered equivalent to the corresponding rectilinear segments *EK*, *FL*, *MN*, ...

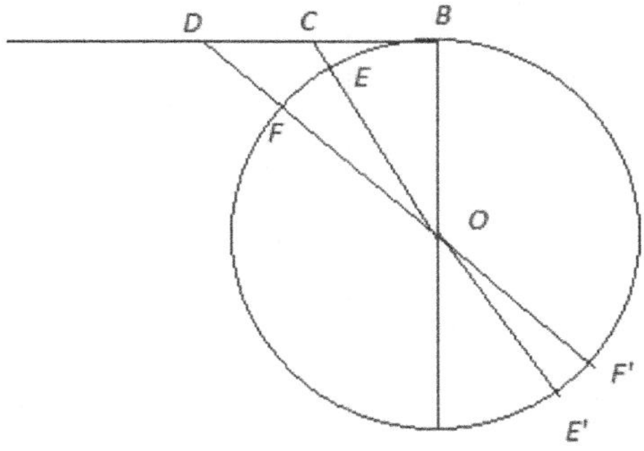

Fig. 6.2 The theorem of tangent and secant. We are considering the particular case in which the chords *EE'* and *FF'* pass through the centre of the circle *O* because it is the only interesting for our analysis. The figure is ours

on the tangent. But, since the motion on the tangent is uniform, the spaces are proportional to the times. Thus, the infinitesimal spaces *CE, DE*, ... are proportional to the squares of the infinitesimal times in which they are completed.

(2) *Second step*: the problem is that the spaces *EK*, *FL*, *MN*,... (Fig. 6.1b) potentially completed under the action of the centrifugal force are not exactly equal to *CE*, *DF*, *MS*, ... (Fig. 6.1b). However, since the movement takes place in an infinitesimal time, the differences *CE-EK*; *DF-FL*; *MN-MS*, ... are insensible.[6] Therefore the segments *CE*, *DF*, *MS*, ... can be assumed with a negligible mistake instead of *EK*, *FL*, *MN*, ... But we have seen that these segments are proportional to the squares of times in which they are completed. On the other hand, it was well known from Galileo's works that in the motion under the action of gravity the spaces are proportional to the squares of the times in which they are completed. Therefore, the effect of the centrifugal force is exactly the same as that of gravity. Hence, the explanation of gravity through the centrifugal forces is certainly correct. This is Huygens' conclusion proved by exact demonstrations.

This conclusion led Huygens to think of a physical mechanism that might explain gravity by means of the centrifugal forces. As we shall see, the step from mathematical to physical considerations is anything but trivial.

Without going into the details of Huygens's demonstrations, we will now focus on three theorems proved by Huygens in *De Vi Centrifuga* that are fundamental for his following considerations on gravity:

1. Proposition II: In the uniform circular motion the centrifugal forces are directly proportional to the squares of the motion's speed.
2. Proposition III: Given two different circles traversed with the same speed, the centrifugal forces are inversely proportional to the circles' radiuses.

[6] One can prove, to use modern terms, that any difference is an infinitesimal of higher degree than the single segments.

These two propositions offer the well-known formula for a centrifugal force F experienced by a body of mass m rotating with a speed v in a circle of radius r.

$$F = \frac{mv^2}{r}$$

3. Proposition XIII: If a body is carried in a circumference and completes its individual revolution in the same time in which a simple pendulum, whose string is as long as the circumference's radius, completes two small oscillations, it will have a centrifugal force equal to its weight.[7]

Commentaries: let us consider once again Huygens' reasoning: in order to explain centrifugal forces, he apparently superimposes a treatment in an inertial reference frame on a treatment in a non-inertial reference frame. But, in fact, this is not the case and, though from a theoretical point of view the distinction between inertial and non-inertial reference frame was not completely clear to Huygens, his comprehension of the phenomena is so good that he made no mistake. As a matter of fact, in reference to Fig. 6.1a, he spoke of the tendency to escape along the tangent in a circular motion, the inertial tendency. It seems that he is referring to an inertial reference frame. On the other hand, while dealing with the explanation of the circular motion he used, in fact, the idea of the man who rotates joined to the wheel. This is clearly a reference frame co-moving with the wheel, a typical non-inertial reference frame. For the man no inertia exists. For him, the ball is pushed out of the trajectory because of a force, the centrifugal force, whose action is not anymore impeded by the rope. We can interpret the situation described by Huygens in this way: in an inertial reference frame, we have the tendency of the body to escape along the tangent; let us now pass to the frame of the man co-moving with the wheel (Fig. 6.2b). For him, when he is in E, the trajectory of the ball will be the curve EC, when he is in F, the trajectory will be FL, when in M the trajectory will be MN.[8] Therefore, the tangent $BCDN$ which is an element extraneous to the non-inertial reference frame, is introduced by Huygens only to facilitate the calculations. In practice, Fig. 6.1b is a comparison between an inertial and a non-inertial reference frame, but the concepts belonging to the two systems are correctly separated by Huygens. Our reading can thus be considered as a Newtonian interpretation of Huygens's treatment. But now the physical question is: Huygens admits inertia as a physical reality, but he also admits the centrifugal forces as a physical reality. To be clearer: Huygens's mathematical treatment is correct because it can be interpreted in terms of inertial and non-inertial reference frames, but, at all appearances, Huygens admitted two intrinsic, physical, ontological properties of circular motion: inertia and centrifugal forces as if they were both present. But, in

[7] A pendulum can be used to measure time if its oscillations take place in the same interval independently of the oscillation-angle. Huygens was well aware that a circular pendulum does not fulfil this condition. However, insofar as its angle of oscillation becomes smaller and smaller, the oscillations tend to become isochronous. Therefore, while referring to "small oscillations", Huygens meant oscillations which can be considered isochronous with a negligible error.

[8] Huygens was also able to describe how the whole trajectory of the ball appears to the observer co-moving with the wheel: see Huygens (1659a), OH XVI, p. 262.

fact, they are not because from the point of view of an inertial reference frame the centrifugal forces do not exist, while from the point of view of a co-moving reference frame, inertia does not exist. Thence, though the mathematical treatment is correct, Huygens' vision of physical reality—and specifically of circular motion—is not yet completely clear, as we shall explain in Sect. 6.1.3.

This issue is extremely interesting from an epistemological point of view: if a physicist like Huygens has a profound empirical knowledge of phenomena and is a very good mathematician, then he is able to arrange things in such a way that some important phenomena—as gravity—are explained. This might occur even though his theory lacks some of the fundamental concepts—such as that of centripetal force—that enable one to construct a satisfying theoretical structure.

There is a further element that should be taken into account: Huygens arrived at defining the centrifugal forces by considering the weight of a body as the quantity of matter which enters its composition[9]—in this case the difference between mass and weight is not decisive—, but all the most important demonstrations of *De Vi Centrifuga* are, actually, kinematical demonstrations. It is true that the concept of weight is fundamental in the structure of *De Vi Centrifuga*. But weight is considered in these propositions as an element whose nature has to be explained by means of kinematical effects, it is not considered as the basis to explain those kinematical effects. To simplify: in Huygens, kinematics explains dynamics, whereas in Newton dynamics explains kinematics.

The treatment of centrifugal forces is connected to two fundamental topics: (a) the nature of gravity; (b) the nature of motion and inertia. Thus, in Sects. 6.1.2 and 6.1.3 we will address these two subjects.

6.1.2 Gravity and Centrifugal Forces

The main problem—of which Huygens was perfectly aware—arising if we want to explain gravity in terms of the centrifugal force of an ethereal matter rotating around the Earth is that, if a heavy body rotates within a vessel containing a lighter fluid, the body will be pushed towards the walls of the vessel, not towards the centre. Therefore, if a vortex of ethereal matter would rotate around the Earth, the stones and all the other bodies, should go to the external regions of the vortex, not towards the centre. The reason why this happens is easy to grasp: in Huygens's formula $F = \frac{mv^2}{r}$, let us suppose that a body of mass m is rotating in a layer of a fluid vortex with velocity v and the distance from the centre is r. Each particle of the vortex has a mass less than m. This means that the centrifugal force of the body is bigger than that of any part of the vortex. Hence, the body will be pushed outside.

The Cartesian vortex theory presents a further problem: since a gravitational vortex rotates around its axis circularly and uniformly, it pushes the bodies towards

[9] Huygens wrote: "[...] conformément a nostre theorie le poids de chaque corps suit precisement la quantité de la matiere que entre dans sa composition". Huygens (1669), in OH XIX, p. 637.

the axis of rotation, but not towards the Earth's centre. To clarify: imagine that the vortex rotates around the terrestrial axis and consider a body afloat in such a vortex; it would be pushed towards the axis along the perpendicular line from the body to the axis. Huygens wrote:

> If one wanted the celestial matter to rotate in the same direction but more rapidly than the Earth, it would follow that this quick motion of a matter moving continuously in the same direction would be felt and transport all the bodies which lie on the Earth [...]. Besides this, however, this circular movement around the terrestrial axis could only push the bodies, which do not follow the same motion, towards the axis itself. Thus, we would not see the heavy bodies to fall perpendicularly to the horizon, but along lines perpendicular to the world's axis. This is contradicted by experience.[10]

To overcome these two problems, Huygens, first of all, devised an experiment in which a certain quantity of water was included in a discoid vessel of a given height and base-surface rotating parallel to the horizon.[11] Some heavier and compact material is set inside water, for example some pieces of sealing wax. When the motion of the disc begins, the pieces of sealing wax are pushed towards the periphery of the vessel, but if the rotation is blocked, water continues to rotate, while the pieces of sealing wax go to the bottom of the disc because of their weight, and they diminish their speed because of the friction with the base, hence they are pushed towards the centre of the base by the centrifugal force of the surrounding fluid. Since the centrifugal force depends on the square of the speed, an increment or a diminution of the speed implies an enormous increment or decrement of the centrifugal force. What Huygens meant with this experiment is that, under particular conditions, it is possible that in a liquid rotating vortex the heaviest bodies are pushed towards the centre of rotation, not towards the external layers. The experiment showed only a model of gravity. It cannot be an explanation, because the weight of the sealing wax's pieces was exactly the problem to solve. Thus, the problem was to find a mechanism that could reduce the speed of the body afloat in the vortex without diminishing that of the particles composing the vortex, and without resorting to the concept of weight.

If all the particles of the vortex rotate in the same direction this mechanism cannot exist. However, Huygens imagined the following hypothesis, which represents a radical modification of Descartes's vortices: suppose an ethereal vortex rotating around the Earth in which the vortex's particles rotate in a plurality of different directions under the condition that the resultant field of velocities is not null with respect to the vortex's centre. It can happen that a single particle colliding with other particles modifies its velocity and, that, for a very brief period, is at rest; however, on

---

[10] Huygens (1690b), in OH XXI, p. 455: "Que si l'on vouloit que la matiere celeste tournast du mesme costé que la Terre, mais avec beaucoup plus de vitesse, il s'ensuivroit que ce mouvement rapide, d'une matiere qui se mouvroit continuellement & toute d'un mesme costé, se feroit sentir, & qu'elle emporteroit avec elle les corps qui sont sur la Terre; [...]. Mais outre cela, ce mouvement circulaire, autour de l'axe de la Terre, ne pourroit en tout cas chasser les corps, qui ne suivent pas le mesme mouvement, que vers ce mesme axe, de sorte que nous ne verrions pas les corps pesants tomber perpendiculairement à l'horizon, mais par des lignes perpendiculaires à l'axe du monde, ce qui est encore contre l'experience".

[11] Huygens (1669, 1690b), in OH XXI, pp. 452–458.

average, the entire mass of the particles will have a certain mean velocity different from 0 with respect to the centre of the vortex. Furthermore, every particle is so little that the resultant of all the impulses acting on it and deriving from the movement of the other particles can hardly make its speed equal to 0 if not for a negligible time. Now, suppose that, for any reason, some of the particles join together and form bigger bodies. It is possible that the impulses—which have different directions—exerted by the particles surrounding the just formed bodies mutually annihilate or, at least, reduce the speed of the bodies until they become slow enough in comparison to the vortex's layer surrounding them so to be pushed towards the centre of the vortex. Huygens claimed that the situation in which the impulses deriving from the surrounding particles with different velocities are not enough to reduce the speed of single particles, but are sufficient to reduce that of a bigger body is physically plausible. Thence, these bodies, having lost a part of their speed, will be pushed towards the centre of the vortex (OH XIX, pp. 635–636). Therefore: *this modified version of the vortex theory would be suitable to explain gravity by means of merely mechanistic devices* (sizes, shapes and movements), which are the only ones scientifically acceptable, as clearly stated at the beginning of the *Discours de la cause de la pesanteur* (Huygens, 1690b, in OH XXI, p. 451).

Through a very brilliant reasoning, Huygens was also able to determine the average speed of the vortex at the equator for a body not to be flown away by its own centrifugal force. We have seen that proposition XIII of *De Vi Centrifuga* states that if a body is carried in a circumference and it completes its individual revolution in the same time in which a simple pendulum, whose string is as long as the circumference's radius, completes two small oscillations, that body will have a centrifugal force equal to its weight.

If a body is at the equator, the radius is the Earth equatorial radius. This means that the pendulum to consider must have its string as long as the Earth equatorial radius. The problem is to find the interval of time in which a pendulum with such a string completes two little oscillations. Huygens was able to determine this interval. It is 1 h 24 m 30 s.[12] Therefore, the speed of the matter rotating around the Earth for a body at the equator not to be flown away by its centrifugal force, must be equal to that of a body rotating around the Earth in 1 h 24 m 30 s. After having developed some easy calculations, the result is that such a speed has to be 17 times the rotational speed of the Earth at the equator. If the speed is higher than this, the body will fly away under the action of the centrifugal force.

Huygens obtained even more significant results: he understood that the whole quantity of matter of a body participates to its gravity and gave a mechanical explanation of this fact:

> There is another remarkable property of gravity, which can also be explained through the free movement of matter across the spaces that are among the bodies' parts, while other different explanations are difficult: all the internal parts of a solid body contribute to its gravity in proportion to their size. The reason why this happens is easy to understand: consider the fluid matter which passes freely through all the bodies' interstices. In this manner, it acts on

[12] Huygens (1669), in OH XIX, p. 639 and Huygens (1690b), in OH XXI, pp. 462–463.

> all the particles composing the body and pushes them towards the Earth's centre. Actually, if such a matter could not freely pass through bodies, it would press them only on the upper part of their surface. In this case, for example, an empty glass ball would weigh as much as another ball of the same size but full of matter, because the matter causing gravity would press on one as much as on the other.[13]

Huygens' idea that the whole quantity of matter of a body determines its gravity is very close to Newton's, who wrote in the General Scholium that gravity penetrates to the centres of the Sun and the planets, acting according to the quantity of solid matter a body contains, and not according to the extension of its surface. Unlike Newton, though, Huygens, like Descartes, ascribed gravity to the action of an external agent. However, unlike Descartes, he did not think that a body might be reduced to its extension. Quantity of matter is for Huygens the fundamental element in order to study the interaction of a body with other bodies, be this interaction due to gravity, to collision or to any other physical phenomenon. Huygens clearly understood that specific weight is an intrinsic property of bodies and that it depends on the quantity of matter distributed in a given volume. Therefore, the weight of a body *B* is proportional to the quantity of fluid matter moved by *B*, according to Archimedes' principle: when the particles of *B* are dense, that is when there is a little quantity of ethereal matter in their interstices, *B* moves a great quantity of ethereal matter and this makes it heavy.[14]

Huygens' reaction to Newton's masterpiece is significant to grasp his way of thinking: he acknowledged that Newton's concept of centripetal force and the mathematical apparatus developed in the *Principia* enabled Newton to offer a proper theory of gravity, to prove that the law by which the bodies fall on the Earth is the same that rules the motion of the planets around the Sun, and also allowed him to provide predictions that fit almost perfectly with all the available data. However, Huygens was unsatisfied with the general picture of the centripetal forces deriving from the *Principia*.[15] Even though it is a well-known fact that, as we shall see in the next section, in some of his works Newton seems to deny that bodies possess the inherent capacity to attract directly other bodies, this is precisely what can be deduced from the *Principia*. It is enough to look at the mathematical form of the

---

[13] OH XIX, p. 637: "Il y a une autre proprieté tres remarquable de la pesanteur qui s'explique encore par ce mouvement libre de la matiere a travers les espaces qui sont entre les parties des corps et qui malaisément le peut estre sans cela. C'est que toutes les parties du dedans d'un corps solide contribuent a sa pesanteur a proportion de leur grandeur; dequoy la raison est maintenant facile, quand on conçoit que la matiere fluide en passant librement par tous les endroicts de ce Corps agit par ce moyen sur toutes les petites parcelles qui le composent et les pousse vers le centre de la Terre, au lieu que si elle n'avoit pas la liberté de ce passage elle ne ce presseroit les corps que par la surface qui regarde vers en haut, ce qui feroit qu'un boule de verre par exemple qui seroit creuse en dedans, peseroit tout autant qu'une de pareille grosseur toute plaine & massive, parce que la matiere qui cause la pesanteur auroit autant de prise sur l'une que sur l'autre".

[14] This concept is well explained by Snelders (1989, p. 213). The author refers this conception to the *Discours sur la cause de la pesanteur* (which is correct), but the same ideas are also present in the paper read by Huygens at the Académie in 1669.

[15] Though dated, an interesting contribution concerning Huygens' reaction to Newton's theory of centripetal forces is Koyré's Appendix A to his *Newtonian Studies* (Koyré 1965).

general gravitation law to see that the masses hold a fundamental role. Therefore, in a Newtonian perspective, the problem was to find the reason why the masses mutually attract each other. However, Newton did not investigate this issue. In other terms: in Newton the centripetal forces, the mutual attraction among two or more masses, is a physical reality, which cannot be reduced to mechanical causes, as is made clear in the General Scholium to the *Principia*. It is exactly this aspect that Huygens did not share. He thought that any movement should be explained by other movements, not by attractions. Huygens claimed to have nothing against Newton's centripetal forces, but it is clear that he regarded them merely as a mathematical device that allowed Newton to find the most impressive physical law: the inverse square law. Therefore, what is true for Huygens is the inverse square law, not the existence of centripetal forces as physical entities. As he wrote:

> I would have never thought of this regular decrement of heaviness, namely that it is as the reciprocal ratio of the distances' squares from the centre. This is a new and very remarkable property of heaviness, of which it is worth to find the reason.[16]

Huygens's aim was to find the reason of the inverse square law. According to Newton, this law depends on the existence of centripetal forces and on the physical–mathematical apparatus deriving from his definitions and his three laws. But for Huygens things are far more complicated: Newton had proved that Descartes's vortices are not stable. Thus, in order to explain the rotation of the planets around the Sun by means of the centrifugal force of an ethereal matter, he developed a new kind of vortex, and he admitted the existence of void (or better, the existence of a very tenuous matter, as we will see) among the different vortices surrounding the Sun and the planets. This led him to develop a series of ad hoc devices in order to explain through the vortex theory a part (a small part) of what Newton had explained by centripetal forces.

Commentaries. First of all, it seems appropriate to raise a question: in what sense the vortex theory and Newton's theory give an explanation of gravity?

In Newton's explanation the theory of movements, namely kinematics, must be explained ontologically at the level of the forces studied by dynamics. This level is more profound than kinematics from a physical point of view since, by means of relatively few properties of the forces, one is able to find the common explanation for a plurality of movements.

Huygens's mechanistic explanations of movements do not contemplate the existence of forces separated from the movements themselves. Such a more profound level does not exist. A priori there is not anything wrong in such a view. It seems reasonable to simplify the explanatory apparatus. The problem is that the physical realization of this idea encounters enormous problems. Let us consider the case of gravity: Huygens was forced to introduce different kinds of vortices to make his theory coherent with Newton's discovery, he had to resort to strange separations

---

[16] Cf. Huygens (1690b), OH XXI, p. 472: "Je n'avois pas pensé non plus à cette diminution reglée de la pesanteur, sçavoir qu'elle estoit en raison reciproque des quarrez des distances du centre; qui est une nouvelle & fort remarquable proprieté de la pesanteur, dont il vaut bien la peine de chercher la raison".

between empty and full regions of the universe; even considering only gravity on Earth, before the publication of Newton's *Principia*, in order to explain gravity through centrifugal forces, Huygens had to postulate strange ad hoc movements of the particles which compose the vortex surrounding the Earth. Furthermore—and this is particularly important—through a construction like that of Huygens, it would have been impossible to reach the inverse square law. Only after Newton had determined such law through completely different principles, Huygens (and Leibniz, too)[17] was able to imagine a series of vortices whose existence was coherent with Newton's inverse square law. Moreover: Huygens' mechanistic theory is simpler because no dynamical level exists. However, it populates the universe of entities, such as the particles composing the vortices, which no one has ever seen. Their existence is hypothetical and their effects are explained more clearly and more "economically" by Newton's theory. In our view, this is the true reason why the treatment in terms of centripetal forces is far superior to the mechanistic one.

The vortex theory of gravity had been criticized long before Newton set out his ideas. In August 1669 the already mentioned discussion on gravity at the Académie took place.[18] On the 7th Roberval presented his Mémoire (OH XIX, pp. 628–630) and distinguished three schools which held three different theses on gravity: according to the first, gravity is a property inherent only to heavy bodies; according to the second, gravity is a mutual attraction among all the bodies; finally, other scholars, such as Descartes, thought that gravity is due to the movement of a particular fluid. Roberval pointed out that all three theses are highly conjectural because they are not based on evident and clear principles. He thought Descartes' view to be particularly implausible because it postulated the existence of an invisible ether. Among the three theses, Roberval preferred the second. The memoires by Frenicle de Bessy and Jacques Buot were read respectively on the 14th and 21st August (summarized in OH XIX, pp. 630–631): Frenicle adhered to the first view, whereas Bout, denying action at a distance, was a Cartesian. On the 28th August Huygens read his Mémoire (OH XIX, pp. 631–640), by far the most profound and innovative in that debate. It is interesting to analyse some of the eleven objections addressed by Roberval and Mariotte to Huygens in the session of the 4th September (OH XIX, pp. 640–642). First of all, they claimed that Huygens excludes without any proof the attraction between bodies, and admits without any foundation that gravity has to be explained by size, shapes and movement. Secondly, if every movement has to be explained by a further movement, there is a *regressus* ad infinitum, which is even more difficult to explain than the attractive qualities themselves. The third objection is the most profound: the only natural (we say inertial) motion is the [uniform and] rectilinear one. Circular motion cannot be obtained as a composition of rectilinear motions. Thus, mechanical causes are not sufficient to produce it. A quality is necessary, which is intrinsically not mechanical (ibid., p. 641). In the ninth objection, Roberval and Mariotte claimed that colliding particles will change their circular movements in

[17] With regard to Leibniz's efforts to frame the inverse square law within the context of vortex theory, see Bussotti (2015).

[18] On Huygens' theory of gravity, see Snelders (1989).

rectilinear ones along their instantaneous tangents. The circular movement of a vortex cannot be maintained since its particles will fly away rectilineraly. Furthermore, particles which move in different manners and enter the interstices of solid bodies should cause an irregular trajectory in the falling bodies themselves. To summarize: the most important objections concern the fact that the existence of the ether and of its complex movements is purely conjectural and that, according to the inertial principle, all the motions would be transformed into uniform rectilinear ones, unless there is a principle that opposes the inertial tendency, and that should be another quality inherent to the bodies.

On 23rd October Huygens answered the objections of Roberval and Mariotte (OH XIX, pp. 642–644). The answer to the first objection is a clear evidence of how irreconcilable are the views of mechanistic physicists and those of thinkers who, instead, admitted the existence of qualities other than size, shape and movement. Indeed, Huygens claimed:

> To the first objection I answer that I exclude attractive and expulsive qualities from nature because I search for an intelligible cause of gravity. It seems to me that ascribing the reason why heavy bodies fall towards the Earth to an attractive quality of the Earth or of these bodies is like to say nothing. In contrast to this, I do not understand how one can say that I introduced movement, shape and size with no foundation since the senses let us know that these things exist in nature.[19]

In this case, Huygens' approach seems to be rather dogmatic: it is true that ascribing gravity to an attractive power is not an explanation and that it recalls the unscientific conceptions of traditional philosophies of nature based upon the confused notions of sympathies and virtues. On the other hand, there was no reason to exclude a priori that gravity is a fundamental phenomenon for which no other more basic cause exists. Anyway, even if Mariotte, Roberval or Frenicle detected the weaknesses of Huygens' theory, their alternative views did not offer any quantitative achievement, while Huygens was able to calculate some particular effects from his theory of gravity.

To the second objection, Huygens answered that, while dealing with motion, it is not necessary to look for its first causes. This answer is fully justified: if one were to search for the causes of every physical phenomenon, no physics would exist.

Answering the third objection, Huygens claimed that he did not consider circular motion to be natural, but since in the world such a motion exists, he regarded it as a matter of fact. This is an unsatisfactory reply and does not touch the core of the problem raised by Roberval and Mariotte. One might think that Huygens explained circular motion *à la* Descartes, but in the answer to the third objection no hypothesis on the origin of circular motion is proposed and Descartes is not mentioned.

[19] OH XIX, p. 642: "A la premiere obiection ie respond que i'exclus de la nature les qualitez attractives et expulsives parce que ie cherche une cause intelligible de la pesanteur, car il me semble que ce seroit dire autant que rien que d'attribuer la cause pourquoy les corps pesants descendent vers la terre, a quelque qualité attractive de la terre ou de ces corps mesmes, mais pour le mouvements la figure et les grandeurs des corps ie ne vois pas comment on peut dire que ie les introduicts sans fondement puisque les sens nous font connoistre que ces choses sont dans la nature".

Regarding the objection that inertia would transform all the circular movements into rectilinear motions, Huygens wrote:

> As for the further objection that the colliding parts of matter would change their circular movement into a straight one along the tangents, I say that it is true that they would do so not only as a consequence of a collision but simply because of the circular motion as well. However, as these particles do not find any place where to go unless, at the same time, other particles approach towards the centre [of the vortex], this rectilinear movement can take place only if there are parts which do not have the centrifugal tendency because they lack circular movement. In such condition, these parts cede without difficulty to those having the centrifugal tendency. I claim that gravity consists exactly in this.[20]

Without a mathematical treatment of vortices' stability, it is impossible to claim whether the kind of vortex envisioned by Huygens might be stable.

Huygens' theory of gravity was the best one available before the publication of Newton's *Principia*[21]: it offered a plausible mechanical explanation, it was partly mathematized and, above all, its adversaries did not elaborate an alternative theory sufficiently developed from a mathematical point of view. The situation is completely different in 1690 when Huygens published the *Discours de la cause de la pesanteur*. This work explains clearly the mechanistic conception, proposes a notion of vortex different from Descartes'—that Huygens thought had been refuted by Newton—and, which is fundamental, offers mathematical demonstrations of several propositions concerning gravity theory. Most of them had already been proved by Newton, but Huygens gave different demonstrations that were consistent with his principles. Therefore, the *Discours* was the most advanced treatise on the vortex theory of gravity at the time, by far superior to Leibniz's speculations which remained at a qualitative level. However, besides being incomparably less complete than Newton's theory, Huygens' views on gravity give the clear impression of being merely a defensive answer to Newton's *Principia*. Huygens recognized the validity of Newton's law of universal gravitation, and aimed at obtaining it within his own vortex theory. Newton's results can be, in part, deduced by creating a vortex with appropriate features, but it is impossible to achieve them starting from a generic vortex because the potential constraint conditions are so numerous that it is impossible, as a matter of fact, to determine them a priori. Instead, on the basis of Newton's principles, the gravity law follows without requiring any further condition beside the three laws of motion. In this respect, Huygens' gravity theory is like Leibniz's planetary theory: Huygens and Leibniz obtain Newton's main results through principles consistent with vortex theory, but could not have conceived an appropriate vortex without

---

[20] OH XIX, p. 643: "Et pour ce qui est adiouté que les parties de la matiere en se choquant changeroient le mouvement circulaire en mouvement droict suivant les tangentes, ie dis qu'il est vray qu'elles le feroient non pas seulement par le chocq mais aussi par le seul mouvement circulaire, mais ne trouvant pas de place ou se mettre si ce n'est qu'en mesme temps d'autres parties s'approchent vers le centre, ce mouvement droit ne peut avoir effect que lors qu'il y a des parties qui n'ont point de tendence du centre faute d'avoir de mouvement circulaire lesquelles alors cedent sans difficulté a celles qui l'ont, et c'est en quoy i'ai dict que consiste la pesanteur".

[21] See Aiton (1972), p. 81: "Of the theories presented in the course of the debate in the Académie, only that of Huygens was a true development of Descartes' theory".

already knowing those results. A further noteworthy problem concerns the fact that Newton's universal gravitation is an interaction between two masses. Earth attracts bodies, but bodies attract Earth too. The vortex theory of gravity can explain how Earth, by means of its vortex of small particles, attracts bodies, but cannot explain the attraction exerted by bodies on Earth. By which mechanism the vortex of each body should act on Earth? This seems an almost insurmountable problem for a complete and coherent theory of vortex gravity.

Despite these difficulties, several remarkable observations can be found in the *Discours*. In particular, we find a passage in which Huygens seems to guess the difference between gravitational and inertial mass:

> For this I will point out what happens in the collision of two bodies, when they meet with a horizontal movement. It is certain that the resistance of bodies, like a ball of marble or lead placed on a smooth table, to being moved horizontally is not caused by their weight towards the Earth, since the lateral movement does not tend to move them away from Earth, and thus it is in no way contrary to the action of gravity, that pushes them down.[22]

Although Huygens' words should not be overestimated because they represent only a hint, here Huygens seems to state that inertial mass is the quantity involved in collisions while gravitational mass concerns the determination of weight. It is noteworthy that on this issue a vortex theory of gravity faces the same problem of Newton's theory: though gravitational and inertial mass are conceptually distinct, they coincide as a matter of fact in every calculation, but this factual coincidence remains theoretically unexplained.

Finally, some brief observations on Huygens's concepts of inertia and centrifugal force: as we have seen, he clearly understood that any motion continues uniformly in a straight line unless some external action occurs. Unlike anyone before him (and specifically unlike Descartes) he worked out a good theory concerning both impact and circular motion. Inertia principle played an important role within his theory. He correctly guessed that the centrifugal force arises because a body is prevented from escaping along the tangent in a curvilinear motion. However, it is significant that he developed no theory on the force which produces the circular motion. Certainly, he was perfectly aware that if a stone is attached to a sling and receives—for example—a horizontal impulse, it begins to rotate because the stone is kept inside the sling by the strenght of the hand. But, as we have seen, he was interested in examining the force experienced by the stone (that he called centrifugal) and not the force exerted by the hand itself. If we calculate the centrifugal force in a reference frame co-moving with the stone and the centripetal force in an inertial reference frame, we find the same value. But the perspective is completely different. And the perspective of the inertial reference frame in which the centripetal force and the inertial tendency are clearly

[22] Huygens (1690b). Cfr. OH XXI, p. 458: "Je feray remarquer pour cela ce qui arrive dans le choc de deux corps, quand ils se rencontrent d'un mouvement horizontal. Il est certain que la resistence que font les corps à estre mûs horizontalement, comme seroit une boule de marbre ou de plomb posée sur une table bien unie, n'est pas causée par leur poids vers la Terre, puisque le mouvement lateral ne tend pas à les eloigner de la Terre, & qu'ainsi il n'est nullement contraire à l'action de la pesanteur, qui les pousse en bas".

separated allows us to obtain a general and unitary picture which is difficult to obtain in the other case. In particular, the perspective of the inertial reference frame allows us to examine the action which is the cause of the curvilinear motion, that is, in our example, the action of the hand.

As we have seen, Huygens was a sort of a tightrope walker who, thanks to his mathematical skills and his clear interpretation of physical phenomena, avoided to overlap mistakenly the concepts valid in inertial reference frames with those valid in the non-inertial ones. However, treating the problems only in terms of centrifugal forces, it is difficult to find a place for inertia: as a matter of fact, in a system where centrifugal forces exist, inertia does not exist. It is necessary to introduce centripetal forces to grasp the physical meaning of inertia. This means that, though Huygens made no error, his mechanics can hardly be framed in a unitary picture. However, this does not prevent us from asserting that his work was an enormous progress in comparison to that of his predecessors and that he was one of the greatest scientists ever.

6.1.3 Inertia and the Nature of Motion

Huygens did not publish his elaborated ideas on the nature of motion and on the problem of inertia. His conception can be found in the posthumous work *De Motu Corporum ex Percussione* (1703), and in letters and notes which were published only in the twentieth century.[23] This probably indicates that Huygens did not find his views on motion fully satisfactory. Scholars disagree on this point: Barbour claims that Huygens never achieved a coherent view of motion and of its physical causes (Barbour, 2001, p. 672), while other authors hold that Huygens treatment of motion and inertia is satisfactory and that he developed an almost complete theory of motion (the standard reference for this thesis is Stan 2016). In any case, Huygens' ideas are of the highest interest and trace a line of thought which is in opposition to Newton's view.

This subsection is divided into two parts. In the first we reconstruct the development of Huygens' train of thought on motion and inertia.[24] We then evaluate the relations between the kinematical and the dynamical aspects of Huygens' ideas on motion and we advance the thesis that his ideas are fully developed from a kinematical point of view, but do not offer an adequate dynamical explanation.

[23] An important contribution on this topic is Mormino (1993) who edited integrally and enriched with new material those texts that had been partially published in various volumes of OH.

[24] On the development of Huygens' ideas on motion see: Dugas (1988) (original 1955), p. 198 (a brief observation); Dugas (1954); Stein (1977); Barbour (2001, pp. 672–675). Mormino offers the most complete analysis in the introductory essay to his edition of Huygens' *Codex 7A*. With regard to the genesis of the concept of inertia in Huygens, see Chareix (2006), who discusses the topic in detail in two long chapters dedicated to Galileo's and Descartes' influence on Huygens.

6.1.3.1 Huygens' Ideas on Inertia and Motion

Huygens' ideas on motion evolved across three phases: in the first one, around 1652–1669, he thought every motion to be relative. Afterwards, from 1669 to 1687, he believed that circular motion and centrifugal forces might offer a criterion to distinguish absolute from relative motion. Finally, after the publication of Newton's *Principia*, Huygens rethought the whole problem of motion and came to the conclusion that every motion is relative. The third phase is by far the most interesting and original.

The work *De Motu Corporum ex Percussione* is the first fundamental source to understand Huygens' ideas on motion. The presentation of this treatise is rather modern because Huygens poses three hypotheses that function as axioms and deduces from them his whole theory. The first hypothesis is the principle of inertia; the second states that if two equal bodies collide with the same velocity, each of them will maintain the same velocity it had before the impact. The third hypothesis is the principle of indifference of the common motion: it states that if two bodies participate to the same uniform rectilinear motion, collisions and other phenomena concerning their kinetic state take place as if that motion did not exist.

Huygens states the principle of inertia in these terms:

> When a body is in motion, if nothing prevents its movement, it continues to move perpetually with the same velocity and along a straight line.[25]

By means of these three hypotheses Huygens proved the first two propositions of his treatise. He then added a fourth hypothesis stating that if a big body A hits a smaller one B which is at rest, A gives up to B a part of its movement and, thus, A loses this amount of movement (OH XVI, p. 39). Having proved a further proposition, Huygens introduced also a fifth and last hypothesis, which is the principle of the momentum conservation.[26]

Huygens conceived, hence, one of the first treatises of what might be called "modern physics", in which a set of initial axioms is clearly stated and the propositions are deduced from them and from other previously proved propositions. Huygens' ideas on motion are clear: the law of inertia is valid, the principle of indifference of the common rectilinear motion holds, as well as the principle of momentum conservation. This is enough to claim that what today we call inertial reference frames are all equivalent and that the consideration regarding the bodies to consider at rest is purely arbitrary so that motions are relative. However, a conceptual problem remained open: did Huygens think that, at least in principle, an absolute motion (perhaps the circular one) might exist? In the published version of *De Motu*

---

[25] Huygens (1656, 1703, OH XVI, p. 31): "Corpus quodlibet semel motum, si nihil obstet, pergere moveri eadem perpetuo celeritate & secundum lineam rectam".

[26] Ibid., p. 41. Huygens fully understood the vector character of momentum and criticized the principle of conservation of the quantity of motion in the scalar form given by Descartes. See Mormino (1993, pp. 36–42). A further proposition used by Huygens as an axiom, though not stated under the title "hypotheses" , is the Torricelli principle. It is claimed along with the demonstration of Proposition VIII (OH XVI, p. 57).

Corporum ex Percussione this problem is not faced. Nonetheless, in a series of interesting notes (published in OH XVI as appendices to *De Motu*) which date back to the period 1652–1654, he states that all the motions are relative and that no absolute reference frame exists to which one could refer—at least in principle—the motion of bodies. Huygens claimed:

> It is to be considered that what is moved, is moved with respect to other bodies from which it changes its distance and place. Analogously, what is at rest is at rest with respect to those bodies from which it maintains the same distance and position.[27]

This means that only relative motion among bodies can be considered and it implies that only relative motions exist since "it is impossible to show a body or a point in the universe to which it would be necessary to refer the motions of all the other bodies".[28]

In the manuscripts edited as Appendices to *De Motu* Huygens criticizes those who consider the Earth to be at absolute rest; his objection can be referred also to those who think of a universal space at rest. Huygens holds that there is no criterion to establish the kinetic state of anything in absolute sense. Rest and motion are relational and, hence, relative states:

> If they claim that Earth is really at rest, I will ask them what is to be really at rest. They will answer that it means to maintain constantly the same place. Thus, the definition of the place the Earth holds depends on some points of the sky or on a surface which is really at rest. But once again it is manifest that the state of rest has to be defined on the basis of "some other place" or of the Earth itself.[29]

Therefore, the assertion that a place or a body is in absolute rest falls under either a *regressus* ad infinitum or a circular reasoning. Huygens' conclusion is that an absolute kinetic state does not exist.

This argument is less convincing than it appears *prima facie*. If we consider Newton's bucket experiment, we can certainly claim that the existence of an immobile object to which motions are referred is only postulated or indirectly proved: however, in Newton's case, there is no *regressus* ad infinitum nor circular reasoning. The problem can be interpreted in these terms: in our daily experience there are only relative motions and we are able to develop the whole physics on such a basis from

---

[27] OH XVI, p. 103: "Quod autem movetur aliorum corporum respectu moveri intelligitur quibuscum distantiam situmve commutat. Itemque quod quiescit eorum respectu quiescere quibuscum distantiam situmve servat eundem". See also Mormino (1993, p. 13). There is no doubt that Cartesian relativism influenced Huygens profoundly. However, it has also to be remarked that, given a body, Huygens did not determine his state of motion or rest only in relation to the neighbouring bodies, as Descartes did. For Huygens the relative motion of each system of bodies can be determined independently of their distance. This makes Huygens' relativism a far more acceptable and physically fruitful view than Descartes'. On this subject, see, e.g., Mormino (1993, pp. 17–18).

[28] Ibid., p. 103: "Neque enim corpus aliquod aut punctum in universo ostendi potest ad quod omnium reliquorum motus referre necesse sit". See also Mormino (1993, p. 13).

[29] OH XVI, p. 111: "Si dicant igitur terram revera quiescere rogabo quid sit revera quiescere, dicent locum eundem constanter obtinere; loci igitur quem terra, tenet definitio ab aliquibus celi punctis pendet aut superficie quae revera quiescat. sed hujus rursus quietem apparet aliunde, aut a terra ipsa definiri debere". See also Mormino, p. 19.

a kinematic point of view. Newton himself would not have anything to object to this approach. However, if we shift from this phenomenal level to the foundational one, then the question of how to determine inertia arises. It is in this further context that the search for an absolute motion makes sense.

Starting from around 1669, Huygens changed opinion with regard to the nature of motion and arrived at the idea that a criterion to distinguish absolute from relative motion can be found in circular movement. Such a criterion is the centrifugal tendency. After his work on *De Vi Centrifuga* (1659), Huygens' new interest in circular motion was probably motivated by his research on gravity (Mormino, 1993, pp. 60–61). The sources to reconstruct Huygens' thought on the problem of motion in the period 1669–1687 are scarce, but the conclusion that he identified the centrifugal force as a criterion of absolute motion is not conjectural. Indeed, in 1694 Leibniz wrote to Huygens: "It seems to me, however, that you yourself, Sir, in the past had the same mind as Mr. Newton about circular movement."[30] Huygens confirmed Leibniz's assertion in a letter dated 24 August 1694: "As to absolute and relative motion, I admire your memory because you have remembered that in the past I had the same opinion as Mr. Newton with regard to circular motion".[31] In several passages of *Codex 7A* Huygens admitted that he entertained the idea that centrifugal forces offered the criterion to distinguish absolute from relative motion: "For a long time, I believed that a criterion of real motion existed in circular motion, from the centrifugal force".[32]

Huygens exemplifies the centrifugal forces-criterion by means of an ideal experiment similar to the one described by Newton at the end of the scholium to the *Definitions* in his *Principia*: given two globes connected by a rope, suppose them to be moved around their gravity centre. The tension of the rope allows us to know the centrifugal force and, hence, the real movement of the globes.

Why did Huygens change opinion after the publication of Newton's *Principia*? Probably the answer is that Newton had shown the absolute motion to be recognizable only in relation to an infinite, immobile mundane space. This connection reminded Huygens that he had claimed two theses: (1) it makes no sense to speak of a mundane space at rest; (2) centrifugal forces represent a criterion of absolute motion. Newton had proved that absolute motion has to be referred to a mundane space at rest. Huygens probably realised that his two theses were contradictory, and decided to maintain the first and to dismiss the second.

To illustrate Huygens' new view, we will focus on two issues: (a) impossibility to give a meaning to the concept of a kinetic state of the mundane space; (b) relation

---

[30] Letter by Leibniz to Huygens dated 22 June 1994. OH X, pp. 645–646. See also Mormino (1993, p. 62): "Il me semble cependant que vous même, Monsieur, estiés autres fois du sentiment de M. Newton à l'egard du muovement circulaire".

[31] OH X, pp. 669–670. See also Mormino (1993, p. 62): "Pour ce qui est du mouvement absolu et relatif, j'ay admirè vostre memoire, de ce que vous vous estes souvenu, qu'autrefois j'estois du sentiment de Mr. Newton, en ce qui regarde le mouvement circulaire".

[32] Mormino (1993, p. 63): "Diu putavi in circulari motu haberi veri motus κριτηριον, ex vi centrifuga".

between uniform rectilinear motion and uniform circular motion. The source we use is *Codex Hugeniorum 7A*.

With regard to item (a) the arguments expounded by Huygens are similar to those he had already presented in his early observations in the manuscripts now published as appendices to *De Motu*. Huygens reflected for a long time on the problem of absolute motion referred to a mundane space, and eventually he denied its existence:

> I kept asking myself if anything distinguished the true motions from those referred to something else. I saw that all who wrote on this topic said that a true and proper motion of a body is different from a relative motion. The true and physical motion takes place when a body moves from a place to another of mundane space. Thus, I say that they conceive these places as really immobile. However, what can we say to be truly immobile, if we are still looking for the true motion? [...] They will say: [motion or rest] are referred to the world's immobile space. Thus, this is what the whole question is about, whether we can say and conceive that the world's infinitely extended space is immobile. This notion seems to me to be false.[33]

Huygens did not deny the existence of an infinite mundane space, but he considered meaningless the idea to ascribe any kinetic state to it. Indeed, he judged the idea of rest and motion to be purely relative (i.e. it cannot be referred to absolute entities such as Newtonian space and time) and relational (concerning only the mutual kinetic state of more than one body). Thus, it is possible to determine the kinetic states of an object only with respect to another object (ibid., p. 140).

It is impossible to ascribe a kinetic state to space before having defined the notion of movement. But, as soon as we define motion as a relational concept, we discover that neither motion nor rest can be applied to space:

> Motion cannot be understood but relatively. To what place? Maybe the immobile space? But what can we consider immobile, if we are still looking for the definition of movement?[34]

Huygens remarks that, probably, the idea that space is at rest came from the thought that the the concept of the motion of space is absurd. In truth, neither motion nor rest can be predicated of space (ibid., p. 141), since rest and motion are properties concerning bodies, not space (ibid., p. 232). As a corollary of this conception, if in the world only one body existed, the question whether this body is in motion or at rest would be meaningless, because movement is a relational concept among bodies (ibid., p. 194). Huygens examines a further interesting problem: suppose a sphere A and imagine that a person P gives an impulse to A. Everyone agrees on the existence of a relative motion between P and A. Then, suppose that P disappears. The inertial

---

[33] *Codex Hugeniorum 7A*, in Mormino (1993, pp. 138 and 140): "Pergebam tamen mecum inquirere nunquidnam inter veros motus ac eos qui referuntur ad aliud interesset. Videbam vulgo omnibus qui de motu scripsere haec diversa dici, alium nempe motum corporis verum ac proprium alium esse respectu alterius cujuspiam. Motum verum ac Physicum esse cum transfertur corpus ab uno ad alium locum spatij mundani. Itaque inquam locos istos tanquam revera immotos concipiunt. quid autem revera immobile esse dicemus, cum adhuc quid sit verus motus quaeramus? [...] spatij dicent mundani immoti. itaque in hoc tota quaestio vertitur an spatium mundi infinite extensi, immotum sit ac dici et intelligi possit. Visum est autem mihi falsam hanc esse notionem".

[34] Ibid., p. 174: "motum non posse intelligi nisi relative. quid locus? an spatium immobile. quid autem immobile dicimus cum adhuc definitionem motus quaeramus?".

motion impressed on the sphere by the initial impulse remains. Therefore, one might claim that it is an absolute motion since it cannot be regarded anymore as a movement in relation to *P*. Huygens replied to this possible objection that the motion between *P* and *A* was purely relative and hence that it makes no sense to speak of the kinetic state of *A* without *P* (ibid., p. 260). Huygens'reply is a mere reassertion of his definition of motion, but in support of Huygens' view we could remark that this ideal experiment does not prove that a movement can be referred to absolute space, for, it might be possible that *A* moved in relation to absolute space before receiving the impulse and that the impulse transformed its motion into a state of rest.

There is, however, a serious candidate for absolute motion: it is circular motion, for dynamic rather than kinematic reasons. Indeed, to every circular motion a centrifugal force is associated that indicates inertia. In the second phase of his thought, Huygens himself regarded centrifugal forces as evidence of absolute motion. Let us see how he tackled the problem of circular (or more generally, curvilinear) motion in the third and last stage of his scientific career. Our thesis is that the weakness of his conception lies exactly in the inability to pass from the kinematic to the dynamic level.

First of all, Huygens offers the criteria to identify circular motion: an inertial observer recognizes circular motion by direct observation. However, this criterion is inapplicable to an observer who participates to circular motion, because the distance between two bodies rotating circularly around their gravity centre does not change even if the two bodies are not in relative rest. There is, however, another criterion: it is the presence of a centrifugal force, which exists in any circular motion. The centrifugal force has various manifestations: if two bodies rotating around their gravity centre are linked by a rope, it is displayed by the tension of the rope; if a body is solid and rotates (for example, a disc), and you set something on top of it, this thing will fly away; if water rotates in a vessel, the water will rise on the vessel's walls.[35]

Huygens is not convinced that centrifugal forces are a sufficient criterion to determine absolute motion. He resorts to two arguments: firstly, he argues that even the circular motions of the parts of a rotating body are purely relative to the parts themselves. Furthermore, he states that a body, in addition to being animated by rotational motion, could also have a uniform rectilinear motion. But this kind of motion, by admission of the Newtonians themselves, can only be relative:

> Given a single body or more joined bodies moving around a centre, it is possible to calculate through the centrifugal force how much circular velocity they received. However, this velocity is also relative to these bodies or to the parts of a single body. It is, instead, impossible to know how much true motion is in the single parts, namely, in reference to that space that they imagine to be immobile. According to their own judgement, a rotating body, be it solid or composed of joined parts, can at the same time proceed rectilinearly with a movement common to all its parts. They admit that it cannot be recognized by any clue to what extent this rectilinear motion is true.[36]

---

[35] Ibid., p. 268: Almost identical concepts are expressed in ibid., pp. 186, 222–224, 226.

[36] Ibid., p. 186: "dato corpore uno, vel pluribus cohaerentibus quae circa centrum aliquod moveantur, potest colligi, ex vi centrifuga, quantam celeritatem circularem acceperint. sed et haec relativa est inter corpora vel inter unius partes. non tamen cognoscetur quantum veri motus partibus singulis

Huygens' new idea is that circular motion can be understood as the relative motion of the parts of a rotating body which are pushed in contrary directions.[37] Their motion cannot become rectilinear because a constraint or a link prevents it. A stone placed on a disc participates to the instantaneous motion of that part of the disc on which it is set, and, of course, the direction of its motion is along the tangent to the disc. Since the stone is not constrained, it flows away:

> But this effect [the stone which flies away] only shows that, because of the impulse exerted on the circumference, the wheel's parts are pushed in different directions, with a relative motion of the ones with respect to the others. Thence, circular motion is only the relative motion of the parts which are pushed in opposite directions, when they are prevented by a constraint or a link.[38]

Huygens concluded that every motion is relative.

Two further elements enrich his analysis:

(1) Huygens did not take for granted what a straight line is and proposed a definition connected to inertia. While the usual idea is that two points are sufficient to define and construct a straight line, Huygens thought, instead, that the feature which distinguishes a straight line from any other line connecting two points can only be found in a physical context. For this purpose, three points are necessary. Thus, be given three points *A, B, C* mutually at rest and on which no force acts; give an impulsive force to *C*; a straight line is the trajectory of *C* as seen from *A* and *B* (ibid., p. 134).

---

insit, hoc est respectu spatij ut fingunt, immoti. nam ex illorum arbitrio, quod circulariter movetur simul et recto motu totum, sive ex pluribus junctum progredi potest quo itaque et partes feruntur: qui rectus motus quatenus sit verus, nullo signo disccerni posse fatentur".

[37] To be pushed in contrary directions means that, given every particle A, B, ... of the disc, there is a particle A′, B′,... such that the instantaneous velocities in A and A′, B and B′, ... are equal and have opposite directions. See figure (the figure is ours).

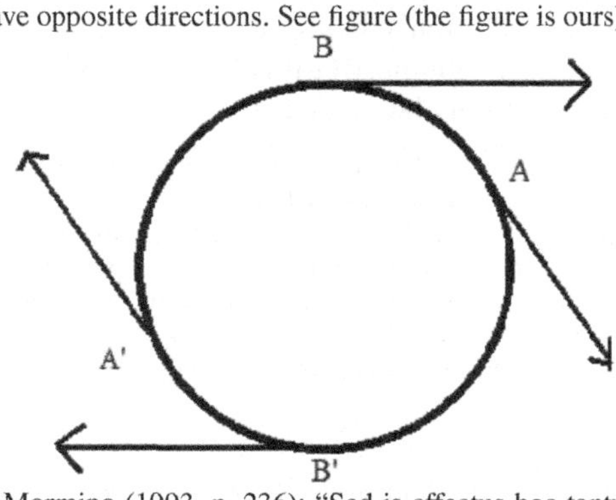

[38] Mormino (1993, p. 236): "Sed is effectus hoc tantummodo declarat impressione in circumferentiam facta partes rotae motu relativo ad se invicem in partes diversas impulsas fuisse. ut motus circularis sit relativus partium in partes contrarias concitatarum sed cohibitus propter vinculum aut connexum".

This vision is fully coherent with Huygens' view of infinite space as a completely unstructured entity. Space is a mere receptacle of bodies, so that nothing can be determined in relation to such a featureless being. To define a property or an entity, something other than space is always necessary, such as the push in the example we have mentioned. In this respect, Huygens' space is different from Euclid's and Newton's. It is true that for both of them space is a receptacle of objects (in Euclid's case of "abstract" objects, in Newton's of physical objects), but space also has minimal intrinsic properties: for example, due to the nature of space, it is possible to determine the distance between two points and to claim that a straight line is the shortest one connecting them. Huygens seems, instead, to think of space, be it geometrical or physical, as a pure container with no other property than that of potentially receiving bodies. Space acquires determinations only when it is filled with objects and when physical principles, such that of inertia, are applied. This is a confirmation that, according to Huygens, it is not possible to define inertia with respect to space. On the contrary, a structure can be given to space only on the basis of an inertial motion. This conception, however interesting, raises some doubts which we discuss in the next subsection.

(2) Huygens describes an ideal experiment that clarifies his conception of motion and of physics in general: consider two bodies, A and C, close to each other and moving with uniform motion in opposite senses along two parallel lines. Let's imagine that when they reach their minimal distance we join them with a rope that has two hooks at its extremities. The two bodies will be hooked at the rope and their uniform rectilinear motion will be transformed into a uniform circular one. Huygens does not admit that this simple operation can transform a relative motion into an absolute one:

> Move the body A along the straight line AB and the body C along the straight line CD, parallel to AB. [...] Lay out a rope BD perpendicularly to AB and CD so that A happens to be in B and C in D at the same instant. Be in B and D hooks to which the bodies hook up. Their rectilinear motion will be transformed into a circular one and the rope BD will reveal this motion through its tension. Since, before the bodies A and B met these hooks we ignore their real motion, maybe we know it now, after they have met them? Could we define how quick they move with respect to that infinite and immobile space? Certainly, this cannot be said. Who considers correctly this question understands that only the previous relative motion remains. Nothing else happens than that motion which previously occurred along parallel lines, while now occurs along the opposite parts of the circumference (which can be defined as parallel to each other), and the distance that before changed continuously now remains unmodified.[39]

[39] Ibid., pp. 190 and 192: "Moveatur corpus A in recta AB; corpus vero C in linea CD ipsi AB parallela [...]. Ponamus jam filum BD utrique AB, CD perpendiculare, atque eo loci positum, ut eodem momento A incidat in B et C in D; sintque unci quidam in B et D, quibus haereant corpora utraque, quo fiet ut rectus eorum motus in circularem vertatur, quem tensione sua filum BD prodet. cum igitur, antequam in uncos istos corpora A, B inciderent, ignoraretur quantum vero motu moverentur; nunc postquam inciderunt, hoc sciemus, ac definiemus quantum respectu spatij illius infiniti et immoti concitetur. Hoc certe dici non potest. Recte vero rem perpendenti, manet tantum motus respectivus qui fuit antea. nihilque aliud accidit nisi quod cum antea esset in lineis rectis parallelis, nunc sit in partibus circumferentiae oppositis quae similiter inter se parallelae dici

This quotation makes Huygens' conception of physics clear. As is well known, it is a mechanistic conception. One might add, more specifically, that it is also an engineering conception. Huygens elaborates his ideas and solves problems step by step, from below. He decomposes events as if they were pieces of a mechanism. He arrives at his most abstract conclusions starting from descriptions which have always a direct mechanical reference, where threads, impacts, men holding ropes in their hand, etc. are involved. You can see this attitude in the examples he produced to show a mechanism analogous (not homologous) to gravity, in his theory of centrifugal forces, in his studies on impact and in his theory of the evolutes so brilliantly conceived and expounded in the *Horologium oscillatorium*.[40] Nature is a machine and circular motion derives from a mechanism or a series of mechanisms which certainly cannot transform a movement from relative into absolute. The principle of indifference of the common motion also relies on the mechanical example of a ship moving rectilinearly and uniformly with respect to the shore and on how movements are seen from a person on the ship and another on the shore (see, e.g., Mormino, 1993, pp. 10–18). In the natural machine every movement is relative. In the second phase of his reflections on motion, Huygens understood that centrifugal forces might imply a non-local reference. This could have been a problem for his relativistic conception of motion, but finally, he preferred to opt for the relative character of circular motion as well.

6.1.3.2 Huygen's View of Inertia in the Secondary Literature

In the third period of his scientific career, Huygens came back to the idea that every motion is relative. We saw that he criticized the idea of absolute motion, namely motion with respect to absolute space, for several reasons: first of all, a rotating body can be also animated by any uniform rectilinear motion, which, according to the principles of Newton's theory, is always relative. Therefore, the absolute kinetic state of a rotating body cannot be established. This criticism is correct: anyone who maintains the existence of absolute motions in the Newtonian sense must admit that they can be determined up to a uniform translation. Therefore, it is not possible to identify an absolute space, but, at best, the set of all inertial reference frames.

Furthermore, Huygens claimed that no kinetic state can be ascribed to absolute space because movement is always a relation between bodies and hence no kinetic state can be assigned to empty space. This is a plausible argument, but it presupposes a relativistic view of motion. From a contrary point of view, one might assume that absolute motion is determinable with respect to the set of inertial reference frames, if not with respect to an absolute space at rest. However, since Newton did not mention

possunt, distantia vero corporum quae prius mutabatur continue, nunc invariata maneat, propter vinculum".

[40] A classical text on Huygens' *Horologium oscillatorium* is Yoder (1988). The author explains the conceptual itinerary which led Huygens to conceive the theory of evolutes and its application to construct an isochronous pendulum.

the set of inertial reference frames in the scholium to the *Definitions*, Huygens' criticism was historically justified.

Huygens' arguments against absolute motion are correct, but they miss the point in an important sense, since Newton introduced absolute space for a precise physical-dynamical reason. The reason is that, given a body, its inertia cannot be referred to neighbouring bodies, as the bucket experiment proves. Huygens regarded the velocities of the parts of a rotating body as reciprocal relations among these parts (see Mormino, 1993, p. 186), but the centrifugal force of a part, namely its inertia, cannot be referred to other parts.

Obviously, Huygens recognized that the rotation of a certain quantity of water in a vessel causes the rise of water along the vessel's walls (see, e.g., Mormino p. 268) and understood that centrifugal forces are a manifestation of inertia. Since one of his fundamental tenets was the idea of the inertial tendency of a body to escape along the tangent, he was obviously aware that centrifugal forces are inertial forces. However, though he understood the single properties of inertia, we think that his concept of inertia as a physical-dynamical notion was not adequately developed and that he did not reach a complete view of this complex phenomenon.

By reading the scholium to the *Definitions* Huygens was led to reflect on Newton's conception. Nonetheless, it seems that he did not realise that the question of the reference of inertial motions is a real physical problem that cannot be evaded. In our opinion this is due to Huygens' mechanistic approach: if every phenomenon has to be reduced to mechanical causes, forces themselves are produced only by impacts and movements. Therefore, they are derived entities in physics, not primitive ones. Every force, be it centripetal or centrifugal, is produced by impacts and movements. Thence, centrifugal forces are quantities through which phenomena can be explained. The same phenomena can also be explained through centripetal forces. This is the meaning of Huygens' much quoted statement:

> Thus, I have nothing against the *centripetal force*, as Mr. Newton names it, through which he makes the planets weigh towards the Sun, and the Moon towards the Earth. On the contrary, I agree without difficulty. Indeed, we know through experience that not only this kind of attraction or impulse exists in nature, but also that it can be explained by means of the laws of movement, as we saw in what I wrote above on gravity.[41]

This means that the explanation of phenomena through centrifugal forces, as in Huygens' gravity, or through centripetal forces, as in Newton's gravity, are equivalent. Once again this is true from a descriptive point of view because the same phenomenon can be described in an inertial or in a non-inertial reference frame. However, if you admit that the phenomenon of inertia is a real one—as Huygens undoubtedly admitted—, the equivalence of the two hypotheses does not hold: the true forces are centripetal, while the centrifugal ones are apparent forces due to inertia. Huygens

---

[41] Huygens (1690b). Cfr. OH XXI, p. 472: "Je n'ay donc rien contre la *Vis Centripeta*, comme Mr. Newton l'appelle, par la quelle il fait peser les Planetes vers le Soleil, & la Lune vers la Terre, mais j'en demeure d'accord sans difficulté: parce que non seulement on sçait par experience qu'il y a une telle maniere d'attraction ou d'impulsion dans la nature, mais qu'aussi elle s'explique par les loix du mouvement, comme on a vû dans ce que j'ay écrit cy dessus de la pesanteur".

missed this point and his conception of inertia remained unclear. This is due to the fact that the whole apparatus of Newton's centripetal forces relied on attraction as action at a distance, while Huygens refused this idea in the name of his mechanistic conception. In this sense, Huygens worked along the line of Cartesian physics, though he was extremely critical of Descartes' interpretation of almost every single phenomenon and of Descartes' fictional physics.

A fundamental study on the concept of inertia in Huygens is Stan, (2016). Despite the originality and conceptual richness of this work, Stan's analysis seems debatable under many respects and we do not agree with his conclusions. Stan's general interpretation can be summarised in the thesis that Huygens referred inertia to the set of all the inertial reference frames that he defined correctly.[42]

Stan distinguishes two ways in which a motion can be called absolute: the first, named by him "Completism", is the idea that motion and rest are real, proper, physical kinetic states of objects; the second is called "Absolutism" and means that absolute motion is a change of position with respect to absolute space (Stan, 2016, p. 279). The relativistic conceptions of motion are also divided into two classes by Stan: the former is the T-Relationism. It does not deny the existence of a true motion: true motion exists and consists of kinematic changes (which can be actual or latent) with respect to some preferred material reference frames, e.g., the Earth, the fixed stars, and so on. The second class is defined by Stan 'Relativism' which is the denial of Completism (ibid., p. 279). Stan claims that Completism was the usual view of astronomers and physicists in the seventeenth century. He mentions Riccioli, Boulliau, Fabri, Newton, Leibniz. Newton falls also under the category of Absolutism together with Borelli. Stan argues that Newton was a completist:

> Newton endorsed Completism even before the *Principia*: "[a body's] centrifugal endeavor is always certain and determinate; which proves [*arguit*] that there is some *certain* and *determinate quantity* of *real motion* in *single* bodies [*motus realis in singulis corporibus*], a motion wholly independent from relations to other bodies, which are as innumerable as the relative motions they constitute (1684–85, 25r)".[43]

The problem with Stan's reading is simply that 'real' and 'absolute' motion are synonymous in Newton, as the scholium to the Definitions clarifies beyond any doubt. It is preposterous to apply to Newton the conceptual distinction between Completism and Absolutism and then to state that he was both a completist and an absolutist: this distinction collapses entirely in Newton's texts in which real motion is motion with respect to absolute space (see Newton, 1726, p. 7).

With regard to Huygens, Stan claims:

> Huygens denied Absolutism vehemently, which has led many to think his target was Newton's absolute space. But, that misses the point: what Huygens rejected is really Completism. (Stan, 2016, p. 281)

---

[42] Stan's main source is Huygens' *Codex 7A*.

[43] Stan (2016, p. 280) (italics and brackets in the text). The reference to Newton concerns the manuscript *De motu corporum in mediis regulariter cedentibus*, see Stan (2016, p. 298).

There is no doubt that Huygens denied both Completism and Absolutism. To prove that Huygens' main target was Completism, Stan mentions two passages of *Codex 7A*. However, it has to be pointed out that the passages where Huygens attacks Absolutism are incomparably more numerous than those in which he attacks Completism. Therefore, to claim that the real target of Huygens is Completism is simply a distortion of textual evidence.

Now Stan enunciates his main thesis:

> The real philosophical value of his [Huygens'] notes, I submit, is in their ability to cope with the two hardest problems for Relativism: giving an account of inertial frames and explaining how rotation is relative motion. It is to these that I turn next, to show that Huygens succeeds. (ibid., p. 282)

To reconstruct Huygens' view of inertia, Stan enumerates three conditions which, according to Huygens, determined the mutual rest of two bodies. The first two are unproblematic: (1) bodies must be kinematically free; (2) they have to maintain the same reciprocal positions and distances (ibid., p. 283). The third condition is by far more complicated: bodies do not accelerate. The fact is that claiming to know whether two bodies accelerate or not already means having solved the problem of inertia. The fundamental question is with respect to what they accelerate or not. For, it is clear that the first two conditions can be satisfied even admitting that an equal acceleration acts on all the bodies under consideration. We agree with Stan that Huygens regarded bodies at reciprocal rest as subject to no acceleration. The problem is: was Huygens entitled to make this assumption? Stan argues that he was:

> Objection: bodies remain at mutual rest even as they receive equal and parallel accelerations. Answer: that scenario is disallowed by Huygens's impact mechanics. An accelerated stream dragging along—by collision—a set *G* of stationary bodies will disturb their mutual rest, save in two cases, and both can be ruled out. One is to require that *G*'s members have equal masses, congruent shapes, and parallel placements in the stream, but of course Huygens does not allow such arbitrary restrictions. His account is fully general, not meant to depend on good luck. The other is a physical impossibility: having single jets in the stream self-adjust their momentum to the mass of the particular *G*-member they carry along, or else relative accelerations arise to disturb mutual rest. (ibid., pp. 284–285)

This might appear a convincing argument, but, in fact, it is not because a general theory of inertia implies that one is able to recognize in any possible case whether two or more bodies accelerate. Thus, the case in which "*G*'s members have equal masses, congruent shapes, and parallel placements in the stream" cannot be excluded. Not even the second case can be excluded, if a general theory of inertia is to be offered. Furthermore, there remains the usual difficulty of determining with respect to what reference frame bodies at mutual rest do not accelerate, since the concept of acceleration is relative.

Assuming that this problem is solved is tantamount to admitting the thesis that Stan wants to prove. For this reason, Stan claims that for Huygens bodies at mutual rest do not rotate and represent an inertial reference frame: "[...] for Huygens the Law of Inertia is true relative to preferred set of frames", which Stan names H-frames (ibid., p. 285). The H-frames can be at mutual rest or moving uniformly and rectilinearly, but none of them can be said to be truly at rest or to have a true velocity. Huygens' H-frames are compatible with Newton's statements in Corollary Five to the Laws of Motion (ibid., p. 286).

Three considerations have to be taken into account: (1) Huygens never formulated his ideas in precise quantitative terms. Therefore, it is an overreading to claim that he developed the concept of inertial reference frame and based his physics on it. Probably he was going in that direction, but he never arrived at a settled theory. His uncertainties are testified by *De Vi Centrifuga* and by the fact that in *Codex 7A* he returned many times on this same topic. The fact that he did not publish the *Codex 7A* is a further clue to his uncertainty. (2) Probably Huygens realised that the problem of acceleration had not been solved. How to be sure that a common acceleration among the bodies composing a H-system did not exist? The non existence of such acceleration can only be assumed and cannot be demonstrated. (3) Huygens' arguments analysed by Stan aim to compare the uniform rectilinear motion with the circular one. But are we sure that, from Huygens' point of view, this was equivalent to address the problem of inertia? In our previous observations we have argued that this thesis is at least dubious.

Granted the existence of H-frames, Stan addresses the main problem: "Huygens must explain how purely relative motion can cause dynamically distinguished centrifugal forces" (ibid., p. 287). Stan states that Huygens' aim was to prove that rotation is relational, and not a true motion, so as to defeat Completism and that "he fulfils his task with mixed success" (ibid., p. 287). Huygens offered two accounts, one referred to H-frames and one without resorting to them.

With regard to the first account, Stan mentions those passages in which Huygens explains circular motion to be a relative motion along parallel straight lines whose direction continuously changes, while the distance does not. This circular motion is determined with respect to a point assumed at rest (i.e. the centre of motion). The velocity of the rotating body is measured with respect to bodies assumed at reciprocal rest: they represent the H-frame (ibid., p. 288). Huygens had already proved in *De Vi Centrifuga* that the endeavour to recede from the axis of rotation is proportional to v^2/r. Therefore, the circular motion is due to a constant change of direction with respect to a H-frame, so that the concepts of angular speed and centripetal acceleration are achieved by Huygens (ibid., pp. 288–289).

We consider this interpretation a modernising overreading. Actually, the concept of centripetal acceleration does not belong to Huygens' physics who always talked only of centrifugal forces. Even if we cannot exclude that he was trying to obtain an explanation in terms of H-frames or that he intuited something like "the class of all inertial reference frames", it must be noted that, at best, his ideas were still in the making and were, in any case, far less clear than those explained by Newton in the Corollary Five. Stan's interpretation is, thence, highly conjectural, while the one we offer fits better to Huygens' texts.

Huygens' second account of circular motion considers non-inertial reference frames. Stan introduces the notion of latent motion (L-motion), a motion that is not actual because some constraints prevent it. Citing a passage from *Codex 7A* (ibid., 292), Stan holds that L-motion is an important element in Huygens' physics: the parts of a rotating body have a centrifugal tendency which cannot be actualized because of existing constraints. The intensity of their L-motion is proportional to their centrifugal force. "The direction is that in which they would move if the constraint was dissolved" (ibid., p. 292). Since Huygens' second account is given with reference to a rotating frame, Stan holds that Huygens gave an account of circular motion both in inertial and non inertial reference frames: "And yet, Huygens asserts both [accounts] nearly in one breath. His keen, exquisite physical insight makes it incredible that he missed their drastic differences" (ibid., p. 293). What is highly debatable in this interpretation is the assumption that Huygens had a very clear idea of inertial and non-inertial reference frames. We believe it is likely that Huygens did not grasp all the implications of this distinction.

A brief contribution which, though dated, highlights important aspects of Huygens' thought is Dugas, (1954). The author claims that for Huygens centrifugal force is real.[44] This is true and we saw that Huygens measured centrifugal forces with respect to what we call now a rotating reference frame. However, three interpretations of this fact are possible: (1) centrifugal force is a real property of phenomena, that is measured in rotating reference frames, but exists independently of our measure, just as inertia is a property that is measured in inertial reference frames, but subsists independently of our measure; (2) Huygens understood that centrifugal forces are the same phenomenon as inertia, but the former take place in rotating reference frames, while the latter in inertial ones; (3) Huygens guessed that centrifugal forces and inertia are connected, but was not able to have a clear idea of this connection. This last is our interpretation, while the second is an interpretation *à la* Stan.

Dugas points out to another important aspect of Huygens' conception of motion: Huygens denied Newton's absolute motion without accepting Leibniz's idea of absolute force:

> But if he [Huygens] rejects Newton's absolutes, he does not do so in order to accept Leibniz's *absolute* force: "You claim – Huygens tells Leibniz – that when several bodies have relative motions, each of them has a degree of real movement or of force, on which I do not agree with you". Huygens remains absolutely closed to Leibniz's metaphysics.[45]

As we shall see, Leibniz had a relativistic conception of space and time, but thought that real motion exists and is grounded in the activity of substances. In physical terms real motion is the manifestation of the *vis viva* of a body. To use Stan's vocabulary, Newton was an absolutist, Leibniz was a completist, but not an absolutist, while Huygens was neither a completist nor an absolutist.

A further interesting contribution is Barbour (2001), who proposes a comparison between Einstein and Huygens. Both of them denied the existence of any action at

---

[44] Dugas (1954, p. 25): "Dans son esprit, la force centrifuge, ainsi mesurée dans le repère tournant, est une force réelle, comparable à la pesanteur." See also Dugas (1988, (original 1955), p. 197).

[45] Dugas (1954, p. 28): « Mais s'il rejette les absolus newtoniens, ce n'est pas pour accepter l'*absolu* leibnizien de la force "Vous voulez", dit-il à Leibniz, "que lorsque plusieurs corps ont entre eux du mouvement relatif, ils aient chacun un degré de mouvement véritable ou de force, en quoi je ne suis point de votre avis". Huygens reste absolument fermé à la métaphysique de Leibniz». The quotation comes from a letter by Huygens to Leibniz on 24 August 1694 (see OH X, p. 670, not p. 669 as Dugas writes).

a distance. Barbour claims that Huygens recognized that the centrifugal forces are locally indistinguishable from gravity, as Einstein did. Barbour draws this conclusion by analysing the results obtained by Huygens in *De Vi Centrifuga* that we have examined in section 6.1.1 (Barbour, 2001, p. 487).

He also argues that Huygens fully grasped the significance of infinitesimal concepts in physics and came close to understanding the notion of covariance:

> In *De Motu Corporum*, the transformations were restricted to transitions between frames of reference in uniform motion relative to each other. In *De Vi Centrifuga*, Huygens went a step further and viewed phenomena from frames in a state of uniform rotation relative to the original frame of reference. This revealed an interesting fact that Einstein was to exploit and significantly generalize many years later. Namely, that if one makes a restriction to considering phenomena over only infinitesimal distances and times the basic concepts of dynamics – above all force – are generally covariant, i.e., the rules for defining and using these concepts are the same in any frame of reference. (ibid., p. 494)

According to Barbour, Huygens saw unambiguously that centrifugal forces are connected to inertia (ibid., p. 493). On the other hand, Barbour does not deny the difficulties behind Huygens' treatment of inertia. While examining Huygens' analysis of impact, Barbour claims that, perhaps, Huygens did not have a full insight of the status of inertial systems (ibid. p. 464).

In Sect. 9.6. (ibid., pp. 476–478), entitled "The enigma of relativity", Barbour claims that all our experiences, including the kinetic ones, are relative. Thus, it is perfectly reasonable to construct a relativistic physics and Huygens had the great merit to be the first to concretely develop it by using mathematics extensively. On the other hand, the principle of inertia, which represents one of the bases of Huygens' physics, is not connected to any contingent empirical fact. This is in contradiction with Huygens' relativistic view, but we face here the "eternal" problem of inertia: it is extremely difficult, in classical physics, to understand its origin and meaning within a relativistic context. The contrast between Huygens' relativistic approach and his admission of inertia is, according to Barbour, so strident that "when Huygens died, about 40 years after completing *De Motu Corporum ex Percussione*, he was still impaled on the horns of this dilemma: *is motion absolute or relative*?" (ibid., p. 478). Since Barbour recognizes that Huygens believed in the relativity of any motion, by claiming that Huygens was "still impaled on the horns of the dilemma" he simply means that Huygens' views on motion were not complete: "...[Huygens] never appears to have arrived at a coherent standpoint [on the nature of motion]" (ibid., p. 672).

We agree with Barbour that Huygens, like Einstein, opposed action at a distance, identified locally centrifugal forces with gravitation and had a relational view of motion. This allowed him to develop the most complete relativistic physics which could be achieved in the seventeenth century, according to the state of knowledge of the time. Nonetheless, his theory has several shortcomings, of which the most relevant is that it cannot offer a satisfying treatment of inertia and, consequently, a complete treatment of motion.[46] In conclusion, we agree with Barbour in considering Huygens the second greatest scientist of his epoch.[47]

6.2 Epistemological and Philosophical Implications of Huygens' Physics

The main role played by Huygens in the history of natural philosophy lies in his awareness of the deficiencies of Cartesian physics and in his attempts to develop and refine the mechanistic view of nature. A letter he wrote to Pierre Bayle on his reading of Descartes' *Principia* is a significant document in the history of mechanistic philosophy and is worth being quoted here:

> M. Descartes had found the way to have his conjectures and fictions taken for truths. And to those who read his Principles of Philosophy something happened like that which happens to those who read the novels which please and make the same impressions as true stories. The novelty of the images of his little particles and vortices are most agreeable. When I read the book of Principles for the first time, it seemed to me that everything proceeded perfectly; and when I found some difficulty, I believed it was my fault in not fully understanding his thoughts. I was only fifteen or sixteen years old. But since then, having discovered in it from time to time things that are obviously false and others that are very improbable, I have rid myself entirely of the prepossession I had conceived, and now I find almost nothing in all his physics that I can accept as true, nor in his metaphysics and in his meteorology.[48]

At the same time, in this very letter, Huygens recognized the fundamental change impressed by Descartes' mechanistic theory to the understanding of natural phenomena. Instead of the empty meaningless jargon of the scholastic tradition (stuffed with 'qualities', 'intentional species' and 'substantial forms'), Descartes offered an explanation of natural phenomena by means of simple intelligible causes: the figures of matter particles and their motions.[49] Huygens captured the essential reason that made corpuscular mechanicism so successful from Descartes onwards:

---

[46] For further insights on the problem of inertia and connected questions in Huygens see: Elena (1983b), Elzinga (1972), Stein (1977).

[47] Barbour entitled the Sect. 9.9. of his work "Why Huygens failed to win the greatest prize" (Barbour 2001, pp. 495–497).

[48] Huygens à P. Bayle, 26 Février 1693, English translations by Westfall (1971b, p. 185). Cf. OH, X: *Correspondance 1691–1695*, p. 403: "Mr. des Cartes avoit trouvè la maniere de faire prendre ses conjectures et fictions pour des veritez. Et il arrivoit a ceux qui lisoient ses Principes de Philosophie quelque chose de semblable qu'a ceux qui lisent des Romans qui plaisent et font la mesme impression que des histoires veritables. La nouveautè des figures de ses petites particules et des tourbillons y font un grand agrement. Il me sembloit lorsque je lus ce livre des Principes la premiere fois que tout alloit le mieux du monde, et je croiois, quand j'y trouvois quelque difficultè, que c'etoit ma faute de ne pas bien comprendre sa pensée. Je n'avois que 15 à 16 ans. Mais y ayant du depuis decouvert de temps en temps des choses visiblement fausses, et d'autres très peu vraisemblables je fuis fort revenu de la preoccupation ou j'avois estè, et à l'heure qu'il est je ne trouve presque rien que je puisse approuver comme vray dans toute la physique ni metaphysique, ni meteores".

[49] C. Huygens à P. Bayle, 26 Février 1693, in OH, X: *Correspondance 1691–1695*, p. 403: "Ce qui a fort plu dans le commencement quand cette philosophie à commencè de paroitre, c'est

its intelligibility that satisfied, at least in principle, the needs of clarity and precision of scientific minds.

To examine Huygens' contribution to cosmology from a philosophical and epistemological perspective, we will consider the *Traité de la lumière* (1690) and the *Cosmotheoros* (written in 1695, but published posthumous in 1698). While the first is a scientific treatise on the nature of light, on its propagation and on various luminous phenomena, the second is a piece of conjectural speculation on the existence of intelligent beings living on other planets. Another topic we will examine is the general structure of Huygens' universe.

6.2.1 The Philosophy of the Treatise of Light

We will illustrate the features of the *Treatise on Light* which have a philosophical relevance without dwelling on scientific details. Huygens' wave theory of light represents the last stage of the mechanistic conception of light which had been outlined by Hobbes and Descartes. More specifically, Huygens' theory is the best formulation of the "kinematic tradition in the continuum theory of light" (Shapiro, 1973, p. 137). Huygens wrote his treatise in the late seventies, and published it only after Newton's *Principia* had appeared in 1687 and after he met Newton in London at a meeting of the Royal Society on 12 June 1689. Huygens' theory of light is, in fact, opposed to the corpuscular theory devised by Newton. Huygens' approach to light theory was based on these principles which he regarded as proved by experiments: (a) identity of the angle of incidence with the angle of reflection; (b) acceptance of Descartes-Snell law of refraction; (c) propagation of light in waves; d) finite speed of light. The so-called Huygens's principle in optics states that "light spreads in circles and not in an instant".[50]

Though Huygens was a mechanicist thinker, he refused Descartes' view that light was not a true motion, but a vibration of the little spheres composing the second material element. Huygens objected that light had to be conceived as a motion, "at least in the true Philosophy, in which one conceives the causes of all natural effects in terms of mechanical motions".[51] That light is motion is not asserted merely as a deduction from a coherent mechanical view of nature, but also as an inference from the experimental considerations that light is generated by heat, and that heat is caused by a quick motion of matter particles. As a consequence of the idea that

qu'on entendoit ce que disoit M. des Cartes, au lieu que les autres philosophes nous donnoient des paroles qui ne faisoient rien comprendre, comme ces qualitez, formes substantielles, especes intentionnelles, etc. Il a rejettè plus universellement que personne auparavant cet impertinent fatras. Mais ce qui a surtout recommandè sa philosophie, c'est qu'il n'est pas demeurè à donner du degout pour l'ancienne, mais qu'il a osè substituer des causes qu'on peut comprendre de tout ce qu'il y a dans la nature".

[50] See Shapiro (1973, p. 219). For a detailed explanation of 'Huygens' principle' cf. Shapiro (1989, p. 230f).

[51] Huygens (1690a), translation Huygens (1912, p. 3).

light is motion, Huygens was convinced that light had a finite speed, dissenting from Descartes who believed that the transmission of light was instantaneous. Huygens rightly guessed that the speed of light was so high that no experiment on Earth could detect it. His conjecture was verified in 1676, when the Danish astronomer Ole Christensen Rømer calculated the speed of light thanks to his observations of the occultations of Jupiter's satellites. Huygens remarked that the very high value of the speed of light was physically possible, because "it is not a question of the transport of a body with so great a speed, but of a successive movement which is passed on from some bodies to other" (ibid., p. 7).

Both the high speed of light and the unobstructed intermingling of luminous rays coming from every directions of space were evidence that light was not a particular body transported from its source to our eyes. To understand how light is diffused in space, Huygens believed that we had to rely on the analogy with sound:

> We know that by means of the air, which is an invisible and impalpable body, Sound spreads around the spot where it has been produced, by a movement which is passed on successively from one part of the air to another; and that the spreading of this movement, taking place equally rapidly on all sides, ought to form spherical surfaces ever enlarging and which strike our ears.[52]

Though the analogy with the transmission of sound made the wave theory of light plausible and understandable, Huygens knew well that it was a limited analogy. In fact, Boyle had demonstrated with his pneumatic experiments that when air is removed from a glass container, sound cannot propagate anymore inside the empty space. However, light is not affected by the production of void. Therefore, light is not transmitted through air like the sound, but through another physical medium, the ether, which Huygens believed was by far thinner than air.

Huygens conceived ether as composed of elastic spherical particles which had two properties: they conveyed the light impulse always at the same speed and transmitted simultaneously various impulses, which explained why light rays intertwine in space without obstructing each other. Huygens' ether was conceived to support an undulatory theory within the mechanical view, as it propagates light with its oscillations without any translatory motion. However, to fulfil its role, ether was endowed with peculiar contrasting features: it had to be material and yet extremely thin, not to obstruct the motion of heavenly bodies; it had to be elastic and yet rigid to explain the very high speed of propagation of light. Notwithstanding its problematic nature, ether was maintained as an indispensable scientific notion well beyond the end of a mechanistic view of light; even in the nineteenth century, when light was discovered to be an electromagnetic wave, ether was deemed as a necessary medium to explain the propagation of electromagnetic waves. Only in 1905 ether was rendered superfluous

---

[52] Huygens (1912, p. 4). Cf. Huygens (1690a), in OH XIX, p. 463: "Nous sçavons que par le moyen de l'air, qui est un corps invisible & impalpable, le Son s'étend tout à l'entour du lieu où il a esté produit, par un mouvement qui passe successivement d'une partie de l'air à l'autre, & que l'extension de ce mouvement se faisant egalement viste de tous costez, il se doit former comme des surfaces spheriques, qui s'elargissent tousjours, & qui viennent frapper nostre oreille".

by Einstein's special theory of relativity and his explanation of the photoelectric effect.[53]

Huygens' theory of light showed how he developed Cartesian mechanicism in a rigorous scientific way with a critical attitude towards Descartes himself. The analogy with sound is employed productively and with care, as Huygens was aware of its limitations. Finally, mechanicism was developed in a way that connected mathematics to experiments since Huygens provided in his *Treatise on Light* a precise geometrical description of light waves and confirmed his theory with a crucial experiment on the double refractions of the Iceland crystal. In his Preface to the *Treatise* Huygens showed a lucid awareness of the epistemic status of a physical theory, pointing out both the distinction and the connection between mathematics and physics: in physical investigations geometrical principles must match phenomena and by means of this correspondence hypotheses are confirmed and a predictive theory is validated.

> There will be seen in it [sc. in this writing] demonstrations of those kinds which do not produce as great a certitude as those of Geometry, and which even differ much therefrom, since whereas the Geometers prove their Propositions by fixed and incontestable Principles, here the Principles are verified by the conclusions to be drawn from them; the nature of these things not allowing of this being done otherwise. It is always possible to attain thereby to a degree of probability which very often is scarcely less than complete proof. To wit, when things which have been demonstrated by the Principles that have been assumed correspond perfectly to the phenomena which experiment has brought under observation; especially when there are a great number of them, and further, principally, when one can imagine and foresee new phenomena which ought to follow from the hypotheses which one employs, and when one finds that therein the fact corresponds to our prevision.[54]

While mathematics is the realm of rigorous demonstrative certainty, physics provides a probabilistic knowledge of phenomena, but the probability is so high that it almost coincides with certainty. Furthermore, the logic of science is different from formal logic because in science the truth of premises is confirmed by the truth

---

[53] In truth, as Einstein himself clarified, ether did not disappear from the arsenal of physical concepts, but was conceived in a new way within his general theory of relativity: "we may say that according to the general theory of relativity space is endowed with physical qualities; in this sense, therefore, there exists an ether. According to the general theory of relativity space without ether is unthinkable; for in such space there not only would be no propagation of light, but also no possibility of existence for standards of space and time (measuring-rods and clocks), nor therefore any space–time intervals in the physical sense. But this ether may not be thought of as endowed with the quality characteristic of ponderable media, as consisting of parts which may be tracked through time. The idea of motion may not be applied to it" (Einstein 2007 [1922], p. 1542).

[54] Huygens (1912, p. vi). Cf Huygens (1690a). Cf. OH XIX, p. 454: "On y verra de ces sortes de demonstrations, qui ne produisent pas une certitude aussi grande que celles de Geometrie, & qui mesme en different beaucoup, puisque au lieu que les Geometres prouvent leurs Propositions par des Principes certains & incontestables, icy les Principes se verisient par les conclusions qu'on en tire; la nature de ces choses ne souffrant pas que cela se fasse autrement. Il est possible toutefois d'y arriver à un degré de vraisemblance, qui bien fouvent ne cede guere à une evidence entiere. Sçavoir lors que les choses, qu'on a demontrées par ces Principes supposez, se raportent parfaitement aux phenomenes que l'experience a fait remarquer; sur tout quand il y en a grand nombre, & encore principalement quand on se forme & prevoit des phenomenes nouveaux, qui doivent suivre des hypotheses qu'on employe, & qu'on trouve qu'en cela l'effet repond à nostre attente". Cf. Shapiro (1989, p. 224f): "Huygens' description of his method has a surprisingly modern ring".

of conclusions: "here the Principles are verified by the conclusions to be drawn from them", Huygens wrote.[55] This view is very close to that held by Kepler in his *Mysterium* (see above pp. 65) and had been already stated by Descartes in the *Discourse on the Method*.[56]

6.2.2 An Overview of the Structure of Huygens' Cosmos

Huygens thinks that gravity and light are transmitted by an ether which is condensed in vortices. The most important questions about the structure of vortices concern four topics: (1) how are the vortices causing gravity distributed in the cosmos? (2) What is the form of such vortices? (3) What is the relation between the ether responsible for gravity and that on which the transmission of light depends? (4) What is the composition of the particles composing the vortices?

We have already answered the second question: Huygens' vortices are spherical. In this subsection the other three will be addressed.

Descartes' universe is composed of vortices which fill the whole space and each vortex is full of matter. No vacuum exists. Before the publication of Newton's *Principia* Huygens accepted Descartes' denial of the void. In the second book of his masterpiece, Newton demonstrated that the friction caused by vortices would slow the motion of planets and comets and, furthermore, that Descartes' vortices are unstable. Finally, because of universal gravitation, a vortex theorist was compelled to find a unitary explanation for gravity on Earth and for the rotation of the planets around the Sun. Huygens did not lose heart: a short note dating from 1688 seems to anticipate the research project developed two years later in the *Discours de la cause de la pesanteur*:

> Vortices destroyed by Newton. Vortices of spherical motion in their place.
>
> To rectify the idea of vortices.
>
> Vortices necessary; [without them] the earth would run away from the sun; but very distant the one from the other, and not, like those of M. Des Cartes, touching each other.[57]

---

[55] On these features of Huygens' epistemology see Bell (1947, pp. 178–179); Elzinga (1972, pp. 18–19).

[56] Descartes (2006, p. 61f). (AT VI, p. 76): "If some of the things I speak about in the beginning of the *Dioptrics* and the *Meteorology* shock readers at first sight, because I call them 'suppositions' and do not seem to want to prove them, I ask them to have the patience to read the whole of the treatise attentively, and I hope that they will be satisfied by it. For it seems to me that my arguments follow each other in such a way that, if the last are proven by the first, which are their causes, the first are reciprocally proved by the last, which are their effects. And one must not imagine that I am committing here the fallacy that logicians call a 'circle'; for as empirical evidence clearly confirms most of these effects, the causes from which I deduce them do not so much serve to prove them as to explain them; conversely, it is the causes that are proved by the effects".

[57] Translation from Koyré (1965, p. 117). Cf. OH XXI, p. 437 and p. 439 note 47: "Tourbillons detruits par Newton. Tourbillons de mouvements spheriques a la place. Rectifier l'idee des tourbillons et Tourbillons necessaires, la terre s'en fuiroit de soleil. mais fort distans l'une de l'autre et non pas comme ceux de M. des Cartes, se touchants".

Huygens acknowledged that, before the publication of Newton's *Principia*, he did not think that the motion of the planets was due to the gravitational action of the Sun. Neither he had thought of the inverse square law. These properties cannot be explained supposing, as Descartes did, the existence of a sky full of matter. Furthermore, such matter would have retarded the movement of celestial bodies (Huygens, 1690b, OH XXI, pp. 472–473).

In reference to the general structure of the universe, Huygens remarked that while Descartes supposed it to be too full, Newton was wrong for the opposite reason as his universe is too empty. If skies are empty, light and gravity cannot be transmitted (Huygens, 1690b, Cf. OH XXI, p. 473).

Huygens supposed that void could exist in two ways: (1) vortices are surrounded by a space completely devoid of matter, namely there is a discontinuity among vortices; (2) there is no discontinuity, but vacuum is inside each particle that makes up the vortex. Each particle might be imagined as a geometrical figure whose surface is composed of matter, while the interior is empty. Huygens opted for this second hypothesis because in this way there is a continuous texture of the universe which becomes denser in the proximity of planets to explain gravity and which is dense and rigid enough in the whole space to permit light's transmission. The hardness and shape of any particle transmitting light as well as the ratio between their empty and full parts can be arranged so that light speed coincides with the enormous, but finite value discovered by Rømer in 1676. He wrote:

> [...] I claim that ethereal matter can be considered sparse in two ways: either there is a distance among the particles which are separate by some vacuum; or they mutually touch, but the texture of every particle is sparse and intermingled with many little empty spaces within. [...] However, if one regards matter to be sparse in the first sense, I do not see how to explain gravity, and, as to light, it seems to me impossible to explain its prodigious speed by such kind of vacuum [...]. It appears more appropriate to conceive vacuum in the other sense: particles can touch, as I supposed in the Traité [de la lumière], and, nonetheless, they oppose almost no resistance to planets' movement, because of the lightness of their texture.[58]

There are two problems with regard to the structure of Huygens' universe: (1) since Newton discovered that the planets rotate around the Sun because of the Sun's gravity, can we offer an explanation of Sun's gravity analogous to the one valid for terrestrial gravity? (2) What is the relation between the ether responsible for gravity and that for light's transmission?

---

[58] Ibid., p. 473: "[...] je dis que la matiere etherée peut ester censée rare de deux manieres, sçavoir ou que ses particules soient distantes entre elles, avec beaucoup de vuide entre deux; ou qu'elles se touchent, mais que le tissu de chacune soit rare, & entre-meslé de beaucoup de petits espaces vuides. [...] Cependant à considerer la rareté de la premiere, je ne vois pas comment alors on pourroit rendre raison de la Pesanteur; & quant à la Lumiere, il me semble entierement impossible, avec de tels vuides, d'expliquer sa prodigieuse vitesse [...]. Il y a plus d'apparence de la concevoir de l'autre façon; parce que les particules s'y peuvent toucher, comme je les ay supposées au dit Traité, & toutefois, à cause de la legereté de leur tissu, resister fort peu au mouvement des Planetes". Koyré (1965, p. 122) points out that Huygens accepted the *vacuum interspersum* or *disseminatum*, but rejected the *vacuum separatum*, according to traditional terms. See also Elena (1983, pp. 11–12); Snelders (1989, p. 217).

Huygens answered the first question in the affirmative and calculated the Sun's gravity with the same technique he had adopted to measure the Earth's gravity (ibid., p. 478). As a result, he determined the speed of the particles composing the gravitational vortex of the Sun. Thus, Huygens believed that all the phenomena described by Newton could be explained by his own theory if one assumes the appropriate density and velocity of the matter composing the gravitational vortices. Huygens did not explain how vortices are distributed in the skies. The vortex-distribution is a complex issue because it has to explain: (A) that the planets do not find any resistance in their motion; (B) that the inverse square law holds. As we have pointed out, Huygens admitted the *vacuum interspersum* to explain gravity and to avoid the enormous density of Descartes' vortices. However, he did not address how vortex density is related to the inverse square law. The only supposition that can make Huygen's construction coherent with Newton's law of the inverse square is that each celestial body has its own gravitational vortex which is denser in the proximity of the body, and becomes progressively rarer so as to satisfy the inverse square law. What does "denser" and "rarer" mean? Again, we do not find an answer in Huygens' texts, but it is reasonable to guess that the particles of ether contain more void and less matter as one moves away from the centre of the vortex. It might appear reasonable to claim that the gravitational vortex of each celestial body is confined to a region not too far from the body itself. In fact, we read in the manuscripts written immediately before the publication of his *Discours*:

> I think that it [the gravitational vortex] is like a small vortex in the middle of the water in any big pond, which is far from extending its effects to the banks of the pond. And so, I consider the fixed stars' numerous vortices in the heavens like numerous little vortices in a lake [...].[59]

This view is confirmed in the *Cosmotheoros* (Cf. OH XXI, p. 820). Scholars consider these statements as evidence that Huygens thought of (relatively) little vortices confined to the regions near a celestial body.[60] This is a plausible interpretation which catches the core of Huygens' view. However, a specification is needed: taken literally, this disposition of the vortices is not coherent with Newton's universal gravitation because the gravity of the Earth, for example, diminishes quickly far from the Earth, but is never null. If Huygens' vortices have to be coherent with the inverse square law, he has to admit that the density of every gravitational vortex diminishes, but is never 0 (one might say it is asymptotically 0). Therefore, from a physical-structural point of view, the skies, far away from big masses, should be full of the very rarefied particles which compose the texture of the gravitational vortices of stars and planets in regions which are remote from any gravitational source. This is necessary to make Huygens' theory consistent with Newton's acquisitions. Whether Huygens thought exactly in these terms is difficult to say, but we may surmise that he

---

[59] Cf. OH XXI, p. 370: "[...] je tiens plustost qu'il en est comme d'un petit tourbillon au milieu de l'eau de quelque grand etang, qui est bien loin de se faire sentir vers les bords. Et ainsi je considere plusieurs tourbillons des fixes au ciel comme plusieurs petits tourbillons dans un lac [...]".

[60] See, e.g., Elena (1983, p. 13); Koyré (1965, p. 124). A good description of Huygens' vortices was given by Delambre many years before Koyré's work (see Delambre 1821, pp. 578–579).

thought of all the consequences of his vortex theory. Therefore, when he claimed that his vortices are little in comparison with those of Descartes and are confined to the proximity of the celestial bodies, probably he meant that the density of the vortices diminishes quickly far from the celestial bodies, but not that it becomes null.

As a matter of fact, it is difficult to avoid the impression that this structure is contrived in order to make it correspond to laws and phenomena which Newton had proved and described through a completely different procedure with far less ad hoc assumptions.

Huygens was not even explicit about the problem of the relation between the matter which determines gravity and that which causes light's transmission. In the traditional scholarly literature, there are two opposite interpretations on this topic. Bell claims that Huygens postulated the existence of a different medium for each interaction:

> It was this limited [mechanistic] view of what can constitute a causal mechanism that led Huygens to postulate his various "media". For the transmission of light, he required an ether, for magnetic fields, a magnetic medium, and for gravitational effects a "subtle matter". (Bell, 1947, p. 162)

However, Bell recognizes that Huygens is not clear on this question (ibid., p. 178). Koyré seems to think that Huygens admitted only one kind of ether:

> [...] he believed – in opposition to Newton – that light was constituted not of particles, but of waves or pulses of particles and as he did not believe – also in opposition to Newton – in the existence of a luminiferous aether distinct from matter, he had to conclude that a perfect vacuum, or even a vacuum not quite perfect but nearly perfect like Newton's, would not transmit light. And, of course, such a rare medium would not be able to provide a mechanical basis for "gravitational" action. (Koyré, 1965, pp. 122–123)

The more recent studies do not address in depth this argument. The answer to the question whether Huygens believed that the ether for gravity and that for light is the same substance can only be conjectural because he was not explicit. Nonetheless, there are several indications that he thought of two different substances.

The luminiferous ether is probably seen by Huygens as a medium uniformly distributed in the heavens, which certainly is not the case for gravitational ether. This interpretation seems supported by Huygens' hypothesis concerning the diffusion of Sun's light: he imagined that the Sun's particles with their quick rotation hit the particles of luminiferous ether and cause their vibration, which is the mechanism by which light is spread:

> [The particles around the Sun rotate] with a terrible rapidity. This made me wonder about the fact that this phenomenon could be the cause of the striking brightness of the Sun, supposing light to be produced as I have already explained in my work. The solar particles, navigating in a subtler and very agitated matter, collide against the particle of the surrounding ether. Indeed, if the agitation of such matter, with the movement it has here on the Earth, can cause the glow of a candle flame or of the camphor alum, how more intense will be such brightness if the movement is 49 times more violent?[61]

[61] Huygens (1690b), Cf. OH XXI, p. 478: "Voila donc une terrible rapidité; qui m'a fait penser si elle ne pouroit pas bien estre la cause de la lumiere eclatante du Soleil, supposé que la lumiere soit

The subtle matter in which the Sun's particles navigate can only be that composing the Sun's gravitational vortex, while the ether is, in this case, clearly the luminiferous medium. Thus, in this passage Huygens distinguishes two physical entities.

This interpretation seems confirmed by the *Traité de la Lumière*, where the same concept is expressed:

> And I do not believe that this movement can be better explained than by supposing that all those of the luminous bodies which are liquid, such as flames, and apparently the sun and the stars, are composed of particles which float in a much more subtle medium which agitates them with great rapidity, and makes them strike against the particles of the ether which surrounds them, and which are much smaller than they.[62]

Once again, the subtle medium is the gravitational ether, and what here Huygens calls "ether" is the luminiferous substance. They are two media pervading space. The gravitational ether is denser near celestial bodies and surrounds each of them so as to form a spherical shell around them. Its density diminishes rapidly moving far from them towards the celestial spaces, but it is never null. The luminiferous ether is more uniformly distributed in the heavens. The two ethers consist of solid particles which are hollow inside. The particles of gravitational ether are strongly agitated in the proximity of stars and planets. Their movements take place in any direction so as to cause a globally circular (or better, spherical) movement. In each particle, the percentage of matter diminishes whereas the percentage of void increases the more they are distant from celestial bodies. Both particles of gravitational and luminiferous ether have various shapes. With regard to the size of the particles, the quoted passage only claims that the particles composing the stars are bigger than those composing both ethereal media. It seems reasonable to think that the particles of gravitational ether are smaller than those responsible for the transmission of light because the former enter the interstices of all bodies, insofar as gravity is a universal attraction. Thus, in space there is a web of two ethereal substances[63] of variable density, since the density of the gravitational ether is uniformly variable, while the density of the luminiferous ether is constant. This web is so tenuous that it does not exert any

---

produite comme je l'explique dans ce que j'en ay écrit; sçavoir de ce que les particules Solaires, nageant dans une matiere plus subtile & extremement agitée, frappent contre les particules de l'Ether qui les environnent. Car si l'agitation d'une telle matiere, avec le mouvement qu'elle a icy sur la Terre, peut causer la clarté de la flamme d'une chandelle, ou du Camphre allumé, combien plus grande fera t'elle cette clarté par un mouvement 49 fois plus prompt & plus violent?".

[62] Huygens (1912, p. 10). Cf. Huygens (1690a), OH XIX, p. 469: "Et je ne crois pas que ce mouvement se puisse mieux expliquer, qu'en supposant ceux d'entre les corps lumineux qui sont liquides, comme la flame, et apparemment le soleil, & les étoilles, composez de particules qui nagent dans une matiere beaucoup plus subtile, qui les agite avec une grande rapidité, & les fait frapper contre les particules de l'ether, qui les environnent, & qui sont beaucoup moindres qu'elles".

[63] A brief, but interesting paper on the structure of Huygens' universe is Menéndez (2004). The author points out that in Huygens, besides the ether responsible for light and gravity, there is also that responsible for the electric and magnetic phenomena (it is not clear whether Huygens distinguished them), whose particles are bigger than those of gravitational ether but smaller than the luminiferous one's (Menéndez 2004, p. 484). We note that Huygens' references to this third/fourth kind of ether are far less frequent and precise than those concerning the gravitational and luminiferous substance.

detectable action on moving bodies, but it is dense enough to permit the spread of light and the action of gravity.

6.2.3 *The* Cosmotheoros*: The Place of Man and Earth in Huygens' Cosmos*

The *Cosmotheoros*, or to be more precise its first book, belongs to a well-established tradition during the seventeenth century of texts of 'science-fiction' (to use an anachronistic term that is, nonetheless, appropriate) which, following the reception of Copernicanism, played with the idea of the existence of living beings on other planets. Two major examples of this tradition had been Kepler's *Somnium* and Fontenelle's *Entretiens*: Huygens refers to them, albeit dismissively. In contrast to these precedents, he claims to have inquired accurately into the matter of inhabitants of other planets on the basis of "probable Conjectures" (*verisimilibus Conjecturis*).[64] He is aware that there are various degrees of verisimilitude, more or less proximate to the truth, and that the use of a sound judgement consists precisely in assessing these degrees (ibid., p. 689). One cannot say, though, that in his text Huygens followed strictly this methodological precept: one thing is to highlight astronomical analogies between the Earth and the other planets, so that we are allowed to say, according to the Copernican system that he strongly supported, the Earth to be *like* the other planets; a very different thing is to jump to the conclusion that by parity of dignity among the planets it is very likely that the other planets have the same beautiful structure of the Earth and that therefore on them there are plants, animals and reasonable creatures which exist there to be the spectators of the wonders of creation (ibid., pp. 699–703). Huygens claimed that the use of analogy enables us to reason from the visible to the invisible, from our planet, which is under our eyes, to the other planets (ibid., p. 699). Analogy is certainly compatible with a sound scientific method: Newton himself included it in his "Rules for Philosophizing". However, as David Hume will admonish in his *Enquiry concerning Human Understanding* (§ IX), only "where the causes are entirely similar, analogy is perfect, and the inference drawn from it, is regarded as certain and conclusive".[65] In all the other cases, analogical reasoning is full of pitfalls and Huygens did not avoid them. Let's give a single instance of a false conclusion even in a case in which the similarity of causes was indeed perfect, according to the Humean criterion: from the fact that gravity acts on every planet Huygens wrongly inferred that all planets are solid (Huygens, 1698, p. 699). In the first book of *Cosmotheoros* he indulged in many other analogical conjectures, concerning the similarities of the forms of life existing on Earth and supposedly on

[64] We have already mentioned Kepler's and Fontenelle's texts (see above section 3.2 and p. 112 resp.). Huygens cites also Cusanus, Bruno and the ancient Greek satirist Lucian. See Huygens (1698, p. 683). On *Cosmotheoros*'s significance see Knobloch (2002).

[65] Hume (1975, p. 104).

the other planets, which are much less plausible than this wrong deduction of solidity by gravity.

Two arguments uphold the conjecture that on other planets there are not only plants and animals but also living beings endowed with reason ("animantia ratione utentia"). The first is that Earth is not a privileged place: "Since it has been established that the Earth must be regarded as one of the planets and that is not superior to them either in dignity or in beauty, who would dare to say that only on the Earth there are beings who enjoy the magnificent and beautiful show of nature?".[66] The second argument is that unless there were inhabitants, immense deserted regions of the universe would have been created purposelessly. It is reasonable to think that other planets exist in order to be contemplated by those beings who live there. As has been noted, these two arguments rely on two diverging presuppositions: on one hand Huygens thinks that beings endowed with reason and therefore similar to men are the end of creation as they possess the capacity to contemplate it, while, on the other hand, he criticizes anthropocentrism, as he insists that there is no reason to assume that the Earth is a special place in the universe.[67]

Huygens states that the extraterrestrial beings should not be regarded properly as men. In this he adopted—probably out of theological caution—a thesis that went back to Campanella and that had been repeated also by Fontenelle; nonetheless, in his conjectural descriptions, anthropomorphism becomes inevitable and the similarities he dwells upon are numerous and detailed. The inhabitants of other planets have senses and reason like men, they have developed astronomy and all the arts necessary to it, geometry, arithmetic, optics and writing. As for their body, they have hands and feet like men, and it is very likely that they stand upright because this is the more convenient position for the observation of the stars (Huygens, 1698, p. 741). They live in societies very similar to ours and there are good reasons to conclude that they take advantage of herbs, animals and metals as we do. All these analogical and anthropomorphic considerations are presented in a sequence that proceeds from complex to simple: Huygens starts from the assumption that the inhabitants of other planets have reached a high degree of civilization comparable to ours and that they are astronomers (since the very reason of their presence is that this vast universe exists to be observed) and then he asserts all the requirements needed to achieve this high level of development ending with the analogy in the bodily structure (Del Prete, 1998, p. 297).

As for the conditions of life, Huygens knows that water is required for nutrition, and he confidently asserts that the late telescopic observations support the probable conjecture that there is water on the planets.[68] He also takes into account the distances of the planets from the Sun and remarks that water would evaporate in Mercury and Venus, while it would freeze in Jupiter and Saturn; however, this correct inference

[66] Ibid., p. 733.

[67] Ibid., pp. 713–15. See Del Prete (1998, p. 294).

[68] Huygens (1698, p. 705). He refers to the observations of darker shapes on the surface of Jupiter which had been interpreted as clouds and to the observation by Cassini of lighter patches which the Italian astronomer believed to be snow on top of mountains.

does not deter him from believing that there could be life in other planets. To maintain his thesis he supposes that water in those planets is not like our water, but it is still a liquid similar to our water for its use and beauty! (Huygens, 1698, p. 707) The analogical reasoning is so pervasive that Huygens argues that the difference between us and extra-terrestrial living beings should not be too much: since the difference among the planets is due only to their distance from the Sun, it is reasonable to conclude that other living beings might differ from us only for the quantity of matter and not in form. Moreover, God has no reason to display on other planets creatures very different from us, because the planets are isolated from each other and so, if extraordinary creatures existed there, nobody could observe them. More in general, Huygens rejects the idea that we should conclude out of the immensity of the universe that nature displays the utmost variety of beings as if every possible being, even the remotest from us, had been necessarily created.[69]

In *Cosmotheoros* we encounter theological concerns and apologetic statements that were common to many natural philosophers in the second half of the seventeenth century. The author worries about hostile readers who might underline the contrast of his views with the first chapter of the Book of Genesis, where it is stated that everything has been created to the advantage of man. He comments that this Scriptural tenet must not be interpreted literally, since it would be absurd to think that all celestial bodies have been created only for us, because they are so far away that most of them are invisible or only partially visibile with the aid of a telescope; the fact that they are set so far from our sight should lead us to guess the existence of other beings who are in the position to observe and admire them as we do here on Earth for our planet (Huygens, 1698, p. 687). He also claims the utility of his investigation for religion and suggests the usual argument from design: celestial bodies are so well constructed and ordered that they show the wisdom of the Artificer in opposition to the pagan views that the world has no beginning or that planets like Earth have been formed by the casual collisions of atoms.[70] The intelligence of the Architect of the Universe is particularly manifested in the formation of living beings; while atomists and Cartesians may try to explain inorganic phenomena by means of the mere motions of particles, they will never succeed in explaining mechanically plants and animals which are characterized by finalistic contrivance.[71]

In the second book Huygens reports astronomical observations of the planets, remarks on their physical constitution and new calculations concerning the size

[69] Ibid., p. 703. Del Prete (1998, p. 295f) remarks that for Huygens the possible and the real do not overlap, and thus his position contrasts with the thesis uphold by A. O. Lovejoy who, in his classical study *The Great Chain of Being* (1936), linked the idea of the plurality of worlds to the principle of plenitude, according to which all the infinite creative power of God must be realized.

[70] Ibid., p. 689. See also the manuscript *Quod animalium productio , praesertim hominum , praecipium sapientiae intelligentiaeque divinae sit opus* ["That the formation of animals, especially of men, is the most significant work of God's intellect and wisdom"] published in OH, XXI, p. 557.

[71] Huygens (1698, p. 701). On this aspect of Huygens' reformed mechanism, see Chareix 2006, chap. VI, p. 261 ff. See also Mormino (2003, p. 127f). However, as Mormino points out, with the exception of living beings, Huygens upheld that the only appropriate way to acquire physical knowledge was by means of a mechanistic approach: see ibid., pp. 115, 120–122, 130–131, 133.

of the planets, the periods of their orbits and their distances. Huygens opens the book with a sharp criticism of the jesuit Athanasius Kircher who in his *Itinerarium Exstaticum* (1656) advanced unreasonable considerations lacking that verisimilitude that Huygens claims for his writing. Kircher maintained astrological opinions on the influences of the planets on human beings and traditional doctrines such as geocentrism and that heavenly bodies are moved by angelic 'intelligences'.[72] In the last part of the second book Huygens accepts the Newtonian idea that the gravity of the planets towards the Sun is the cause of their orbits and tries to elaborate a new vortex theory: as pointed out in Sect. 6.1, against Descartes' cosmology of plenum, he thinks that vortices are not contiguous but are dispersed at great distances from each other; therefore no vortex can destroy or absorb another vortex (Huygens, 1698, p. 819–821). Huygens is persuaded that the universe is infinitely extended, but this does not entail that the number of the stars is infinite, because we may conceive an infinite space beyond the region of the fixed stars in which God may have created numberless things which escape our knowledge (ibid., p. 817). The work ends with a scathing criticism of Descartes' cosmology and cosmogony:

> All this theory of the origin of the Comets, and of the planets and the world as well, is built by Descartes upon foundations so shaky that I often wondered how he made so much effort to compose fictions of this kind. It seems to me that we will be very advanced if we understand how things existing in nature are like, which now we are very far away from knowing. But to know how they were made and how they began to be what they are now is something that, in my view, the human mind will never be able to imagine nor to reach by its conjectures (ibid., p. 821).

Cosmotheoros should not be dismissed as a fanciful literary invention, if only because Huygens devoted time and care to its composition for many years and dedicated it to his brother Constantin. In a letter to Leibniz Huygens defined it "un petit traitè en matiere philosophique",[73] implying that it was not a scientific work. In fact, the first book is full of imaginative speculations, much more fictional than those which Huygens reproached to Descartes, since they are developed by loose analogies, dependent on a combination of anti-anthropocentric and yet teleological ideas. Due to its free analogical conjectures about extra-terrestrial life, *Cosmotheoros*' first book is unrelated to the scientific results which Huygens obtained and published in his wide and rich scientific production. The second book gathers astronomical data and remarks which are interesting for the history of science, but their connection to the considerations of the first book is purely extrinsic. Huygens repeats in the second book that Earth cannot be the only planet fit for hosting life and that it would be unreasonable to think that God created planets of so great a size as Jupiter and Saturn with the only purpose that their light could be observed and their orbits could be investigated by the little men (*homunculi*) that we are.[74] It is obvious that this conclusion

[72] See Kircker (1656): the work narrates the story of a voyage through the cosmos in the form of a dialogue between a human, Theodidactus, and an angel, called Cosmiel. Kircker adopted the Tychonic astronomical system and upheld that it is God's will that no living being exists on other planets. For Huygen's criticism see Huygens (1698, pp. 765–771).

[73] C. Huygens to G. W. Leibniz, 29 Mai 1694, in OH, X: *Correspondance 1691–1695*, p. 609.

[74] Ibid., p. 781.

is not logically entailed by the astronomical contents which precede it. This writing shows that Huygens has to be ranked among the many natural philosophers who supported natural theology, as it was customary in the protestant countries during the seventeenth and eighteenth centuries. It is also evident that he did not refrain from making use of teleological considerations which depended on the assumption that the universe has been created to be contemplated by intelligent beings. However, it must be noted that these theological and teleological assumptions are confined to this piece of free cosmological speculation while they do not appear in Huygens' scientific works.

References

Aiton, E. J. (1972). *The vortex theory of planetary motions*. New York: American Elsevier Publishing Company.

Barbour, J. B. (2001). *The discovery of mechanics*. Oxford: Oxford University Press.

Bell, A. E. (1947). *Christian Huygens and the development of science in the seventeenth century*. London: Edward Arnold.

Bertoloni Meli, D. (2005). Who is afraid of centrifugal forces. *Early Science and Medicine, X*: 535–543.

Bertoloni Meli, D. (2006). Inherent and centrifugal forces in Newton. *Archive for History of Exact Sciences, 60*, 319–335.

Bussotti, P. (2015). *The complex itinerary of Leibniz's planetary theory: Physical convictions, metaphysical principles and Keplerian inspiration*. Basel: Springer-Birkhäuser.

Chareix, F. (2004). Huygens and mechanics. In K. Fletcher (Ed.), *Proceedings of the international conference "Titan—From discovery to encounter", 13–17 April 2004, ESTEC, Noordwijk, Netherlands*, (pp. 55–65). ESA Publications division.

Chareix, F. (2006). *La philosophie naturelle de Christiaan Huygens*. Paris: Vrin.

Del Prete, A. (1998). *Universo infinito e pluralità dei mondi. Teorie cosmologiche in età moderna*. Napoli: La Città del Sole.

Delambre, J. B. J. (1821). *Histoire de l'astronomie moderne, Tome 1*. Paris: Courcier.

Descartes, R. (2006). *A discourse on the method of correctly conducting one's reason and seeking truth in the sciences*. Translated with an introduction and notes by Ian Mclean. Oxford: Oxford University Press.

Dugas, R. (1988). *A history of mechanics*. New York. Dover. Original edition: *Histoire de la mècanique*, Neuchâtel: Éditions du Griffon, 1955.

Dugas, R. (1954). Sur le Cartésianisme de Huygens. *Revue D'histoire Des Sciences Et De Leurs Applications, 7*(1), 22–33.

Einstein, A. (2007[1922]). Ether and the theory of relativity. In M. Janssen, J.D. Norton, J. Renn, T. Sauer, & J. Stachel (Eds.), *The genesis of general relativity*. Boston studies in the philosophy of science (Vol. 250, pp. 1537–1542). Dordrecht: Springer.

Elena, A. (1983). Huygens y el Cartesianismo (A propósito de la noción de gravedad). *LULL, 5*, 5–16.

Elzinga, A. (1972). Huygens' theory of research and Descartes' theory of knowledge. II. *Zeitschrift für allgemeine Wissenschaftstheorie/Journal for General Philosophy of Science, 3*(1), 9–27.

Hume, D. (1975) *Enquiries concerning human understanding and concerning the principles of morals*. Oxford: Clarendon Press. Reprint from the posthumous edition of 1777 and ed. with Introduction, comparative table of contents, and analytical index by L. A. Selby Bigge, (3rd edn) with Text revised and notes by P. H. Nidditch.

Huygens, C. (1656, posthumous 1703). *De Motu Corporum ex Percussione*, OH XVI (pp. 37–91). English translation by M. S. Mahoney, 1977 reviewed 1995, *On the Motion of Bodies resulting from Impact*, freely available at www.princeton.edu/~hos/mike/texts/huygens/impact/huyimpct.html

Huygens, C. (1659a, posthumous 1703). *De Vi Centrifuga*, OH XVI: 235–301. English translation by M. S. Mahoney, *On centrifugal Force*, freely available at www.princeton.edu/~hos/mike/texts/huygens/centriforce/huyforce.htm.

Huygens, C. (1669) Debat de 1669 à l'Acadèmie sur les causes de la pesanteur, OH XIX (pp. 628–645).

Huygens, C. (1690a). *Traité de la lumière*, OH XIX (pp. 451–537).

Huygens, C. (1690b). *Discours de la cause de la pesanteur*, OH XXI (pp. 443–488).

Huygens, C. (1698, posthumous). *Cosmotheoros*, OH XXI (pp. 633–821).

Huygens, C. (1888-1950). *Oeuvres complètes de Christiaan Huygens publiées par la Société Hollandaise des Sciences*. La Haye: Martinus Nijhoff – Amsterdam: Swets & Zeitlinger (abbreviation OH)

Huygens, C. (1912). *Treatise on light*. Translated by S. P. Thompson. London: Macmillan and Co.

Kircker, A. (1656). *Itinerarium exstaticum quo mundi opificium*. Romae: Typis Vitalis Mascardi.

Knobloch, E. (2002). Vielheit der Welten - extraterrestrische Existenz. In W. Voßkamp (Ed.), *Ideale Akademie - vergangene Zukunft oder konkrete Utopie* (pp. 165–186). Berlin: Akademie.

Koyré, A. (1965). *Newtonian studies*. Harvard (Mass.): Harvard University Press.

Menéndez, V. (2004). Christiaan Huygens y sus conceptos de materia. Claves para entender las difficultades de un genio aferrado a un paradigma. In R. A. Martins, L. A. C. P. Martins, C. C. Silva, & J. M. H. Ferreira (Eds.), *Filosofia e história da ciência no Cone Sul: 3° Encontro* (pp. 479–485). AFHIC.

Mormino, G. (1993). *Penetralia Motus. La fondazione relativistica della meccanica in Christiaan Huygens, con l'edizione del Codex Hugeniorum 7A*. Firenze: La Nuova Italia.

Mormino, G. (2003). Le rôle de Dieu dans l'oeuvre scientifique et philosophique de Christiaan Huygens. *Revue D'histoire Des Sciences, 56*(1), 113–133.

Newton, I. (1726). *Philosophiae Naturalis Principia Mathematica* (3rd edn.) London: Innys.

Shapiro, A. (1973). A study of the wave theory of light in the seventeenth century. *Archive for History of Exact Sciences, 11*, 134–266.

Shapiro, A. (1989). Huygens' *Traité de la Lumière* and Newton's *Opticks*: Pursuing and eschewing hypotheses. *Notes and Records of the Royal Society of London, 43*, 223–247.

Snelders, H. A. M. (1989). Christiaan Huygens and Newton's theory of gravitation. *Notes and Records of the Royal Society of London, 43*(2), 209–222.

Stan, M. (2016). Huygens on inertial structure and relativity. *Philosophy of Science, 83*, 277–298.

Stein, H. (1977). Some philosophical prehistory of general relativity. In J. Earmann, C. Glymour, & J. Stachel, (Eds.), *Minnesota studies in the philosophy of science. Vol VIII: Foundations of space-time theories* (pp. 3–49). Minneapolis: University of Minnesota Press.

Stinner, A. (2001). Linking 'The book of nature' and the 'Book of science': Using circular motion as an *Exemplar* beyond the textbook. *Science and Education, 10*, 323–344.

Westfall, R. S. (1971b). *Force in Newton's physics. The science of dynamics in the seventeenth century*. London-New York: McDonald-Elsevier.

Yoder, J. G. (1988). *Unrolling time. Christiaan Huygens and the mathematization of nature*. Cambridge: Cambridge University Press.

Chapter 7
Newton and His System of the World

Newton is perhaps the greatest scientist of all time. It is difficult to conceive how a single man was able to obtain the results he achieved: Newton invented the infinitesimal calculus (credit for which also goes to Leibniz); he gave fundamental contributions to algebra and to algebraic geometry; as for optics, he devised a corpuscular theory and understood that white light is composed of all the colours of the spectrum. All this would have been more than enough to secure his reputation as a great scientist. However, by far his most important contribution to modern science concerns mechanics where he created the system of classical physics. The masterpiece, in which he expounded his theory, is the celebrated *Philosophiae Naturalis Principia Mathematica* (three Latin editions were published during Newton's lifetime: the first in 1687, the second in 1713, and the third in 1726; the first English translation by Andrew Motte was published in 1729). From a structural point of view, Newton's universe is basically the Copernican-Keplerian universe. However, he was the first to be able to conceive a working *dynamical theory* of the universe.

The very core of Newton's physics is the full understanding of the concept of inertia, from which the following fundamental notions derive: (1) the distinction between mass and weight; (2) the basic concept of dynamics—which he called force—defined as a function of acceleration and not of velocity (more precisely, force is defined as a function of the second derivative of space with respect to time rather than as the first derivative); (3) the clear distinction between inertial and non-inertial reference frames. Moving from these ideas, it can be said that: (a) motions are seen and calculated in the same manner for all the observers at rest or moving rectlinearly and uniformly; (b) centrifugal forces, which were fundamental notions of the physics of that age (see, e.g., Descartes and Huygens), are fictitious; they are inertial forces, which appear only in non-inertial reference frames, for, according to his view, the real forces determining the curved motions are centripetal and not centrifugal. By means of these concepts he was able to develop for the first time in history a complete system of physical causality, of which the most famous result is the law of universal gravitation. In principle, Newtonian theory has an even wider

P. Bussotti and B. Lotti, *Cosmology in the Early Modern Age: A Web of Ideas*, Logic, Epistemology, and the Unity of Science 56, https://doi.org/10.1007/978-3-031-12195-1_7

scope, since it enables us to solve all the questions connected with the decomposition of forces, namely most of the problems dealt with in theoretical and applied mechanics.

Newton's achievements bring to light an important aspect of the development of scientific thought: a scientific concept becomes significant when it is inserted in a complete theorical structure in which it is applied and employed to solve problems and prove theorems. This is precisely the case with Newton's concept of inertia. Although Galileo, Descartes and Huygens formulated propositions that are similar to Newton's first law, it can be safely asserted that only Newton fully understood the concept of inertia and transformed it into a principle that became, the basic structural law of his cosmology, as we shall see.

Besides his scientific works, Newton also addressed several theological subjects. In his published works the most famous example of his theological views is the General Scholium, which is included in the second edition of his *Principia*. In § 7.2 we clarify the relations between Newton's theological and scientific thought, paying attention to distinguish between the objective conceptual links of his scientific theory and the subjective theological and philosophical interpretation that Newton himself gave of it. Several aspects of Newton's work offer a paradigmatic example of how an interdisciplinary approach, in which natural philosophy, epistemology and theology are taken into account, can shed light on the evolution of scientific theories.

7.1 The Successful Synthesis of Newton's Dynamics

In order to comprehend the core of Newtonian dynamics it is first of all necessary to identify its fundamental ideas, which are summarized below. This is particularly important as far as the concept of *force* is concerned (see below, item 4). The main ideas are:

(1) In mechanics, mean quantities have—in practice—no importance. Instead, instantaneous quantities are significant. It follows that in order to grasp the causes and the development of physical phenomena, one has to analyse the situation when time tends to 0. Let us give an example: to understand under which conditions an orbit is an ellipse with given properties, knowing the mean velocity or the mean acceleration of a body is useless, as these quantities are not sufficient to define an elliptical orbit. It is instead paramount to know velocity and, above all, the changes of velocity (acceleration), for infinitesimal parts of a trajectory, which are covered in an infinitesimal time. This is the reason why infinitesimal calculus is one of the bases of modern physics.

(2) Velocity and acceleration are oriented magnitudes. They do not only have an absolute quantity, but also a direction and a sense. Presumably Newton was not the first to grasp this concept, but was certainly the first to discern its consequences fully.

(3) The inertia principle: a body does not change its state of rest or of uniform motion, unless an action is exerted on it (see *Principia*, section "Axioms or Laws of Movement", Law I). This means that the actions (forces) determine the changes of velocity (accelerations), not the velocities. Galileo and Descartes formulated ideas or principles verbally similar to Newton's, but they did not construct a system of physical causality based on this premise.
(4) The items (1), (2), and (3) provide the framework for understanding the fundamental concept defined by Newton at the beginning of his work: the *concept of force.*

7.1.1 Newton's Concept of Force and Mass

Newton introduces the notion of force, which is the cornerstone of his dynamics, through a series of definitions (*Principia*, section "Definitions", defs. III-VIII). In def. III, he describes "Inherent Force" as the disposition of a body to resist any action tending to change its state of rest or uniform motion. In fact, Newton is here defining the inertial mass of a body, a concept which is of vital importance within his dynamics and which, as we will see, has to be kept separated from the gravitational mass.

Definition IV deals with another fundamental notion: that of "Impressed Force", which is defined as an action exerted on a body to change its state of rest or uniform motion. This definition expresses the idea of a physical quantity whose action changes the velocity of a body in a certain time. The "centripetal force" (def. V) is a force impressed on a body that attracts it towards a centre. This force has a threefold determination: "Absolute quantity of a centripetal force" (def. VI); "Accelerative quantity of a centripetal force" (def. VII); "Motive quantity of a centripetal force" (def. VIII). These three definitions express all the features of a centripetal force.

"Absolute quantity" is the power of the centripetal force's source, independently of the effects it exerts on space and bodies. The "accelerative quantity" is the force considered as exerted on a material point located at a certain distance from the body which exerts the force. It is equal for every point of a single circumference whose centre coincides with the geometrical centre of the force (see Fig. 7.1) and decreases the further one goes from the centre. It indicates the force exerted on a single particle (considered as a geometrical point) of a body, not on the whole body. Newton argues that it is proportional to dv/dt.

"Motive quantity" is the force considered as referred to a body in its entirety, not to its single parts. Within this definition, one finds the famous statement that the motive force is equal to the accelerative force multiplied by the quantity of matter.[1]

[1] Particularly significant studies on Newtons' concept of force are: Bertoloni Meli (1993), Boudri (2002), Chandrasekhar (1995), De Gandt (1995), Ellis (1962), Guicciardini (1999), Heimann-McGuire (1971), Westfall (1971a, 1971b, 1981). Chandrasekhar (1995), though not very detailed from a historical point of view, is an excellent book where the reader can find profound mathematical and physical explanations of the main Newtonian concepts, including that of force; Westfall's

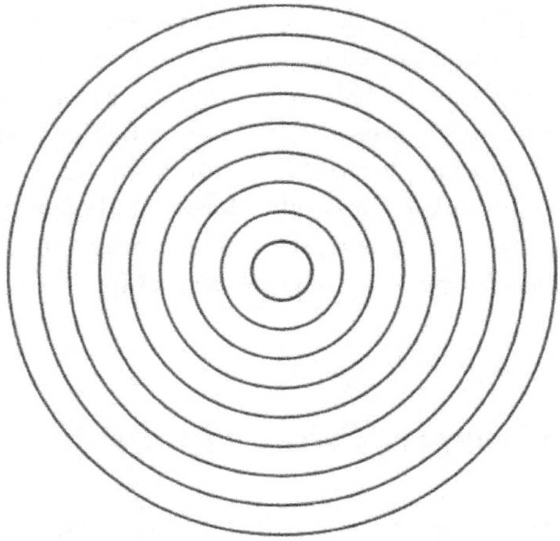

Fig. 7.1 Accelerative force: suppose that the force is located in the geometrical centre of the figure. The accelerative force is the same around each single circumference, but changes from circle to circle. In most cases it is more intense near the centre and weaker the further one goes from the centre. There are, however, cases, as that of the force producing the simple harmonic motion (Newton 1687, Book I, Proposition X, Problem V), in which force is directly proportional to distance. The figure is ours

In the light of this conceptual apparatus, Newton elaborated his physical theory. His objectives were the following:

(I) to find a system of physical causality that offers not only a kinematical and structural model of phenomena but also a dynamical one. In other words: to find the physical causes of motions and of their changes. These causes are the forces.

(II) To develop rational mechanics, namely to find the abstract laws according to which the forces act.

(III) As regards the system of the world, to prove that the force responsible for the movements of the planets and the comets around the Sun, as well as for the movements of the satellites around the planets and the general structure of the whole universe, is nothing other than the same force responsible for the falling of bodies on our Earth. It is *gravity*. This is the reason why Newton was so interested in centripetal forces.

Having defined the notion of centripetal force, Newton offers this illuminating explanation, which manifests the depth and clarity of his order of ideas and can be read as the statement of a work project he would develop in his *Principia*:

> Of this sort is Gravity [centripetal force], by which bodies tend to the centre of the Earth, Magnetism, by which iron tends to the loadstone; and that force, whatever it is, by which

---

works are fundamental for their reconstruction of the development of the most important concepts of Newton's science along his life.

the Planets are perpetually drawn aside from the rectilinear motions, which otherwise they wou'd pursue, and made to revolve in curvilinear orbits. A stone whirled about in a sling endeavours to recede from the hand that turns it; and by that endeavour, distends the sling, and that with so much the greater the force, as it is revolv'd with the greater velocity, and as soon as ever it is let go, flies away. That force which opposes itself to this endeavour, and by which the sling perpetually draws back to the stone towards the hand, and retains it in its orbit, because 'tis directed to the hand as a centre of the orbit, I call the Centripetal force. And the same thing is to be understood of all bodies, revolv'd in any orbits. They all endeavour to recede from the centres of their orbits; and were it not for the opposition of a contrary force which restrains them to, and detains them in their orbits, which I therefore call Centripetal, would fly off in right lines, with an uniform motion. A projectile, if it was not for the force of gravity, would not deviate towards the Earth, but would go off from it in a right line, and that with an uniform motion, if the resistance of the Air was taken away. [...]. And after the same manner, that a projectile, by the force of gravity, may be made to revolve in an orbit, and go round the whole Earth, the Moon also, either by the force of gravity, if it is endued with gravity, or by any other force, that impells it towards the Earth, may be perpetually drawn aside towards the Earth, out of the rectilinear way, which by its innate force it would pursue; and be made to revolve in the orbit which it now describes; nor could the Moon without some such force, be retain'd in its orbit. If this force was too small, it would not sufficiently turn the Moon out of a rectilinear course; if it was too great, it would turn it too much, and draw down the Moon from its orbit towards the Earth.[2]

Starting from Newton's definitions, it can be said that the force moving the celestial bodies in their orbits is a centripetal force, but this does not mean a priori that it should be gravity. Newton proved that such a force is gravity through a process that required two steps: (1) through rational mechanics, he identified all the features of the centripetal forces connected to the movements of the celestial bodies; (2) he determined, by means of specific data and calculations, that, among the possible centripetal forces, gravity is the one responsible for the movements of celestial bodies. These two steps were developed respectively in the first and in the third book of the *Principia*. Clearly, while offering the general theory of central forces in the first book, Newton had in mind the contents of the third book, since he focused particularly on the cases in which the trajectory of the body acted upon by the central force is a conic and the force is located in one of the conic's foci, which is the case of the planets. Thus,

[2] Newton (1729 I, pp. 4–5). Cf. Newton (1726, pp. 3–4), Def. V: "Hujus generis est gravitas, qua corpora tendunt ad centrum terrae; vis magnetica, qua ferrum petit magnetem; & vis illa, quaecunque sit, qua planetae perpetuo retrahuntur a motibus rectilineis, & in lineis curvis revolvi coguntur. Lapis, in funda circumactus, a circumagente manu abire conatur; & conatu suo fundam distendit, eoque fortius quo celerius revolvitur; &, quamprimum dimittitur, avolat. Vim conatui illi contrariam, qua funda lapidem in manum perpetuo retrahit & in orbe retinet, quoniam in manum ceu orbis centrum dirigitur, centripetam appello. Et par est ratio corporum omnium, quae in gyrum aguntur. Conantur ea omnia a centris orbium recedere; & nisi adsit vis aliqua conatui isti contraria, qua cohibeantur & in orbibus retineantur, quamque ideo centripetam appello, abibunt in rectis lineis uniformi cum motu. Projectile, si vi gravitatis destitueretur, non deflecteretur in terram, sed in linea recta abiret in coelos; idque uniformi cum motu, si modo aëris resistentia tolleretur. [...] Et eadem ratione, qua projectile vi gravitatis in orbem flecti posset & terram totam circuire, potest & luna vel vi gravitatis, si modo gravis sit, vel alia quacunque vi, qua in terram urgeatur, retrahi semper a cursu rectilineo terram versus, & in orbem suum flecti: & sine tali vi luna in orbe suo retineri non potest. Haec vis, si justo minor esset, non satis flecteret lunam de cursu rectlilineo: si justo major, plus satis flecteret, ac de orbe terram versus deduceret".

from a logical point of view, the structure of Newton's text is perfect: the first book is dedicated to the general theory of central forces; the third book is dedicated to the planetary movements and contains the proof of the equivalence between gravity and the force which determines such movements. There is a deduction from the general (central forces) to the particular (gravity is the specific central force determining the system of the world). If, however, we go beyond the logical structure of the text and look at the genesis of the concept of central force, it seems legitimate to conclude that a specific problem (why do the planets rotate around the Sun as they do?) led Newton to create a general concept (that of central force) which is applicable also in contexts different from celestial mechanics. As a matter of fact, Newton knew from Kepler's laws and from empirical confirmations that the planets rotate in an ellipse of which the Sun occupies one of the foci. He realized that, without a general theory of central forces, it was impossible to offer a satisfying dynamical explanation of those phenomena. He thus elaborated the whole theory of central forces, which is expounded in the first book. After that, he applied such a theory to the movements of the celestial bodies and reached a series of conclusions (for example concerning the tides, the movements of the Moon and of the comets), which explain Kepler's results and go far beyond them. Among Newton's innumerable achievements three are fundamental:

(A) *The solution of the direct problem of central forces*: given a body rotating in an ellipse, a hyperbola or a parabola (respectively *Principia*, Bk. I, Sect. III, Props. XI, XII, XIII), in which the source of the centripetal force lies in a focus of the curve, how can the law of such a force be found? Newton discovered that the inverse square law holds.

(B) *The solution of the inverse problem of central forces*: given any centripetal force (intensity and direction) and the function linking force and the body's position, how can we determine the body's trajectory moving under the action of this force (*Principia*, Bk. I, Sect. VIII, Prop. XLI)? Newton found that, when the centripetal force acts according to the inverse square law, then the trajectory is a conic section. This is a crucial result because, jointly with item (A), it guarantees that, in a two-body problem, all and the only possible trajectories are conic sections.

(C) Properly speaking, in a concrete physical situation, it cannot be said that a body m rotates around the centre of the force M: the latter one is another body with a certain mass (in general far bigger than m) and both M and m rotate around their barycentre.

Therefore, for the first time in the history of science, Newton found *a physical, dynamical reason* to explain motions under the action of a centripetal force and, more specifically, in the third book, under the action of gravity. This discovery can be regarded as the most important one in the history of physics. Newton's universe is populated by orbits which are conic sections: they are ellipses, hyperbolas or parabolas, according to various parameters determined by the ratio between the masses of two interacting bodies and their respective initial velocities.

Given that gravity is a universal force acting wherever there are masses, Newton understood that the motions of the planets and the structure of the universe cannot be explained only by considering the mutual action of two bodies. For example: in the movements of the planet P around the Sun (or better, around the barycentre of the solar system), not only the mutual action Sun-planet P has to be considered, but also (at least) the actions of the other planets. This is particularly evident when dealing with systems such as Sun-Earth-Moon, where, to determine precisely the Moon's orbit, both the actions of the Earth and of the Sun have to be calculated. Therefore, Newton realised that the complex interactions which were subsequently called the 3-body problem or the n-body problem had to be considered. The physical knowledge as well as the mathematical techniques of his epoch were entirely inadequate for facing such a difficult problem in its general and quantitative terms. Nonetheless, Newton gave some very interesting and correct semi-quantitative solutions for some special cases in the amazing proposition LXVI (*Principia*, Bk. I, Sect. IX) with its 22 corollaries!

Commentary. So far we have focused on mass, but, in fact, two different concepts of mass have been introduced: the first is the inertial mass which coincides with the quantity of matter. The bigger the body's mass is, the bigger the resistance to change its state of rest or uniform motion is. We have also addressed the issue of gravitational attraction and of gravitational mass. However, there is no a priori reason why the tendencies to attract a body or to be attracted by it[3] should be related with its quantity of matter. There is no reason to exclude that something like a "gravitational charge", entirely or partially independent on inertial mass, might exist. This is exactly the case with the electromagnetic attraction/repulsion. The charge of a body does not depend on its mass. While studying a particle whose mass is m and electric charge is q, its motion is given by the system of equations

$$m\ddot{x} = \boldsymbol{F} = q\boldsymbol{E} \rightarrow \ddot{x} = \frac{q}{m}\boldsymbol{E}$$

This means that, in an electromagnetic field, acceleration does not depend only on the field E but also on the ratio between charge and mass of the test particle, which is not constant for any body.

Whereas, if, while dealing with gravity, one introduces the scalar potential $\Phi(x, t)$ of a gravitational field, it holds

$$m\ddot{x} = \boldsymbol{F} = -m\nabla\Phi$$

Thence, it follows $\ddot{x} = -\nabla\Phi$, namely the acceleration of a test particle depends only on the gravitational field (on its potential). But this deduction is based on the supposition that the two values of m in the last equation represent the same concept,

[3] As to the relation between gravitational and inertial mass, see Einstein-Infeld (1938, pp. 34–38). The reader can find useful indications in Sonego (2020, pp. 53–55). Will (1993) is an excellent book on the various principles of equivalence, with a particular focus on the "weak equivalence principle" and on the "Einstein equivalence principle".

which is not the case, because when we write the second Newton law $\boldsymbol{F} = m\ddot{x}$, we are referring to the inertial mass, whereas when we write $\boldsymbol{F} = -m\nabla\Phi$ we are referring to the gravitational mass. Thence, we should indicate the two concepts through two different symbols, for example m_i for the inertial mass and m_g for the gravitational mass so as to obtain.

$$m_i\ddot{x} = \boldsymbol{F} = -m_g\nabla\Phi \rightarrow \ddot{x} = -\frac{m_g}{m_i}\nabla\Phi,$$

An equation which is completely analogous to that for the movement of a particle in an electromagnetic field and which does not depend only on the gravitational field, but also on the ratio between gravitational and inertial mass. Why should one think that $m_i = m_g$, or, at least, that their ratio is constant for any body? In fact, as is well known, it is only the general relativity that provides a satisfactory answer to this question.

We introduced the specific example of the electromagnetic attraction only with the intent to clarify that attraction and inertial mass are not necessarily linked. This example is clearly anachronistic with regard to Newton; nonetheless, he was perfectly aware of the problematic relation between gravitational and inertial mass. In the Proposition VI of the third book of *Principia,* he stated:

> that all bodies gravitate towards every planet; and that the weights of bodies towards any the same planet, at equal distances from the centre of the planet, are proportional to the quantities of matter which they severally contain.[4]

In the demonstration of this proposition, Newton explained his experiments with pendulums composed of different materials (gold, silver, lead, glass, sand, salt, wood, water, and wheat) and having the same length l. He found that the oscillation period is the same for every material. Things work like this: the equation of motion for a pendulum of length l subject to a gravitational acceleration g and whose oscillations form a very small angle θ with the perpendicular is[5]:

$$m_i l\ddot{\theta} + m_g g\theta = 0$$

This corresponds to harmonic oscillations whose period T is:

[4] Newton (1729 II, p. 220). Cf. Newton (1726, p. 400): "Corpora omnia in planetas singulos gravitare, & pondera eorum in eundem quemvis planetam, paribus distantiis a centro planetae, proportionalia esse quantitati materiae in singulis".

[5] As we have seen, Newton's equation is of the form $m_i\ddot{x} = \boldsymbol{F}$. In the case of the pendulum, consider the tangential acceleration a_T of the bob. The first member of Newton's equation is $m_i a_T$. The tangential motion is caused by the component of the gravitational acceleration depending on $\sin\theta$, that is on $g\sin\theta$. The gravitational force acts, hence, with the intensity $m_g g\sin\theta$, so that such is the value of the right member of Newton's equation. Since: (1) In any circular motion the tangential acceleration is $l\ddot{\theta}$, where l indicates the radius; (2) we are considering very small angles so that $\sin\theta$ is approximately equal to θ, the equation $m_i l\ddot{\theta} + m_g g\theta = 0$ follows.

$$T = 2\pi\left(\frac{l}{g}\right)^{\frac{1}{2}}\left(\frac{m_i}{m_g}\right)^{\frac{1}{2}}$$

Newton found with a very good precision T to be constant independently on the material composing the pendulum. This implies that the ratio $\frac{m_i}{m_g}$ is a universal constant, which can be posed equal to 1. However, in Newton's physics this equivalence or proportionality is accidental. Einstein's and Infeld's explanation of the link between the gravitational and the inertial mass is worth quoting:

> In our world it happens that they [the two masses] are equal, but we can well imagine that this should not have been the case at all. Another question arises immediately: is this identity of the two kinds of mass purely accidental, or does it have a deeper significance? The answer, from the point of view of classical physics, is: the identity of the two masses is accidental and no deeper significance should be attached to it. The answer of modern physics is just the opposite: the identity of the two masses is fundamental and forms a new and essential clue leading to a more profound understanding. This was, in fact, one of the most important clues from which the so-called general theory of relativity was developed. (Einstein-Infeld 1938, p. 36)

A further observation is also connected to the concept of mass: the third law of Kepler states that $\frac{a^3}{T^2} = K$ where a indicates the half-major axis of the planet's orbit, T its period of revolution around the Sun, and K is a constant. Kepler believed that this law was an exact law of nature. From Newton's law of universal gravitation, it is easy to deduce the value of the constant K according to this formula $\frac{a^3}{T^2} = \frac{GM}{4\pi^2}$, where, M is the mass of the Sun and G is Newton's gravitational constant. However, unlike Kepler, Newton was aware that $\frac{a^3}{T^2}$ is only approximatively a constant because the mass m of a planet influences the gravitational field of the Sun. He also fully understood that, given a system Sun-planet, both of them rotate around their centre of mass. As a matter of fact, the exact expression of Kepler's third law is:

$$\frac{a^3}{T^2} = \frac{G(M+m)}{4\pi^2}$$

The member on the right side of the equation is not a constant at all, but depends on the mass of the planet. This means that it was a lucky circumstance for physics and astronomy that the mass of each planet is so small with respect to the mass of the Sun that it can be neglected without appreciable variations. Had this not been the case, Kepler would have found no constant relation between the cube of the major half axes and the square of periodical times and Newton could have not assumed Kepler's laws as a sure guide to which refer his research. In some circumstances fortuitous factors influence the development of science. This is one of such circumstances.

A particularly important observation concerns the concept of force: so far, we have given for granted that Newton in the *Principia* defined the force as $m\frac{dv}{dt}$ or as ma (the two expressions are equivalent in Newtonian mechanics, but not in the theory of relativity). As a matter of fact, in the commentary to the Definition VIII, Newton wrote:

> Wherefore the Accelerative force will stand in the same relation to the Motive, as celerity does to motion. For the quantity of motion arises from the celerity, drawn into the quantity of matter; and the motive force arises from the accelerative force drawn into the same quantity of matter.[6]

Therefore, indicating the quantity of motion by p and the motive force by f, it should be $p = mv$ and $f = ma$.

On the other hand, Law II reads as follows:

> The alteration of motion is ever proportional to the motive force impress'd; and is made in the direction of the right line in which that force is impress'd.[7]

If one interprets "alteration of motion" and "alteration of the quantity of motion" without any reference to the time in which such an alteration takes place, then, here Newton seems to offer a different definition of the motive force f, that is $f = m\Delta v$, and not $f = ma$.

Westfall interprets in this way the "alteration of motion" and claims that two different conceptions of motive force cohabit in the *Principia.* According to Westfall, force as impulsive action is considered by Newton to be equal to $m\Delta v$, while force as continuous action (as it is the case of centripetal forces) is equal to ma. Westfall argues that one of the main issues studied by Newton at the beginning of his scientific career was impact, where impulsive forces are considered, so that Newton thought of force as $m\Delta v$. In manuscripts dating between the end of the 1660 s and the beginning of the 1670 s, Newton first considered force to be equal to ma (Westfall, 1971b, pp. 345–357). In Westfall's view, Newton maintained this twofold definition of force also in the *Principia*, both in the formulation, and operatively. Newton realized that he was offering two different definitions of force, but did not think that there was any problem of compatibility between them (ibid., p. 357). In the *Principia* Westfall detects the definition of force as $m\Delta v$ in the celebrated first proposition of the *Principia*, the area law—whose proof Newton already held at the end of the 1670s (Westfall, 1971b, pp. 436–437)—, where Newton used the impulsive definition of force, because he considered that the force acts through successive impulses. Westfall comments:

> [...] Newton did not consider the two definitions to be incompatible. As I have been arguing, dynamics in the seventeenth century was bedevilled by dimensionally incompatible definitions of force. However great his contributions to dynamics, Newton did not succeed in escaping from the same dilemma [...]. (ibid., p. 437)

An argument in favour of Westfall's thesis is the famous Corollary I of the Laws of Motion. Here Newton argues that if a body is acted upon by two conjoined forces, it will describe the diagonal of a parallelogram in the same time it would describe the single sides. Clearly Newton is referring to an impulsive force which acts instantaneoulsy and he is composing velocities, not forces. He is considering force as $m\Delta v$.

---

[6] Newton (1729 I, p. 8). Cf. Newton (1726, p. 5): "Est igitur vis acceleratrix ad vim motricem ut celeritas ad motum. Oritur enim quantitas motus ex celeritate & ex quantitate materiae, & vis motrix ex vi acceleratrice & ex quantitate ejusdem materiae conjunctim".

[7] Newton (1729 I, p. 19). Cf. Newton (1726, p. 13): "Mutationem motus proportionalem esse vi motrici impressae, & fieri secundum lineam rectam qua vis illa imprimitur".

This is confirmed by the Corollary 3 of Proposition I where he refers to Corollary I of the Laws of Motion speaking of composition of movements, not of forces. On the other hand, in the proof of Proposition I, he wrote that the force acts with a single but big impulse (*impulsu unico sed magno*).[8] Therefore, here Newton distinguishes also terminologically the force from its impulse. Thus, it seems that Westfall is basically right in pointing out that two different meanings of force cohabit in the *Principia*. However, it can also be assumed that Newton not only guessed the differences of the two meanings, but that, at least on some occasions, he also employed a different terminology to designate them. At the same time, the presence of these two meanings never caused him to make any mistake in his physics equation. It is furthermore to add that, while force as $m\Delta v$ is drawn from the tradition of studies on the impact, force as ma is a completely new Newtonian concept and it is the one which prevails in the *Principia*. Therefore, we can safely say that in some propositions of the *Principia* there is a terminological inexactness and ambiguity as to the use of the term force. However, from a conceptual perspective, the three notions of impulse, quantity of motion and force were completely clear to Newton and, in most cases, force in Newton's masterpiece means mdv/dt.[9]

7.1.2 The Problem of Inertia

A fundamental notion of Newton's physics is the idea of absolute space.[10] His arguments on this topic are extremely illuminating in order to understand his physics, of which the notion of inertia is one of the main concepts.[11] In fact, the reason why Newton dedicated most part of the scholium to the Definitions to the notion of absolute space depends on the concept of inertia. As we have seen, the first law states that a body does not change its state of rest or of uniform motion, unless an action is exerted on it. The first question is: how is it possible to determine the state of rest or uniform motion? Actually, it seems that any motion is relative. For example, Newton was well aware that the motion of a planet is an ellipse with respect to the Sun, while it is not an ellipse with respect to any other reference frame. With respect

[8] With regard to this part of Newton's proof De Gandt (1995, pp. 24–25), writes: "The action of the central force is absolutely punctiform or instantaneous; it is produced in the manner of an impact at point B (*impulsu unico sed magno*), then at point C, at point D, and so on. Immediately after this action, the innate force takes over and the motion is again uniform [...]. The orbit is made up of a series of rectilinear segments, and at each vertex of the polygonal path the body is 'deflected' by an impulsion".

[9] With regard to this discussion on the concept of Newtonian force we are grateful to Francesco Sfiligoi for his valuable considerations. An interesting book on this problem is Torretti (1999) (in particular see pp. 46–49).

[10] Newton introduced the concepts of absolute time and absolute space in the famous scholium to the "Definitions" of *Principia* (Newton 1729, pp. 9–18). For the extensive literature on absolute space in Newton see here below, Sect. 7.1.2.

[11] Newton also believed in the existence of absolute time, but his arguments for the existence of absolute space are by far more interesting, thence we will refer to them.

to the planet itself, the motion of the planet is null and indistinguishable from the planet's state of rest. Descartes had claimed that any motion is relative and that it can be determined only referring it to neighbouring bodies. On the other hand, the first law of Newton does not speak of a state of relative rest or uniform motion: the concept of rest and uniform motion are stated as if they are absolute. From a purely conventional and operative point of view we could argue that the notions of physics do not describe exactly the structures of the world, but are our conceptual models of the world; therefore, the concept itself of inertia may be regarded as relative. For the purpose of studying the motions on the Earth we are free to consider the Earth at rest; if we study the motions of the planets in the solar system, we are free to consider the Sun at rest. All the bodies moving uniformly with respect either to the Earth or to the Sun (in the two cases) will be considered moving uniformly. Though we assume that no motion is absolutely uniform with respect to any possible body, there are motions which, according to specific physical situations, can be considered inertial without any detectable mistake. In other terms: we might adopt an instrumentalist conception of the physical laws and claim that inertia is an abstract orientative idea, a limiting concept that does not mirror exactly a physical state. Physical states approximate, more or less accurately, this limiting concept. This viewpoint, that many contemporary physicists would embrace, was not Newton's one. In his view, the state of rest and/or of uniform motion should and could be determined in absolute terms.

In Newtons's cosmology and physical ontology, absolute space is a fundamental concept because it offers the scene on which the events take place. Therefore, a clear understanding of such an idea is a necessary condition for a complete comprehension of Newton's views on natural philosophy. A more detailed examination of this idea is necessary because absolute space is connected to the concept of inertia and, as Westfall pointed out, Newton was the first to understand fully the implications of the inertia principle for dynamics and conceived the idea that the whole of mechanics is to be founded on such principle (Westfall, 1971b, pp. 344–348).

Given the importance and complexity of the subject, this section will be divided into different subsections. First of all, we will offer a brief view of the structure and logic of the scholium to the Definitions; afterwards, the bucket and the two globes experiments will be presented and explained. Later on, we will focus on some interpretations of Newton's scholium that differ from the one we propose; finally, we will analyse the complex relation between Newton's absolute space and the concept of inertial reference frame.

7.1.2.1 Structure and Logic of the Scholium

The structure of the scholium is very clear, but the logic behind it is not so terse and this is the reason why scholars have offered numerous different readings of it.

First of all, Newton clarifies that there is no need to define time, space, place and motion in the ordinary sense of these terms. In contrast, he felt the need to define absolute or mathematical time and space. Absolute or mathematical time is defined as that entity which, according to its nature, flows uniformly without regard to anything external. Absolute space is that entity which, according to its nature, remains always similar to itself and immovable without reference to anything external. On absolute time, Newton added few considerations, whereas, he dwelled longer on absolute space: he explained the mistake of confusing relative for absolute space and expounded the notion of absolute and relative place.

A first fundamental question arises: when Newton offers the definition of absolute time and space, is he defining two "objects" whose existence he takes for granted or, as it is the case in mathematics, does he think that, given a definition, it is necessary to prove the definition to be well posed and that, hence, the defined objects exist only from a mathematical point of view? At the end of this subsection, we will try to answer this question.

The scholium continues defining the absolute motion as a translation of a body from an absolute place to another. Newton is very careful in determining the relation between absolute and relative motions. The example of the ship offers an interesting view on Newton's order of ideas. Suppose, he claims, that a certain object is moving on a ship which is sailing the sea. Suppose the Earth to be in absolute rest. Then, the absolute movement of the object will be composed of the movement of the object with respect to the ship and of the ship with respect to the Earth. Thence, we can interpret Newton's thought like this: the absolute motion of a body P can never be determined through a composition of its relative motions because, first of all, P might be potentially included in an infinity of other objects animated by relative motions; furthermore, even if the inclusions were in a finite number, we cannot be sure of the state of motion of the body including all the others in which P is included. Thence, the absolute motion cannot be reconstructed inductively through a series of relative motions. An absolute frame of reference is necessary.

In this case too, we face the same problem concerning the relation between definition and existence: did Newton define the absolute motion as a state whose existence he took for granted or did he want to prove the existence of absolute motion? Like for absolute space, we answer this question at the end of this subsection.

Newton continues his analysis by examining the methods by which the astronomers attempt to achieve a measure of time and space as close as possible to the true, absolute duration and distance between two or more events. Here the situation is very different for time and space: for, though it is possible that no body exists whose movement is so regular to offer an accurate measurement of the duration between two events, the time equation—namely the difference between mean solar time and apparent solar time—, with some appropriate corrections, offers a good approximation of absolute time. Instead, as far as the absolute space intervals and the absolute motions are concerned, there exists no body from whose state of rest or motion we can deduce, albeit approximately, the absolute state of motions of other bodies.

With regard to the immutability of the parts of absolute time and space, Newton writes:

> As the order of the parts of Time is immutable, so also is the order of the parts of Space. Suppose those parts to be mov'd out of their places, and they will be moved (if the expression may be allowed) out of themselves. For times and spaces are, as it were, the Places as well

> of themselves as of all other things. All things are placed in Time as to order of Succession; and in Space as to order of Situation. It is from their essence or nature that they are Places; and that the primary places of things should be moveable, is absurd. These are therefore the absolute places; and translations out of those places are the only Absolute Motions.[12]

This means that the places where things are located are immobile and, thence, coincide with parts of absolute space.

Newton, then, focuses on the difference between absolute and relative rest and motion. Since the absolute motion of a body cannot be reconstructed through the series of relative motions of which it might be thought to be composed of, how can we recognize it? As Newton explains, we can recognize it through its properties, causes and effects. Here, he clearly explains that the absolute kinetic state of a body cannot be recognized in relation to the kinetic state (in particular, rest) of the surrounding bodies because we cannot be certain of the absolute kinetic state of these bodies. If in the attempt to determine the absolute motion of a body, we would proceed by referring to particular bodies, we might fall in a *regressus* ad infinitum without any chance to reach a definitive conclusion. We know that absolute motions depend on physical causes and that these causes are the impressed forces. But since, for us, it is difficult to detect an absolute movement starting from its causes, because we hardly have a direct experience of the forces which cause absolute motion, then, we must consider their effects and through them we must try to argue the features of the absolute motions. The effect of a true force is an apparent force, an inertial force and, in particular, the centrifugal force. Thence, it is through the centrifugal forces that we can infer the absolute motion of a body. To this aim Newton presents the bucket experiment. He claimed that to detect the absolute motion of a body, starting from its causes rather than from its effects, is difficult, but not impossible. To illustrate this possibility Newton presents the two globes experiment.

This is the structure of the scholium. As for its underlying logic and for its purpose, one can wonder whether Newton wanted to prove the existence of absolute space and motion or if he took their existence for granted and aimed at something narrower, namely to prove that Descartes' conception of relative motion is wrong and that the inertia of a body cannot be referred to the neighbouring bodies. A third, intermediate position should also be considered: Newton might have taken for granted the existence of absolute time and space, while trying to demonstrate the existence of absolute motion. We posit that in the scholium Newton intended to prove the existence of absolute motion. With regard to absolute space, instead, things are more complicated because his line of thought is not straightforward. One has the impression that Newton was convinced that the bucket and the two globes experiments offered arguments in favour of the absolute space's existence, but that, in fact, he does not explicitly conclude from them to the existence of absolute space. The text

[12] Newton (1729 I, p. 12). Cf. Newton (1726, p. 8): "Ut ordo partium temporis est immutabilis, sic etiam ordo partium spatii. Moveantur hae de locis suis, & movebuntur (ut ita dicam) de seipsis. Nam tempora & spatia sunt sui ipsorum & rerum omnium quasi loca. In tempore quoad ordinem successionis, in spatio quoad ordinem situs, locantur universa. De illorum essentia est ut sint loca: & loca primaria moveri absurdum est. Haec sunt igitur absoluta loca; & solae translationes de his locis sunt absoluti motus".

of the scholium is ambiguous in that it offers something more than a mere existential assumption, but falls short of a proof. However, if we consider the whole of Newton's production and, in particular the *De gravitatione et aequipondio fluidorum*, it seems that Newton thought that a proof of the absolute space's existence was possible.

7.1.2.2 The Bucket Experiment

To prove that inertial motions cannot be referred to the neighbouring bodies, Newton devised the bucket experiment, one of the most ingenious ideations ever conceived of by a human mind. The experiment runs like this:

- Let us consider a bucket hung to a chord. Be the chord rotated until it becomes completely rigid.
- Let us fill the bucket with water. Obviously the water's surface will be flat.
- Let us free the chord and let us move the bucket circularly in the opposite sense to that with which the chord has been winded. At the beginning the surface of the water will remain flat.
- After a certain time, the bucket is going to communicate its motion to the water and the water will begin to rotate and to assume a concave form, ascending along the walls of the bucket. The rotation of the water will continue until the water will be at rest with respect to the bucket (which means that the water moves with the same speed of the bucket).

For a clearer understanding of the experiment, it is useful to illustrate it with the addition of the two following phases of which Newton did not speak explicitly:

- Let us now block suddenly the bucket's rotation. The bucket will be at rest with respect to the neighbouring bodies, but the water will continue to rotate and to maintain the concave form for a certain period.
- After a certain time the water will stop, reassuming the flat form that it had at the beginning.

This sequence of events can be summarized in four different stages (see the following Fig. 7.2):

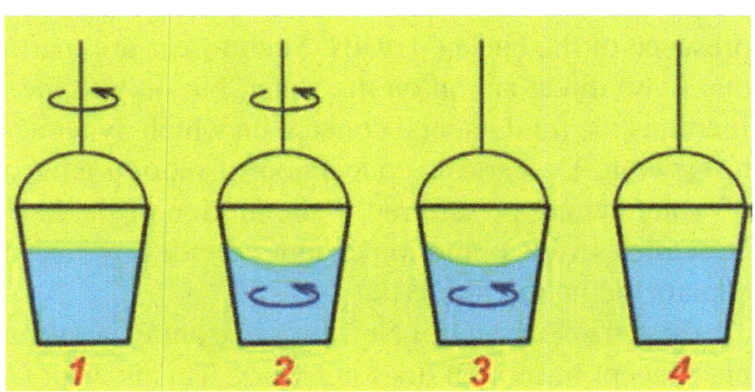

Fig. 7.2 The different phases of the bucket experiment

Now, if we analyse the movement and the position of the bucket and of the water, we will see that it is impossible to claim that the motion of the water can be determined in relation to that of the bucket. At the beginning, when the bucket rotates the water is impelled by no force, because its position is flat (phase 1 of the figure). In the second phase, when the water is at rest with respect to the bucket, its rotational movement, indicated by the ascent of the water along the bucket's wall, is maximal; in the third phase, when the bucket is at rest (as it has been stopped), the water continues its rotational movement. At the end (fourth phase) both the bucket and the water are at rest. The following table summarizes the situation.

	Relative motion water/bucket	Surface of the water	Rotation of the water
1	Maximal	Flat	No
2	Absent	Curved	Yes
3	Maximal	Curved	Yes
4	Absent	Flat	No

The experiment shows clearly that there is no connection between the relative motion bucket-water and the rotational motion of the water, because when there is no relative motion (phases 2 and 4) the water rotates in the phase 2, but does not rotate in the phase 4; when the relative motion is maximal (phases 1 and 3), the water does not rotate in the phase 1, but rotates in the phase 3. This is an indisputable observation, it is not an interpretation. Nonetheless, we see that there is a perfect connection between the flat-curved form of the water's surface and the absence-presence of its rotational motion.

What did this argument prove according to Newton? One answer is indubitable: the inertial motion of a body cannot be referred to the neighbouring bodies. For, it is clear that the forces of receding from the axis of circular motion, the centrifugal forces, are due to the inertia of the water's particles; they are not true forces. Let us suppose that, at a certain instant a part of the bucket's wall breaks. Then, according to the inertial principle, the particles of the water near the gap will fly away along their instantaneous tangent rectilinearly and uniformly. Since the walls of the bucket are solid, the particles cannot escape along their instantaneous tangent and, hence, they accumulate on the walls. Ergo, the centrifugal forces indicate the presence of an inertial tendency which cannot be actuated because of contingent reasons—the presence of the bucket's walls. Such forces are inertial or fictitious forces: they are due to no direct action on the water, but only to the inertial tendency of the water. Inertia is the fundamental concept on which dynamics is built. Therefore, according to Newton, Descartes's conception of motion is erroneous because the movement of water cannot be referred to the motion of the bucket. Furthermore, in principle, Descartes's view of motion cannot provide any basis for dynamics, since it does not explain the nature of inertia.

The *pars destruens* of Newton's reasoning is so completed: Descartes was wrong; his concept of motion does not work. This is what Newton thought to have proved and, in fact, proved correctly. Now the question arises whether there is also a *pars construens* in the expounded argument. Did Newton think he had identified the object in reference to which inertial motion could be determined? We think that this question can be answered in the affirmative.

Let us start from absolute motion and only afterwards we will deal with absolute space. Newton's argument can be posed in an apagogic form: let us suppose Descartes to be right and that any motion is relative, so that it can be referred to the neighbouring bodies. Let us try to verify this supposition through the bucket experiment. You find a motion which cannot be referred to the neighbouring bodies. Thence, not every motion is relative. Therefore, absolute motion exists. This is a simple application of *modus tollens*. Thus, the physical entity—absolute motion–, defined in the previous sections of the scholium is not a fictitious abstract concept, but it has a physical existence. This does not seem to be an overreading of Newton's reasoning. Let us continue: according to Newton *every motion has to take place in a physical environment which pre-exists to the physical objects*. If every motion were relative, then such environment would be the relative space; since absolute motion exists its environment has to be the absolute space. Therefore, it remains for us now to analyse the proposition *every motion has to take place in a physical environment which pre-exists to the physical objects*. Is this proposition an assumption or did Newton think it to be provable?

The scholium to the Definitions does not enable us to give an indubitable answer, but we think that the *De gravitatione et aequipondio fluidorum* does. In this complex text Newton offered several arguments against Descartes's physical conceptions.[13] The most important argument is the one we are going to examine now. Let us consider all of Descartes's presuppositions. Let's admit all Descartes' presuppositions: (a) the motion of a body can be determined only in relation to the kinetic state of the surrounding bodies; (b) space is identical with extension, i.e. with matter; (c) no empty space, separate from the bodies, exists; (d) there is a permanent circulation of curvilinear motion due to the mutual contact of all the particles of the universe; then, given these presuppositions, it is not possible to establish any reference frame because there is no certainty that the points of the reference frame are determined once and forever with respect to the frame itself. Suppose Descartes to be right: given a body *B* the position of the surrounding bodies changes. Often (as in the case of the impact) this happens because such a position is influenced by the motion of *B*. If it is not possible to separate the space from the bodies, how is it possible to establish a reference frame which allows us to study the physical phenomena and to foresee their development? The points themselves of the reference frame would move in a given time not only with respect to other reference frames, but also with respect to the initially chosen frame. Following Newton's reasoning, it would not even be possible to determine a reference frame relative to any observer whatsoever. The observer has to be considered at rest, but if its kinetic state changes

[13] The *De Gravitatione et Aequipondio Fluidorum* was not published during Newton's lifetime. Most scholars dates it at (1666). Some of the most interesting works on *De Gravitatione* are Henry (2011), Hill (2003), Kochiras (2012) ; Slowik (2002). Slowik's book is particularly remarkable. For the topics of our discussion, Sect. 7.1.2. ("Newton's Arguments Against Cartesian Relationalism", pp. 12–17) and the entire Chap. 2 on the structure of Newton's space are particularly relevant.

under the influence of the neighbouring bodies and if space is not separate from the bodies, how is it possible to consider an observer at rest? Let us suppose, as Newton wrote, that we want to consider the idea that Jupiter occupies, after one year, the same position it held exactly one year before. He claims that Descartes's theory would not even be able to describe this situation. As a matter of fact, it would be impossible to determine Jupiter's position with respect to the ether's particles because such particles could have changed their position with respect to Jupiter and, hence, we would be compelled to determine their position with respect to some other bodies, thus creating a *regressum* ad infinitum. It would not be possible to refer the position to the fixed stars or to the Sun since the vortices surrounding them would influence with their motions both the size and the position of these celestial objects. For instance, a comet traversing the solar vortex would influence the kinetic state of the particles composing the solar vortex and, thence, the kinetic state of the Sun would also be modified. Therefore, Jupiter's position would be relative to a Sun moving in unpredictable manners. Ergo, if you claim that the motion of bodies is determined with reference to the positions of neighbouring bodies, and if you identify space and matter, then you are not even allowed to determine the relative motions of the bodies (exactly what Descartes intended to establish) because you are not allowed to construct a reference frame relative to the observer. Newton wrote:

> But that this may be clear, it is first of all to be shown that when a certain motion is finished it is impossible, according to Descartes, to assign a place in which the body was at the beginning of the motion; it cannot be said whence the body moved. And the reason is that according to Descartes the place cannot be defined or assigned except by the position of the surrounding bodies, and after the completion of a certain motion the position of the surrounding bodies no longer stays the same as it was before. For example, if the place of the planet Jupiter a year ago be sought, by what reason, I ask, can the Cartesian philosopher define it? Not by the positions of the particles of the fluid matter, for the positions of these particles have greatly changed since a year ago. Nor can he define it by the positions of the Sun and fixed stars. For the unequal influx of subtle matter through the poles of the vortices towards the central stars (Part III. Art. 104), the undulation (Art. 114), inflation (Art. 111) and absorption of the vortices, and other more true causes, such as the rotation of the Sun and stars around their own centres, the generation of spots, and the passage of comets through the heavens, change both the magnitude and positions of the stars so much that perhaps they are only adequate to designate the place sought with an error of several miles; and still less can the place be accurately defined and determined by their help, as a Geometer would require. Truly there are no bodies in the world whose relative positions remain unchanged with the passage of time, and certainly none which do not move in the Cartesian sense: that is, which are neither transported from the vicinity of contiguous bodies nor are parts of other bodies so transferred. And thus there is no basis from which we can at the present pick out a place which was in the past, or say that such a place is any longer discoverable in nature. For since, according to Descartes, place is nothing but the surface of surrounding bodies or position among some other more distant bodies, it is impossible (according to his doctrine) that it should exist in nature any longer than those bodies maintain the same positions from which he takes the individual designation. And so, reasoning as in the question of Jupiter's position a year ago, it is clear that if one follows Cartesian doctrine, not even God himself could define the past position of any moving body accurately and geometrically now that a fresh state of things prevails, since in fact, due to the changed positions of the bodies, the place does not exist in nature any longer. Now as it is impossible to pick out the place in which a motion began (that is, the beginning of the space passed over), for this place no

> longer exists after the motion is completed, so the space passed over, having no beginning, can have no length; and hence, since velocity depends upon the distance passed over in a given time, it follows that the moving body can have no velocity, just as I wished to prove at first. Moreover, what was said of the beginning of the space passed over should be applied to all intermediate points too; and thus as the space has no beginning nor intermediate parts it follows that there was no space passed over and thus no determinate motion, which was my second point. It follows indubitably that Cartesian motion is not motion, for it has no velocity, no definition, and there is no space or distance traversed by it. So it is necessary that the definition of places, and hence of local motion, be referred to some motionless thing such as extension alone or space in so far as it is seen to be truly distinct from bodies.[14]

However, on the basis of our astronomical theories, we are able to predict the position of Jupiter in one year, independently of the positions of any other celestial body. Thus, Newton thought that this possibility revealed the existence of an absolute reference frame independent of the bodies' positions. The possibility of such a

[14] Newton (1962, pp. 129–131). Cf. Newton (1666, in Newton 1962, pp. 97-98): "Sed ut hæc pateant, imprimis ostendendum est quod post motum aliquem peractum nullus potest assignari locus juxta Cartesium in quo corpus erat sub initio motus peracti, sive dici non potest unde corpus movebat. Et ratio est quod juxta Cartesium locus non definiri et assignari potest nisi ex positione circumjacentium corporum, et quod post motum aliquem peractum positio corporum circumjacentium non amplius manet eadem quæ fuit ante. Verbi gratia si Jovis Planetæ locus ubi erat ante annum jam peractum quæratur; qua ratione, quæso, Philosophus Cartesianus describet? Non per positiones particularum fluidæ materiæ, siquidem istæ particulæ positiones quas ante annum habuere, quam maxime mutaverint. Neque describet per positiones solis et fixarum stellarum, quoniam inæqualis influxus materiæ subtilis per polos vorticum in sidera centralia, (Part 3 Art 104), Vorticum undulatio (Art 114,) inflatio (Art 111,) et absorptio, aliæque veriores causæ, ut solis et astrorum circa propria centra gyratio, generatio macularum, et cometarum per cælos trajectio, satis mutant et magnitudines siderum, et positiones, ut forte non sufficiant ad locum quæsitum sine aliquot miliarium errore designandum, et multo minus ut ipsarum ope locus accurate describi ac determinari possit, quemadmodum Geometra describi postularet. Nulla equidem in mundo reperiuntur corpora quorum positiones inter se diuturnitate temporis non mutantur et multo minus, quæ non moventur in sensu Cartesij hoc est vel quatenus transferuntur e vicinia contiguorum corporum vel quatenus sunt partes aliorum corporum sic translatorum: Et proinde nullum datur fundamentum quo locus qui fuit in tempore præterito, jam in præsentia designari possit, vel unde possumus dicere talem locum jam amplius in rerum natura reperiri. Nam cum locus juxta Cartesium nihil aliud sit quam superficies corporum ambientium vel positio inter alia quælibet remotiora corpora: impossibile est ex ejus doctrina ut in rerum natura diutius existat quam manent eædem illæ corporum positiones ex quibus individuam denominationem sumpsit. Et proinde de loco Jovis quem ante annum habuit, parique ratione de præterito loco cujuslibet mobilis manifestum est juxta Cartesij doctrinam, quod ne quidem Deus ipse (stante rerum novato statu) possit accurate et in sensu Geometrico describere, quippe cum propter mutatas corporum positiones, non amplius in rerum natura existit. Jam itaque post motum aliquem completum cum locus in quo inchoabatur hoc est initium trajecti spatij non assignari potest nec amplius esse: illius trajecti spatij non habentis initium nulla potest esse longitudo; et proinde cum velocitas pendet ex longitudine spatij in dato tempore transacti, sequitur quod moventis nulla potest esse velocitas; quemadmodum volui primo ostenderere. Præterea quod de initio spatij transacti dicitur de omnibus intermedijs locis similiter debet intelligi; adeoque cum spatium nec habet initium nec partes intermedias sequitur nullum fuisse spatium transactum et proinde motus nullam determinationem, quod volui secundo indicare. Quin imo sequitur motum Cartesianum non esse motum, utpote cujus nulla est velocitas, nulla determinatio et quo nullum spatium, distantia nulla trajicitur. Necesse est itaque ut locorum determinatio adeoque motus localis ad ens aliquod immobile referatur quale est sola extensio vel spatium quatenus ut quid a corporibus revera distinctum spectatur".

frame implies that space can be identified neither with matter universally nor with the relative positions of bodies locally. Furthermore, since not all motions can be explained as relative motions (as it has been proved by the bucket experiment) and since any motion has to be referred to space, then absolute space, to which absolute motions are referred, exists. It is through these complex and often indirect, reasonings, both physical and epistemological, that Newton believed he had proved the existence of absolute space. Of course, this means neither that his proof works (as we will see in a few pages) nor that he arrived to the concept of absolute space through a demonstration. He certainly endorsed this concept because of physical and metaphysical convictions which are a priori and independent of any demonstration. However, from a formal point of view, it cannot be said that absolute space was assumed without demonstrating its existence. Newton was convinced that he had proved such existence.

The two globes experiment. In the bucket experiment absolute motion has been inferred by an apparent force, namely the centrifugal endeavour of water's particles, which is an effect of the true motion of water. Newton also wondered whether it was possible to infer the absolute motion both from its causes and effects and not only from its effects. In other terms: is it possible to find a real, not an apparent, force, which is the exactly identifiable cause of an absolute motion? Newton claimed this to be difficult because the parts of absolute space in which the absolute motions take place do not fall under our senses. However, a solution can be offered. In an ideal experiment, let us consider two globes tied with a rope. Make them rotate around their common centre of gravity. Each globe will tend to escape along the instantaneous tangent to its trajectory because of its inertia. This produces the tension of the rope, which is a real force—not an apparent one—that attracts the globes towards their common centre of gravity. Furthermore, if during the rotation we apply equal forces to the opposite faces of the bodies, the increment or the decrement of the rope's tension will indicate the increment or the decrement of the absolute motion as well as its direction, or determination (*determinatio* is the word used by Newton). The tension of the rope is an absolute quantity; this means that the experiment would also work in an absolute vacuum where only the two globes and the rope exist. Newton perfectly understood the difficulty to determine absolute motion. As he argued: suppose we add the fixed stars to our universe composed only by the two globes and the rope linking them. From the movement of the globes with respect to the fixed stars, it is impossible to determine whether absolute movement belongs either to the globes or to the stars or to both of them. However, the tension of the rope will indicate the absolute motion of the globes and thus we will also be able to deduce the absolute kinetic state of the fixed stars.

Commentary. The leitmotif of the scholium is inertia. This final comment to the "Definitions" offers a highly refined treatment of the problems connected to inertia. Since, given a body *B*, its inertia cannot be determined in relation to any surrounding body, it has to be referred to something else. This is absolute space. This does not mean that absolute space is causally connected to the inertia of a body, in the sense that it produces it. Inertia is an inherent property of matter which depends only on matter itself. In the commentary to the third of his *Regulae Philosophandi* Newton wrote thus:

> Not that I affirm gravity to be essential to bodies: by their *vis insita* I mean nothing but their *vis inertiae*. This is immutable. Their gravity is diminished as they recede from the Earth.[15]

Inertia is referred to absolute space in the sense that, given a certain body, the inertial effects of its motion—such as the ascending of the water on the bucket's walls or the tension of the rope in the two globes experiment—do not depend on and are not measurable through the existence of other bodies. If the inertia of a body were any function of its interrelations with other bodies, it would not be an immutable quantity. Thence, it is to be referred to and measured in accordance with an absolute and immutable entity: absolute space.

Absolute space and inertial forces also exert a specific cosmological role within Newton's physics because they explain the forms of celestial objects. If a body has a real rotational movement with respect to absolute space, its form cannot be spherical since the centrifugal forces at the equator are more intense than those at higher latitudes as the velocity at the equator is the maximal one. The intense velocity at the equator and in its proximity induces the particles of the bodies to recede quickly from the axis's movement (exactly like in the bucket experiment) and, thence, this endeavour produces an ellipsoidal form. As a consequence, a rotating body can be perfectly spherical only if it is at rest with respect to absolute space: in this case, no receding endeavour from the axis's movement exists.

This is not a mere theoretical precept. Newton measured the deviation of the form of a planet from that of a perfect sphere and ascribed it to the inertial effects. In the Proposition XVIII, Theorem XVI of the third book of the *Principia*, Newton wrote:

> *That the axes of the Planets are less than the diameters drawn perpendicular to the axes.* The equal gravitation of the parts on all sides would give a spherical figure to the Planets, if it was not for their diurnal revolution in a circle. By that circular motion it comes to pass that the parts receding from the axe endeavour to ascend about the equator. And therefore if the matter is in a fluid state, by its ascent towards the equator it will enlarge the diameters there, and by its descent towards the poles it will shorten the axe. So the diameter of Jupiter (by the concurring observations of astronomers) is found shorter betwixt pole and pole than from east to west. And, by the same argument, if our Earth was not higher about the equator than at the poles, the Seas would subside about the poles and, rising towards the equator, would lay all things there under water.[16]

---

[15] Newton (1729 II, pp. 204–205). Cf. Newton (1726, p. 389): "Attamen gravitatem corporibus essentialem esse minime affirmo. Per vim insitam intelligo solam vim inertiae. Haec immutabilis est. Gravitas recedendo a terra, diminuitur".

[16] Newton (1729, p. 239). Cf. Newton (1726, p. 412): "*Axes planetarum diametris quae ad eosdem axes normaliter ducuntur minores esse.* Planetae sublato omni motu circulari diurno figuram sphaericam, ob aequalem undique partium gravitatem, affectare deberent. Per motum illum circularem fit ut partes ab axe recedentes juxta aequatorem ascendere conentur. Ideoque materia si fluida sit ascenfu suo ad aequatorem diametros adaugebit, axem vero descensu suo ad polos diminuet. Sic jovis diameter (consentientibus astronomorum observationibus) brevior deprehenditur inter polos quam ab oriente in occidentem. Eodem argumento, nisi terra nostra paulo altior esset sub aequatore quam ad polos, maria ad polos subsiderent, & juxta aquatorem ascendendo, ibi omnia inundarent".

The following Proposition XIX (Problem III) where Newton determined the form of the Earth with a good approximation is one of the longest and most beautiful of the *Prinicipia.*[17] The theoretical presupposition of the entire reasoning is that the centrifugal force increases descending from regions with higher latitude towards the equatorial regions. Therefore, Newton evaluated the effects of inertial forces on the shapes of celestial bodies. Inertial forces are referred to absolute space. This means that he distinguished the physical effects which, in his opinion, can be referred to absolute space from those that are purely relative to the reciprocal kinetic states of a system of bodies.

7.1.2.3 Different Interpretations of the Scholium on Absolute Time and Space

The situation with regard to Newton's ideas on absolute time, and, above all, space, can be summarized, in broad lines, like this: in Newton's time the problem of the relativity or absoluteness of space and time was rather debated; with different aims and arguments, Huygens, Leibniz and Berkeley criticized Newton's conceptions. Afterwards, the successes of Newtonian mechanics made sure that Newton's conception of absolute time and space was also accepted for a long period. Thus, Euler, Maclaurin, Lagrange, Laplace, Poinsot, Poisson (only to mention some of the greatest physicists who lived in the eighteenth and nineteenth century) accepted Newton's ideas. Starting from the 1860s, things began to change: after a phase of great results, the physicists also started to reflect on the conceptual bases of their discipline and realised that one of its fundamental concepts—that of inertia—was anything but clear. Thus, from that period to the beginning of the twentieth century, authors such as Neumann, Lange, Föppl, the Friedlaenders, Frank, Hofmann, Reissner gave seminal contributions to the inquiry on inertia and to the discussion on absolute or relative space/time. The two most celebrated authors who contributed to these discussions were Mach and Poincaré.[18] Some scholars talk of a "Machian Legacy", holding that the reading of the scholium proposed by most authors (be they philosophers as Nagel, Reichenbach, Sklar and Earman, or historians of science as Jammer, Alexander and Koyré) had been profoundly influenced by Mach's thesis that in the sholium Newton intended to prove the existence of absolute time, absolute space and absolute motion.[19]

Of course, Einstein's theories of relativity changed completely the situation and, by resorting to a part of the mathematical apparatus also utilized in general relativity, neo-Newtonian theories of gravity were proposed. The most renowned is that conceived by Cartan in the 1920s. After the second world war a series of studies began to elaborate a machian theory of inertia. In this respect the research of

[17] For an explanation of Newton's reasoning, see Bussotti (2017c, pp. 129–135).

[18] We refer to Föppl (1894, 1904a , 1904b, 1909), Frank-Rothe (1911), Friedlaender-Friedlaender (1896), Hofmann (1904), Lange (1885a, 1885b, 1885c, 1886, 1902), Mach (1883), Neumann (1870), Poincaré (1902), Reissner (1914).

[19] Rynasiewicz (1995-II, pp. 307–312).

Sciama, Assis, Barbour, Barbour-Bertotti are fundamental.[20] This field of research is rich nowadays, too. At the same time, starting with the neo-positivism, more philosophical research was developed on the nature of space and time in connection to Newton's and Einstein's conceptions. Carnap's and Reichenbach's works are pioneering in this sense. More recently, authors such as Sklar, Friedman, Earman, Brown, Barbour,[21] among others, have offered profound philosophical analyses of the concepts of inertia, space and time based upon an excellent knowledge of the whole mathematical apparatus used in modern physics. They provide, so to say, a view from a higher standpoint on Newtonian mechanics. We will refer to some of their ideas and results in the next subsection insofar as such ideas are useful to fully grasp the greatness and the limits of Newton's conception.[22] Thus, the discussions on the concept of inertia and on the meaning of Newtonian physics were carried on both by physicists and by philosophers of physics. There are very close links between the two groups, and authors such as Mach, Poincaré and Barbour belong to both of them. But for sure, Sciama was a physicist and not a philosopher and Friedman and Earman are philosophers and not physicists as it is evident not only from their scientific background, but also from their works. This distinction, hence, makes sense.

However, a third tradition of thought exists: that of the historians of science. How did they interpret inertia, absolute space and time in Newton? How did they evaluate the scholium of the Definitions? The answer is rather interesting: until around the 1970s there were many studies on these problems. Some of them focused on the internal logic of Newton's arguments, some on the context in which Newton worked and was educated. Other studies also explored the influence of his background on his conceptions as well as the criticisms by Leibniz, Huygens and Berkeley. However, it was taken for granted that Newton in the Scholium had the intention to prove the existence of absolute time, space and motion and to identify an entity to which inertia could be referred. Nowadays several authors think that Newton did not aim at proving the existence of absolute entities, but that he pursued other purposes, which are identified differently by various scholars. We believe that the traditional interpretation of Newton's scholium is correct, and for this reason we are going to examine some alternative readings. We will focus on the main arguments that have been advanced against the traditional interpretation that we have supported. Even if the advocates of new interpretations are inclined to present them in contrast to the

---

[20] Among the numerous works written to construct a Machian theory of inertia, we mention Assis (1989, 1999, 2003), Assis-Graneau (1995), Barbour (1975, 1982), Barbour and Bertotti (1982), Berman (2008), Davidson (1957) (a very profound paper in which some of Sciama's conceptions are criticized); Giné (2009), Putz (2019), Raine (1981), Sachs-Roy (Eds.) (2003), Sciama (1953), Sciama (1957), Sciama (1959).

[21] There is a very extensive literature on this subject, but for our aims the following references are sufficient: Barbour and Pfister (1995) (Eds.), Belkind (2012), Earman (1989), Friedman (1983), Graneau and Graneau (2003), Gründbaum (1963), Lucas (1984), Nerlich (1976), Reichenbach (1957) (original 1927), Reichenbach (1960) (original 1920); Sklar (1974).

[22] These authors also deal with special and general relativity, but in our context we are interested in their speculations on Newtonian mechanics.

traditional reading of the Scholium, there are cases in which the conventional and the new interpretations are compatible.

A pioneering and very significant paper on Newton's conception of absolute space is Stein (1967). This essay is not precisely contrary to the interpretation we have supported, but it suggests a different reading of the scholium and of the structure of Newton's argumentations on absolute space. Stein begins his analysis of the scholium from the last sentence, in which Newton claimed that the whole of his treatise has been written to determine the true motions from their causes, effects and apparent differences, and conversely how from apparent or true motions we may come to the knowledge of their causes and effects. Did Newton achieve his aim? Stein's answer is worth noting: if we admit Newton's hypothesis, inserted between the 10th and 11th propositions of the third book, according to which the centre of the system of the world is immovable, then Newton was successful.[23] Stein does not examine the question whether Newton assumed the existence of absolute space or believed to have proved it. However, if we transpose Stein's thesis within this debate, it can be said that Newton assumed the existence of a single fixed point and then proved the existence of absolute space as the space to which that point belongs.

Stein continues examining Newton's statement in the fifth corollary of the Laws of Motion where he claimed that rest and uniform rectilinear motion are indistinguishable. Newton acknowledged that dynamics does not permit to distinguish these two kinetic states. Then why did he not identify them?[24] Stein's answer is that the concepts of absolute time and space belong to Newton's theoretical apparatus with the same logical status and ontological value as those of "attraction" and "gravitation" (ibid., p. 190). Newton's investigations led him to believe that things mutually attract, though a manifest proof of this fact does not exist, and had also prompted him to believe that absolute space exists. The bucket experiment, the two globes experiment, the bulge of the Earth and, after Newton, in the nineteenth century, Foucault pendulum can be interpreted as evidences in favour of the existence of the absolute space, at least if you admit the possibility of such existence. Newton admitted this possibility and interpreted the experiments and the shape of the Earth as proofs of the existence of absolute space. Thus, Stein claims:

> Now, this is obviously in large part a question about words and meanings. If one denies—on whatever grounds—that "absolute" space and motion can be well-defined notions at all, then one obviously cannot accept *any* experimental result as confirming that the earth posseses a certain quantity of "absolute" rotation. But it is surely nonsense to deny the *definability* of "absolute" motion, since prior to definition—or to some stipulation of conditions to be satisfied—the mere constitution of a word confers no logical attributes whatever upon it. One can with reason deny only the intelligibility of some proposed definition or the realizability of some proposed set of conditions. (Stein, 1967, pp. 191–192)

How can Stein's assertions be interpreted within our discussion? These conclusions can be drawn: the definition of absolute space and motion is a legitimate operation from a logical point of view; however, the experiments produced to prove

[23] For the argumentations in favour of this thesis see Stein (1967, pp. 181–182).

[24] Ibid., 183.

their existence are subject to interpretations. If you admit the possibility of their existence, then you can find a series of experiments that can be read as proofs of such existence. Obviously, if you deny in principle the existence of absolute space and motion, you will offer weaker interpretations of the experiments that, according to Newton, proved the existence of absolute space and motion.

An important contribution to the discussions on Newton's bucket experiment and on his conception of absolute space was given in 1978 by Laymon, who stated thus:

> In this paper I will argue that Newton did not intend his bucket experiment to be interpreted as showing either that absolute space exists or that a relativistic mechanics is not possible. The intended function of the bucket experiment is more modest. It is to be used, as Newton clearly states, as an illustration of a case where absolute rotations can be distinguished from relative motions, *given* Newton's dynamical theory. It can also be used, again as Newton noted (although this time not so clearly), to refute Descartes's system of dynamics. [...] I will also challenge the usual analysis of the "logic" of the bucket experiment, which is that Newton showed experimentally that the curvature of the water surface is *independent* of the relative motion of the water with respect to the sides of the bucket. I will show that Newton's argument is that the experiment demonstrates that the degree of surface curvature is *inversely* related to this relative motion. Therefore, relative motion with respect to the bucket sides cannot be true motion since true motion even for Descartes is *directly* proportional to its effects. (Laymon, 1978, p. 399)

In his initial analysis of the scholium up to the bucket experiment, Laymon does not propose any argument in favour of his thesis, but merely observes that Newton did not give any direct proof of the existence of absolute space in the scholium.[25] In another passage, Laymon claims:

> The function of the bucket experiment is not per se to show that absolute space exists. It shows this existence only *indirectly* in the sense that to show that a concept has application is to give inductive support to the claim that the entities presupposed by the concept do exist. (ibid., p. 403)

This idea is substantially correct, in the sense that the bucket experiment is intended to prove the existence of absolute motion and to identify a mark of distinction between absolute and relative motions. What is highly disputable is Laymon's claim that Newton's statement that the centrifugal force cannot be determined with regard to the neighbouring (*ambientium*) bodies in the bucket experiment, refers only to the bodies which immediately surround the water, and not to distant bodies, such as the fixed stars. Authors such as Mach, Reichenbach and Nagel have, instead, interpreted Newton's claim as an assertion of the impossibility to refer the centrifugal forces to any body whatsoever, and this we believe to be the correct reading. Laymon writes that "[his own] interpretation of the bucket experiment, as showing an inverse instead of a direct relationship [between the motion of the bucket's walls and the water's], is somewhat different from the usual interpretation". His analysis is certainly new, but erroneous. There is no inverse relation between the motion of the bucket's walls and that of the water, as can be clearly seen from our presentation of the bucket experiment and as will be discussed while dealing with Rynasiewicz's contribution.

25 On this question, we refer the reader to our analysis of Newton's whole argumentative line.

Newton showed something different and more profound: he proved that there is no relation, either direct or inverse between the two motions; thence, it is impossible to infer the state of motion of the water from that of the bucket.[26] Of course, this remains true if you suppose the bucket to be included in a further container, and this container in a further one, and so on, until including potentially the whole universe. Therefore, Newton demonstrated and thought to have demonstrated that the movement of water cannot be interpreted in reference to any external body, whatever its distance. He used the term *ambiens* only as a direct opposition to Descartes's conception, but his idea does not only refer to the bodies immediately surrounding water, but also to all the bodies of the universe. To conclude: Laymon is right when he states that the bucket experiment and the two globes experiment are not intended per se as a demonstration that absolute space exists; but he is wrong in asserting that Newton did not offer, and did not intend to offer a proof of the existence of absolute space. He did it in the terms we have clarified above.

A text which has become, so to say, the manifesto of those scholars who interpret Newton's scholium in a weak sense is Rynasiewicz (1995). The author claims that in the scholium Newton took for granted the existence of absolute space and absolute motions. He never tried to prove empirically their existence. Two even stronger theses are proposed:

> Nowhere in the scholium does Newton seek to establish that each body has a state of true or absolute motion unique to it (Rynasiewicz, 1995 I, p. 134).

And:

> When Newton asserts that absolute motion and rest are distinguished from their relative counterparts by their properties, causes, and effects, he means to adduce reasons for believing absolute motion and rest are not special instances of the relative motion or rest of a body with respect to other bodies. The arguments that follow thus do not presuppose that absolute motion is by definition motion with respect to absolute space (ibid., p. 134).

In a long introductory section ("Difficulties with the customary view", ibid., pp. 135–139), the author highlights two problems concerning "the customary view": (1) if the aim of the bucket and of the two globes experiments were to prove empirically the existence of absolute motion, why did Newton redundantly propose two experiments aimed at the same purpose? This seems an implausible question. First of all, there is nothing strange in presenting more than one experiment to confirm a thesis. It is almost an ordinary procedure among scientists, even more justified when you are dealing with such an elusive phenomenon as absolute motion. Secondly, the two experiments have different purposes: the bucket experiment focuses on the effects which allow to distinguish the absolute from the relative motions, while the two globes experiment concerns the identification of an absolute force, displayed by the tension of the rope. (2) Rynasiewicz, then, points out that in all the parts of

[26] The Nagel-Reichenbach table referred to by Laymon himself (ibid., p. 405n) is not only correct, but catches Newton's thought. Had Newton believed that an inverse relation between the motion of the water and that of the bucket existed, Newton would have made a serious mistake, which seems extremely implausible.

the scholium preceding the bucket experiment there is no evidence of the empirical existence of absolute motion. Again, this is true, but irrelevant. It depends on the fact that Newton first defined what the properties of absolute motion ought to be if such a motion existed and, afterwards, with the two experiments shows that it exists.

Before beginning his examination of the bucket experiment, Rynasiewicz claims:

> First, in none of them [arguments Newton developed in the scholium before introducing the bucket experiment] is it argued that absolute motion exists. Instead, this is taken for granted. The properties and causes invoked are not adduced as empirical criteria in virtue of which absolute motion can be detected or established to exist. In each case, that absolute motion or rest indeed has the property or causes in question is not something the argument seeks to show, but serves instead as a premise. Nor is it assumed that absolute motion is, by definition, motion with respect to absolute space (ibid., p. 146).

It is true that Newton did not argue for the existence of absolute motion before the exposition of the bucket experiment, but this does not mean that he took its existence for granted. He enumerated a series of properties that absolute motion should have if it existed. The existence of absolute motion is proved by the argument we have developed above (see Section 7.1.2.1.). The properties enumerated by Newton are criteria to distinguish absolute from relative motion. They are not empirical properties since no experiment exists that allows us to verify if a motion fulfils such criteria. With regard to Rynasiewicz's last remark, Newton does not define absolute motion as motion referred to absolute space because this is so obvious that no definition is needed. In Rynasiewicz's opinion, conversely, the scholium was not conceived to prove the existence of absolute space or that of absolute motion, but, rather, to demonstrate that, hypothetically, absolute motion has to be referred to absolute space (Rynasiewicz, 1995 II, pp. 320–321).

In his analysis of the bucket experiment Rynasiewicz highlights that Newton reveals an anti-correlation between the absolute movement of the water and the relative movement of water with respect to the bucket (ibid., p. 148). However, an anti-correlation is indeed a correlation. Therefore, if the movement of water were always opposite to that of the bucket, one would be fully justified in thinking that a causal (or an anti-causal) relation exists between the two motions. Hence it would be possible to infer the absolute motion from the relative one. Newton's argument makes sense only if one supposes that his objective (as it is clear) is to prove that no correlation exists between absolute and relative motion; these are not related either directly or inversely. Newton did not expound on those parts of the experiment we have called (3) and (4), but his thought is clear. Let us consider, for instance, Newton's description of the experiment: the bucket rotates and communicates its motion to the water; when the bucket reaches its maximal rotation, the water has just begun to ascend; the bucket maintains its maximal rotation for a while and, in this period of time, the water changes its state of rotation because it keeps going up for some time. Therefore, the same intensity of the bucket rotation—the maximal one—is associated with different rotation-intensity of the water. This means that there is no correlation between the rotation of the bucket and that of the water; there is no evidence of an anti-correlation. Newton, with his typical brachylogical style, compressed the entire experiment into one exposition. We have expounded it in four phases to better clarify its implications. But, it is clear that he excludes a correlation and is not thinking of an anti-correlation, as Rynasiewicz claims.

Among the authors who offered an interpretation of the scholium that is different from the so called "Mach's legacy" is DiSalle (2002). This author has written extensively on the concept of time and space in Newton.[27] His paper begins with a series of acceptable observations: from a physical standpoint, Newton's mechanics needs absolute time,[28] but not absolute space (DiSalle, 2002, p. 36); according to Newton, experiments allow us to identify an absolute rotation and acceleration, but not an absolute velocity; unlike Leibniz, what characterizes Newton's notion of time is not the distinct individuality of its moments, but its structure: absolute time flows uniformly (ibid., p. 41). Newton postulates a difference between inertial motions, in which equal spaces are traversed in equal times, and non-inertial motions, where this does not occur. Thus, absolute time is well defined with reference to inertial motions and this allows also to distinguish the inertial motions from the non-inertial ones. Of course, as far as absolute space is concerned, the problem is that an immobile space is physically indistinguishable from a space moving rectilinearly and uniformly, an issue of which Newton was aware of (ibid., pp. 42–45). After these initial remarks, DiSalle presents his thesis:

> Newton never claims to *prove* that the centrifugal forces on the water or the globes are caused by rotation relative to absolute space, or claims that any such experiment could demonstrate the existence of absolute space. What he says, instead, is that the centrifugal forces *define* absolute rotation [...]. Thus Newton has not tried to *justify* the causal link between rotation and centrifugal effects, but simply to identify it as definitive of true rotation. Thus he has defined a theoretical quantity, absolute rotation, by exhibiting how it is detected and measured by centrifugal effects. (ibid, p. 47)

DiSalle justifies his interpretation in this way:

> His [Newton's] discussion of the water-bucket makes this explicit: from the endeavour to recede from the axis, "one can find out *and measure* the true and absolute circular motion of the water, which here is the direct opposite of its relative motion".[29] And concerning the globes, he states not only that from the tension on the cord "we might compute the quantity of their circular motions," but also that changes in the tension would provide a measure of the increase or decrease in rotation. "In this way both the quantity and the direction of this circular motion could be found in any immense vacuum, where nothing external or sensible existed with which the balls could be compared." [...] In both cases, we are not inferring a theoretical entity from a phenomenon, but defining a phenomenon as the measure of a theoretical quantity. Newton's argument, in sum, was never an argument from physical

[27] Among DiSalle's contributions we mention DiSalle (1991, 1995, 2006, 2020).

[28] Velocity is the first derivative of space with respect to time and acceleration is the first derivative of velocity with respect to time. Thence, if time did not flow uniformly, this would complicate the whole mathematical apparatus of Newton's physics with the risk of making the solution of physics problems impossible or, at least, extremely intricate. It is, however, to point out that Newtonian mechanics only needs a time flowing uniformly, which, from an ontological standpoint, is a less demanding requirement than absolute time. As we will see, while speaking of Neumann, it is possible to conceive a relationalist idea of time which works in Newtonian physics.

[29] Newton (1687, p. 10): the quotation comes from Newton's scholium to the Definitions and the term 'measure' is emphasized by DiSalle.

> phenomena to metaphysical conclusions about the "absoluteness" of rotation. Instead, it was an argument of a sort that is fundamental to every empirical science: an argument that a novel theoretical concept has a well-defined empirical content. (ibid., pp. 47–48)

We do not agree with this reading, as is clear from our discussion of the whole apagogic demonstrative process through which Newton believed to have proved the existence of absolute motions. Another element confirms our point of view: in the "Definitions", Newton gives clear definition of matter, quantity of motion, various kinds of forces, absolute time and absolute space. Why does he not define absolute motion explicitly? Quite simply the answer is that there is no need to define absolute motion, since it is the motion relative to absolute time and absolute space. In other words: once regarded as a motion with respect to absolute time and space, absolute motion has a provable physical existence. Therefore, the two experiments produce demonstrations, not definitions. Furthermore, why resort to such complicate experiments to define a kind of motion? It would have been enough to define as "absolute" any motion in which centrifugal effects appear. The fact that Newton proved that such effects cannot be understood by referring them to neighbouring bodies implies that he was not looking for a mere definition, but for a true demonstration.

Another original interpretation of the scholium has been advanced by Belkind who holds two theses:

(1) "For Newton the notion 'place' represents the particular volume a body occupies within a larger volume that contains it. Thus, by 'absolute motion', Newton means the body's position relative to an immovable container, not relative to a background space. Relative motion, on the other hand, is the change of position relative to a movable container". (Belkind, 2007, p. 272). Therefore, absolute space is the set of immovable places. The problem with this idea is that the concept of "immovable container" is not clear. For, with respect to what is the container immovable? Obviously, the answer cannot be with respect to the body, because in this case the container would connote a relative determination of the body's place, not an absolute one.

(2) The main purpose of the scholium is to criticize Descartes's conception of quantity of motion. For a correct definition of the quantity of motion, absolute space, considered as the set of immovable places, is required.

Belkind objects to the "classical" readings of the scholium:

> Sklar claims that absolute accelerations 'give rise' to inertial effects. The notion of one property 'giving rise' to another phenomenon intimates the existence of some causal process, but this turn of phrase is ambiguous. Either this means that *the property of absolute acceleration* itself is the cause of the effect; or we take absolute space to be the causal agent. According to the first reading, the function of the space container is merely to provide an idle reference for the *definition* of absolute acceleration, while space itself is causally inert. Newton himself argued that inertial effects are produced by the inherent force of matter, which is 'the power of resisting by which every body, so far as it is able, perseveres in its state either of resting or of moving uniformly straight forward' (Newton, 1999, p. 404). For him inertial effects originate from the inherent force of a body, a force which is 'proportional to the body and does not differ in any way form [from] the inertia of the mass'. [30]

---

[30] Belkind (2007, p. 273). For Belkind's reference to Newton (1999) see our bibliography.

The criticism against Sklar is unjustified, and there is no contradiction at all in Sklar's and Newton's claims. The former takes for granted the existence of the inherent force and is simply stating that, without an external force that produces an acceleration, no inertial effect would exist because the body would maintain its state of rest or of uniform motion. Inherent force alone cannot explain the inertial effects and certainly this was not what Newton thought. Both an external force and the inherent force are necessary to produce inertial effects. What Sklar claims is, hence, very clear. Absolute space is causally inert; it is the theater on which the scene of physics is staged. From a logical and physical point of view, absolute space fulfils the role that we would assign to an inertial system. We will consider the relation between Newton's absolute space and our inertial systems in the next subsection.

Referring to scholarly literature,[31] Belkind claims that "Newton's argument for absolute space is not at all an attempt to explain inertial effects" (ibid., p. 274). In this respect he agrees with the authors he mentions. However, he states that none of them understood the intention with which the scholium was written precisely because none of them grasped thoroughly Newton's doctrine of place (ibid., p. 276). What is worth noting in Belkind's paper is his idea of Newtonian absolute space as a set of immovable places which, however, do not form a space 'backdrop' (ibid., p. 280). This concept does not seem clearly defined. Belkind possibly means: if one studies a physical phenomenon in which several bodies, included one into the other, move, then the most external body *B* and a region of space sufficiently extended and external to *B* have to be considered. Such a region should be regarded as a part of absolute place at rest. The following quotation seems to confirm this reading:

> The change of relative place depends on the additional entities which are included in the composite body. The position of relative place is the position of the region a body occupies relative to the composite whole. Relative rest therefore is the 'continuance' of the body within the same region of the ship. Absolute rest is defined as taking up the same part of immovable space: "true rest is the continuance of a body in the same part of that unmoving space in which the ship itself, along with its interior and all its contents, is moving. (Newton, 1999, 409)". Thus, Newton does not define absolute rest as a body changing its position relative to a space 'backdrop'. Rather, absolute rest is the continuance of the body in the same part of the unmoving space in which the composite body, the ship together with its interior is moving. That is, he imagines absolute space as providing a large enough 'immovable' region that contains the moving part. (ibid., p. 280)

We argue that Belkind's reading of Newton's passages is not correct. As a matter of fact, Newton is not saying that absolute rest is the constant position of a body with respect to a region that is large enough to be regarded as immovable: it is impossible to understand according to which criteria that region should be considered immovable. Newton thinks that absolute rest is the maintenance of a body position exactly with respect to a global spatial background. There is no other possibility if one wants to offer a coherent interpretation of Newton's thought, unless we regard Newton as an operationalist and a conventionalist who claims that we might assume freely a

[31] Belkind mentions Stein (1970), Laymon (1978), Rynasiewicz (1995) and (2004), DiSalle (1994) and (2002).

certain region to be at rest. But this view is so anachronistic and so far from everything Newton wrote that it is not worth being considered.[32]

Absolute space and inertial reference frames. Newton's conception can be summarised as follows:

(1) in physics a privileged class of reference frames exists: the inertial reference frames. The study of natural phenomena is, in general, easier using such frames because the forces assume the form $\boldsymbol{F} = m\,\boldsymbol{a}$, which represents the authentic actions exerted by external bodies on a test body. All the inertial reference frames express, hence, dynamics in the same manner. This can be deduced from the fifth Corollary to the Laws of Motion where we read:

> The motions of bodies included in a given space are the same among themselves, whether that space is at rest, or moves uniformly forwards in a right line without any circular motion.[33]

We think that the most reasonable interpretation of what Newton here defines "a given space" is what we call an inertial reference frame.[34] Newton is stating that, from a dynamical point of view, there is no possibility to distinguish a space at rest from a space moving uniformly in a right line.

(2) Newton was aware that physics might also be developed considering non-inertial reference frames. This is clearly proved, for example, by the Proposition XIX, Problem III of the third book of the *Principia* concerning "the proportion of the axe of a Planet [but specifically the Earth] to the diameters perpendicular thereto" (Newton, 1729 II, p. 239). Here Newton used a reference frame centred in the centre of the Earth and corotating with the Earth and resorted to the notion of centrifugal force, which, of course, appears only in such systems. Therefore, he conceived the existence of non-inertial reference frames and separated their treatment from that of the inertial frames. The true actions among bodies can be detected only in inertial frames, but Newton also used non-inertial frames when these were more convenient.

(3) Absolute space is the objectual background with respect to which the kinetic state of the bodies can, at least in principle, be determined. It is the physical entity which supports absolute motions. This means that, according to Newton, among all reference frames, absolute space is the privileged one, the prototype of all the other reference frames with respect to which the true or absolute

[32] On Newton's scholium to the Definitions and on his concept of inertia, space and time the literature is huge. We have restricted our analysis to those authors and those works who were more directly connected to our interpretation. Among other contributions to this problems we mention: Bussotti (2017c), Cassini (2005), Freudenthal (1986), in particular Chap. 1, pp. 13–28; Gould (1962), Hugett (2008), Kochiras (2012), Mcguire (1978a, 1978b), Mclaughlin (2008), Newburgh (2007), Toulmin (1959), Zylbersztajn (1994).

[33] Newton (1729 I, p. 30). Cf. Newton (1687, p. 20): "Corporum dato spatio inclusorum iidem sunt motus inter se, sive spatium illud quiescat sive moveatur idem uniformiter in directum sine motu circulari".

[34] For a different interpretation which, though interesting, we do not find convincing, see Arthur (1994).

motions of bodies can be determined. Its status is truly peculiar since an inertial reference frame can be considered indifferently at rest or in uniform motion while Newton's absolute space is "immovable" (Newton, 1729 I, p. 9), being the set of all the immovable places, that is of those places "that, from infinity to infinity, do all retain the same given positions one to another" (ibid.). One of the main problems which, starting from the 1870s, gave rise to a long-term discussion on absolute space is that, within Newtonian physics, all inertial references frames are equivalent because there is neither a kinematical nor a dynamical way to identify a privileged reference frame. As the Corollary V to the Laws of motion clarifies, Newton was aware of the impossibility to distinguish among different inertial reference frames on the basis of physical observables. The following pages are, hence, dedicated to the discussions on reference frames between the end of the 19th century and the beginning of the 20th as they throw a retrospective light on the reasons that led Newton to admit absolute space.

In Newton's works a formal definition of the concept of reference frame cannot be found: Newton's ideas are clear, but, as to the notion of reference frame, they are not formalised. This is not surprising from an historical perspective. While Newton's text is extremely precise and formalised if compared to those of the other physicists of his generation, we cannot expect to find in that age a level of formalization comparable to that of later physics. A fundamental question concerns the ontological status of inertial reference frames in Newton: they are abstract entities indispensable to describe the theatre of nature; nonetheless, Newton felt the need to make them dependent on a concrete object—where "concrete" signifies "really existing in the world". This object is absolute space. It is the reference entity of the inertial frames and of the property of inertia itself. As is well known, Newton was criticized by Leibniz, Huygens and Berkeley for his recourse to absolute entities and, in particular, to absolute space. During his scientific career Huygens elaborated a series of ideas on the concept of physical space, but he published almost nothing on this subject and it is not easy to draw a complete picture of a possible dynamics developed on the basis of Huygens' train of thoughts.[35] Leibniz reflected profoundly on the problem of absolute/relative space/time (see Section 8.2.), but he did not elaborate in depth a dynamics alternative to Newton's. This is even clearer in Berkeley. As already highlighted, the extraordinary successes of Newton's dynamics overshadowed the difficulties lying behind the concept of absolute space until the 1870s. In that period discussions arose about the necessity to define inertia as a relative concept, namely as a notion referred to observables and not to an impalpable and perhaps metaphysical concept as the absolute space. Scientists and philosophers felt the need of a precise definition of the concept of inertial reference frame. These discussions belonged to a deep debate on the foundations of physics. Since physics had already reached a highly formalised and refined structure, some scholars began to analyse the conceptual bases of the discipline itself. Furthermore, in that period the discovery of electromagnetism and the philosophical climate created by positivism led to search for a more precise and rigorous conception of the notions of reference frame, inertia, mass

[35] On Huygens's conception of space see Chap. 6.

and force, a conception which should be free from any reference to non-observable or not directly detectable entities. As we will see, several physicists at the end of the nineteenth century clarified many aspects which were emendable in Newton's view, but a completely satisfactory treatment was hardly achieved.

First of all, it became clear that the concept of absolute space is redundant and not necessary in Newtonian physics. The notion of the class of inertial reference frames is sufficient. This is a logical abstract concept without ontological implications. Furthermore, the idea of a privileged reference frame (as the absolute space) is, if not contradictory, at least scarcely compatible with the fact that the differential equations of Newtonian dynamics are second order and not first order equations. In other terms, they permit to determine the acceleration of a body which is the same in every inertial reference frame. However, the choice of the inertial reference frame which we consider at rest is absolutely arbitrary. The choice of a privileged reference frame would imply that the forces determine velocities and not changes of velocities. This would be an Aristotelian physics rather than a Newtonian physics. The physicists who discussed the Newtonian idea of absolute space between the end of the nineteenth century and the beginning of the twentieth century intended to refer the concepts of inertia and mass as well as the notion of reference frame to observables. This was not an easy task. The definition of inertial reference frame is anything but trivial, if you claim to anchor this concept to observables. After all, in the universe there are events and moving bodies, not reference frames! The notion of inertial reference frame implies that not all observers are equal. There is a privileged class of observers who agree on the way in which they see the world and for whom the laws of movement are simpler than for the other observers. A reference frame of this sort is something different from merely establishing a system of coordinates because coordinates are only a map which associates to any point of the n-dimensional space an n-tuple of numbers.

As regards to the definition of the concept of reference frame the most important author is surely Ludwig Lange (1863–1936)[36] who criticized Newton for referring inertia to the "phantom" of absolute space:

> According to his explanations, he [Newton] bases the law [of inertia] on a certain coordinate system that he calls the "absolute, homogeneous, infinite and immovable" space. This "absolute space" is real and not merely something conceived, though admittedly it is not accessible to our imperfect human sensory perception. It consists of indiscernable absolutely fixed points arranged next to each other; and it is only by comparison of the bodies with these "in themselves" fixed points, not by comparison with other matter, that the nature of the positions and motions of the bodies are to be recognized. In brief, Newton's absolute space is a phantom that should never be made the basis of an exact science.[37]

Lange claimed that the fundamental coordinates systems of dynamics should be called "inertial systems", a locution of his invention. He divided his study into two

[36] Among the works dedicated by Lange to the definition of inertial reference frame we refer to Lange (1885a). As to the literature, Barbour (2001), pp. 654–662 and Pfister (2014) clarify Lange's order of ideas. Giulini (2002, pp. 370–371) improves the mathematical part of Lange's work.

[37] Lange (2014, p. 252). We could not check the original German version of Lange (1885a).

parts: in the former he considered only the problem to construct systems in which "points left to themselves", namely on which no external force acts, move on a straight line; in the latter he added the condition that such motion has to be not only rectilinear, but also uniform. Lange begun his discussion with this crucial consideration: if you consider a finite time, it is always possible, given three mobile points *P*, *P'*, *P"* to determine a reference frame (to be precise, an infinite number of reference frames, as Lange points out) with respect to which the motion of the three points is rectilinear.[38] As Barbour highlights: "this means we can always find a coordinate system such that three point particles moved in an entirely arbitrary manner will, for a finite time at least, move along straight lines emanating from one point. (Barbour, 2001, p. 658)".

On the contrary, this operation does not work in general if the points are more than three. Therefore, the very concept of rectilinear trajectory acquires a physical meaning only in presence of more than three points. Otherwise, to establish a rectilinear trajectory is only a matter of convention. Thus, the following characterization of the inertia law can be offered:

> Definition I. An "inertial system" is any coordinate system of the kind that in relation to it *three* points *P*, *P'*, *P"*, projected from the same space point and then left to themselves – which, however, may not lie in one straight line – move on three arbitrary straight lines *G*, *G'*, *G"* (e.g., on the coordinate axes) that meet at one point.
>
> Theorem I [Law of rectilinear trajectory]. In relation to an inertial system the path of an *arbitrary fourth* point, left to itself, is likewise rectilinear.
>
> Definition II. An "inertial timescale" is any timescale in relation to which *one* point, left to itself (e.g., *P*), moves uniformly with respect to an inertial system.
>
> Theorem II [Law of uniform rectilinear trajectory, inertia]. In relation to an inertial timescale *any other* point, left to itself, moves uniformly in its inertial path. (Lange, 1885a, 2014, p. 254)

Lange provides the mathematical demonstrations that the two definitions are well founded and that the two theorems are true.[39] Lange's definition of inertial system and his statement of inertial law deserve to be commented on both for their conceptual and historical relevance.

Conceptually, Lange's definition avoids any reference to absolute space and is only tied to the movement of any three points. Therefore, in this respect, there is no reference to non-observables. Lange's definition also depends on the possibility to establish an "inertial timescale" independently of any reference to absolute time. This important result had been already achieved by Neumann in (1870). He based his definition of uniform flow of time and of identity of two time-intervals on the concept of the motion of a force-free body. Neumann observed that if only one body is considered—that can be regarded as a material point—to claim that this force-free material point travels equal distances in equal times is an assertion completely empty of any empirical content. The situation changes if two material force-free points are taken into account because, in this case, the motions of these two points can be compared. Here Neumann gives his definition of time unit: consider two material

[38] For an intuitive and clear explanation of Lange's idea, see Barbour (2001, pp. 657–658).

[39] Lange's proofs present both redundancies and oddities which, however, can be emended (see Giulini 2002).

force-free points, then equal distances travelled by one point correspond to equal distances travelled by the other point and conversely. Therefore, equal distances travelled by one of them define a unit of time, which is applicable to the motions of the other and, more in general, to all physical circumstances. Ergo, given this unit of time, you can calculate what distance is travelled by each material point. As Barbour points out, this procedure materializes a clock (Barbour, 2001, p. 655), that is it determines a time interval by means of a uniform movement. Neumann wrote:

> A material point left to itself [force-free] travels equal distances in equal time intervals. So the words of this law sound. It seems impossible to give a well determined meaning to these words until we do not know what equal time-lengths mean.
>
> But this is true only apparently! For, when we remove that stone of scandal, when we remove the irrational concept of equal time intervals and of such a proposition, only a well determined residue remains, which sounds:
>
> *Two* force free material points move reciprocally so that equal traversed distances of one always correspond to equal traversed distances of the other one.[40]

Thus, is every problem solved? Have the "phantoms" of Newton's absolute time and space been removed? The answer is not easy, but we think that it should be negative. Neumann's and Lange's definitions—but we should also include Streintz (1883)—are based on the existence of bodies on which no force is applied, something which is simply impossible if the law of universal gravitation is true. It is perhaps possible to eliminate all the other interactions, but it is certainly impossible to eliminate gravity. A body, however distant from another, is acted upon by the gravitational influence of this other body. Quite simply, material points or bodies on which no force acts do not exist. Bertrand Russell, almost at the end of his *Principles of Mathematics*, criticized Streintz, who had developed Neumann's ideas:

> If motion *means* motion relative to fundamental bodies [or inertial reference frames] (and if not, their introduction is no gain from a logical point of view), then the law of gravitation becomes strictly meaningless if taken to be universal – a view which seems impossible to defend. The theory requires that there should be matter not subject to any forces, and this is denied by the law of gravitation. The point is not so much that universal gravitation must be *true* as that it must be significant – whether true or false is an irrelevant question. (Russell, 1903, p. 491. Emphasis in the text)

The conclusion is that the concept of inertial reference frame can be defined by referring it not to absolute time and space but to material points, which is a great

---

[40] Our translation, except for the last sentence transl. by Barbour (2001, p. 656). Cf. Neumann (1870, p. 18): "Ein sich selbst überlassener materieller Punkt durchläuft in gleichen Zeitintervallen gleich grosse Wegabschnitte. So lauten die Worte jenes Gesetzes. Und es scheint unmöglich, mit diesen Worten einen bestimmten Inhalt zu verbinden, so lange wir nicht wissen, was unter gleichen Zeitlängen zu verstehen ist. Aber nur scheinbar! Denn wenn wir jenen Stein des Anstosses bei Seite werfen, jenen irrationalen Begriff der gleich grossen Zeitintervalle abscheiden, so bleibt von dem Satze immerhin noch ein bestimmtes Residuum übrig, welches so lautet: *Zwei* materielle Punkte, von denen jeder sich selbst überlassen ist, bewegen sich in solchen Weise fort, dass gleiche Wegabschnitte des einem immer mit gleichen Wegabschnitte des andern correspondiren". (All the emphases are both in Barbour's and Neumann's texts).

improvement towards the definability in terms of observables. Nonetheless, a metaphysical[41] *residuum* remains; this *residuum* is the idealization of material points on which no force acts. This is perfectly in order if our aim is to solve physical problems: for example, if we want to study motions on the Earth, nothing prevents us from considering Earth as an inertial reference frame. But, if the idea is to found the notion of inertial reference frame on a basis free from any metaphysical assumption, then such a basis cannot be the body on which no force acts because this is a concept whose extension is empty. It seems that this problem cannot be solved within Newtonian physics, or at least its solution is extremely complicated and unachievable with the mathematical and physical tools available not only to Newton, but also to the physicists of the late 19th and the beginning of the 20th century. This is an impressive confirmation of Newton greatness: it is true that his dynamics is based on non-observables, but this seems to be an unavoidable consequence, not a defect of Newton himself. The following historical observations on the concept of inertia are a further confirmation of this state of affairs.

Newton considered absolute space as the entity to which the inertia of bodies must be referred; with Lange's work the notion of reference frame was defined and inertia was, obviously, referred to inertial reference frames. As already pointed out, the first work that influenced the trend of ideas that led to Lange's definition was Neumann (1870). Here the author refers bodies' inertia not to absolute space, but to something that he called *Body Alpha* (*Körper Alpha*). To use a literary metaphor, one might say that absolute space is contracted in a single body, whose position and nature are left undetermined. Julian Barbour summarizes very clearly Neumann's conception:

> The body itself is most mysterious: '*Where* this body is situated and the grounds for according a single body such a distinguished, indeed sovereign position—these are questions to which we receive *no* answer'.
>
> Unperturbed by these mysteries, Neumann states formally:
>
> As *first principle* of the Galileo-Newton theory one should take the statement that at some unknown place in the universe there is an unknown body, indeed, an *absolutely rigid* body, a body whose shape and size are unchanged for all time.
>
> Neumann proposes to call this the *Body Alpha* and says that the *second principle* of the theory is that the motion of any material point left to itself (i.e., subject to no disturbance) is rectilinear with respect to *Body Alpha*.[42]

Having introduced the *Body Alpha*, Neumann shows how to reduce to the smallest possible number the independent parameters necessary to determine the position of a system of n bodies with respect to the *Body Alpha* or, equivalently, to the absolute space. Neumann's *Body Alpha* can be considered as an attempt to materialize an entity such as absolute space. However, it is not less metaphysical than absolute

[41] The term "metaphysical" is polysemous. Here by "metaphysical" we mean, in neopositivistic terms, what is "not referable to anything observable".

[42] Barbour (2001, p. 648). Cf. Neumann (1870, pp. 15–16). With reference to the *Body Alpha*, Barbour says ironically that when he read for the first time Neumann's work, he thought that it was a joke written on the first of April, but after having seen the date 3 November, he was forced to change his mind! (Barbour, 2001, p. 648).

space, with the further epistemological (if not physical and logical) problem that it is far from any intuition of ours. This is not true of absolute space. Barbour's assertion seems a reasonable one:

> If the concept of *Body Alpha* encapsulated superbly the *rigidity* of the Newtonian frames of reference, it was not at all felicitous as a characterization of the Galilean invariance of dynamics. Neumann was forced to admit that *Body Alpha* was subject to a certain lack of definiteness – any particular *Body Alpha* can always be replaced by another provided only the new body is in a state of uniform rectilinear motion relative to the first. (Ibidem, p. 653. Emphasis in the text)

The "successor", so to say, of Neumann's *Body Alpha* is Lange's inertial reference frame, a notion much better defined than the *Body Alpha*, but which would hardly have arisen without the pioneering work of Neumann, who had the merit to pose the problem of the inertia status in Newton's theory.

The notion of inertial reference frame establishes a series of systems with respect to which inertial movement is determined. This means that each inertial frame is perfectly equivalent and that no privileged frame as Newton's absolute space exists. Therefore, from a kinematical point of view, inertia is a completely relative concept; it is relative to any inertial reference frame, for which no prototype such as Newton's absolute space is necessary. But what is the situation with regard to the dynamical aspect of inertia? Newton considered inertia an intrinsic property of each body, proportional to its mass. If in the universe there were only a single body, it would have its own inertia insofar as it has a mass. However, the relativization of inertia at the end of the nineteenth century did not regard only its kinematic aspect, but also its dynamic one. A group of scientists began to realize that inertia is not a property of a single body, but that it depends also on the presence of other bodies. The fundamental source of inspiration for these scientists was Ernst Mach, and particularly his celebrated *The Science of Mechanics*.

In this text, Mach objects against the Newtonian concepts of absolute time and space. As is well known, Mach refused the introduction in physics of any concept not directly detectable through experience. With respect to Newton's concepts of absolute space and time, he wrote:

> It is scarcely necessary to remark that in the reflections here presented Newton has again acted contrary to his expressed intention only to investigate actual facts. No one is competent to predicate things about absolute space and absolute motion; they are pure things of thought, pure mental constructs, that cannot be produced in experience. All our principles of mechanics are, as we have shown in detail, experimental knowledge concerning the relative positions and motions of bodies. Even in the provinces in which they are now recognised as valid, they could not, and were not, admitted without previously being subjected to experimental tests.[43]

[43] Mach (1883, 1919, p. 229). Cf. Mach (1883, 1889, 2nd ed., p. 213): "Dass Newton auch in den eben mitgetheilten Ueberlegungen gegen seine Absicht, nur das Thatsächliche zu untersuchen, handelt, ist kaum nöthig zu bemerken. Ueber den absoluten Raum und die absolute Bewegung kann niemand etwas aussagen, sie sind blosse Gedankendinge, die in der Erfahrung nicht aufgezeigt werden können. Alle unsere Grundsätze der Mechanik sind, wie ausführlich gezeigt worden ist, Erfahrungen über relative Lagen und Bewegungen der Körper. Sie konnten und durften auf den Gebieten, auf welchen man sie heute als gültig betrachtet, nicht ohne Prüfung angenommen werden".

As to the bucket experiment and to Newton's interpretation of it, we read:

> Newton's experiment with the rotating vessel of water simply informs us, that the relative rotation of the water with respect to the sides of the vessel produces no noticeable centrifugal forces, but that such forces are produced by its relative rotation with respect to the mass of the earth and the other celestial bodies. No one is competent to say how the experiment would turn out if the sides of the vessel increased in thickness and mass till they were ultimately several leagues thick.[44]

From this passage it appears legitimate to ascribe to Mach what Einstein called "Mach principle", that is: "the inertia of a body depends on the quantity and distribution of matter in the whole universe".

However, things are more complicated than this: first of all, Mach never stated in general terms the principle that was attributed to him. His closer statement to the Einstein's version is the one we have quoted. Is this enough to ascribe to Mach the so-called Mach principle? The question is not easy to be answered: Mach certainly denied the existence of absolute time and space, and the passage we have mentioned is undoubtedly close to the so-called Mach principle. However, the assertion that inertia has to be referred to the mass distribution of the entire universe is as metaphysical as Newton's reference to absolute space, at least without an experimental proof. Therefore, it is inconsistent with the entire philosophy of natural sciences advocated by Mach. Consistently with his philosophical views, Mach should have formulated his principle as an hypothesis to verify and not as an alternative assumption to the Newtonian concepts. The statement in our last quotation—"[...] such forces are produced by its relative rotation with respect to the mass of the earth and the other celestial bodies. No one is competent to say how the experiment would turn out if the sides of the vessel increased in thickness and mass till they were ultimately several leagues thick"—is not enough to let us understand if Mach considered his idea of inertia as a result of the universe's distribution mass an hypothesis to verify experimentally or an a priori assertion. Therefore, we have a double reason of interpretative uncertainty: (1) did Mach believe in what has been called by Einstein "Mach principle"? (2) In case of affirmative answer, did he conceive this principle as an a priori assumption or as a hypothesis to be proved?

Paul Carus, who entertained a very extensive epistolary exchange with Mach, denied that Mach had positively proposed a new mechanism to explain inertia. Furthermore, when a series of works written by different authors, who ascribed the principle to Mach, were published, Mach's reactions were extremely cautious. He neither explicitly refused nor accepted the paternity of the principle.[45]

---

[44] Mach (1883, 1919, p. 232). Cf. Mach (1883, 1889, 2nd ed., p. 216–217): "Der Versuch Newton's mit dem rotirenden Wassergefäss lehrt nur, dass die Relativdrehung des Wassers gegen die Gefässwände keine merklichen Centrifugalkräfte weckt, dass dieselben aber durch die Relativdrehung gegen die Masse der Erde und die übrigen Himmelskörper geweckt werden. Niemand kann sagen, wie der Versuch verlaufen würde, wenn die Gefässwände immer dicker und massiger, zuletzt mehrere Meilen dick würden".

[45] On all the problems related to Mach principle before Einstein's general relativity see Norton (1995).

However, most of contemporary scientists who were influenced by Mach's ideas, answered affirmatively to the first question and, hence, ascribed to him the principle. As to the second question, they did not answer explicitly, but, believing in the truth of Mach principle, tried to verify its validity experimentally.

A further observation concerns the possibility to transcribe Mach's principle into mathematical terms in order to construct a theory in which the inertia of a body depends on the presence of other bodies. This is an extraordinary difficult task which is continuing nowadays, with significant successes, but also with several difficulties. The fact that 21 slightly different versions of Mach principle have been proposed testifies to the difficulty of this notion (Barbour-Pfister, 1995, p. 530).

If applied to Newtonian physics,[46] Mach principle requires a problematic assumption: the existence of a new interaction among bodies, which depends on their inertial mass. Its effects, added to those of all the forces considered by Newton, and in particular to gravity, should be compatible with Newton's physics and coherent with the observations. Most of the scholars who, at the turn of the nineteenth and twentieth centuries, thought that the inertia of a body is also influenced by the mass distribution of all the other bodies, worked experimentally. Considering certain phenomena, they attempted to determine slight deviations from the predictions of Newtonian mechanics which might be explained as the result of the mass distribution.

Immanuel and Benedict Friedlaender explicitly stated what Mach had never clearly asserted:

> However, it seems to me that the correct form of the law of inertia will only then have been found when *relative inertia* as an effect of masses on each other and *gravitation*, which is also an effect of masses on each other, have been derived on the basis of a *unified law*.[47]

The two brothers argued to have reached this conclusion independently of Mach. Interestingly enough, they also speculated on the possible way in which a mass might influence the inertia of another mass:

> A body that approaches or recedes from a second body would have no influence on it as long as the velocity of approach, which is to be taken as positive or negative, remains unchanged; in contrast, any change of the velocity would have the effect previously shown [a potentially detectable inertial effect].[48]

[46] We do not discuss general relativity or even subsequent theories of gravitation.

[47] Barbour and Pfister (Eds.) (1995, p. 310). Barbour and Pfister (eds.) (1995) offers a partial English translation of Friedlaenders's work. The first part by Benedict is partially translated at pp. 114–119; the second by Immanuel is partially translated at pp. 309–311. Cf. Friedlaender-Friedlaender (1896, p. 17): "Mir will aber scheinen, dass die richtige Fassung des Gesegtzes der Trägheit erst dann gefunden ist, wenn die relative Trägheit als eine Wirkung von Massen auf einander und die Gravitation, die ja auch eine Wirkung von Massen auf einander ist, auf ein einheitliches Gesetz zurüchgeführt sein werden".

[48] Barbour and Pfister (eds.) (1995, p. 117). Cf. Friedlaender-Friedlaender (1896, p. 30): "Ein Körper, der sich einem zweiten nähert oder von ihm entfernt, würde ohne Einfluss auf diesen sein, solange die positiv oder negativ zu nehmende Annährungsgeschwindigkeit unverändert bleibt; jede Aenderung der Geschwindigkeit hingegen würde die vorher gezeigte Wirkung ausüben".

Given this theoretical hypothesis, the Friedlaenders devised an experiment involving a flywheel and a torsion balance. The aim of the experiment was to show the foreseen inertial effects, but no effect was detected.

August Föppl thought that, by studying appropriate systems of celestial bodies, it would be possible to detect slight variations of the Newtonian gravitational effects. These variations ought to be ascribed to the inertial influence of other masses on the masses of the system. In particular, he thought that the inertial forces might depend on the relative velocities among bodies:

> These forces, which for brevity I shall in what follows simply call 'velocity forces,' have nothing to do with gravitational forces, which arise concurrently with them, and specifically they can – and probably will – follow a quite different law as regards the distance dependence compared with the gravitational forces.[49]

The concept of velocity forces might also offer the possibility to determine a privileged inertial reference frame, something similar to Newton's absolute space, without giving up the idea that all motions are relative:

> We are now in the position – and on this I put considerable value – to specify a condition meeting our requirement for causality that must be satisfied by the true inertial system required by the law of inertia. Namely, the true inertial system is the reference system for which all the velocity-dependent forces that arise from the individual bodies of the universe are in balance at the test point. Even if in practice it is clear that we have not gained very much through this statement, it does appear to me that we have thereby obtained a very suitable basis for forming a clear concept of what is known as absolute motion in the framework of mechanics. There is at the least a prospect opened up of a way of determining the inertial system once the law that establishes the velocity dependent forces has been found. In other words, it will be possible to construct the absolute space that appears in the law of inertia without having to sacrifice the notion that ultimately all motions are merely relative.[50]

Föppl argued that the possible velocity forces might be directly calculated by a physicist who lived on a planet *P* rotating around a star *S* under the condition that *P* is the only planet rotating around *S*. Obviously this consideration has no

---

[49] Barbour and Pfister (eds.) (1995, p. 124). (This volume contains at pp. 120–127 a partial English translation of Föppl (1904b). Cf. Föppl (1904b, p. 388): "Diese Kräfte, die ich der Kürze halber weiterhin einfach „Geschwindigkeitskräfte" nennen will, haben nichts mit den Gravitationskräften zu tun, die neben ihnen auftreten, und sie können namentlich und werden auch vermutlich ein ganz anderes Gesetz in Bezug auf die gegenseitige Entfernung befolgen, als diese".

[50] Barbour and Pfister (eds.) (1995, p. 123). Cf. Föppl (1904b, p. 387–388): "Hiermit sind wir nun auch, worauf ich besonderen Wert lege, in den Stand gesetzt, eine unser Kausalitätsbedürfnis befriedigende Bedingung anzugeben, der das von dem Trägheitsgesetze geforderte wahre Inertialsystem genügen muss. Es ist nämlich jenes Bezugssystem, für das sich alle von den Geschwindigkeiten abhängigen Kräfte, die von den einzelnen Weltkörpern ausgehen, an dem Aufpunkte im Gleichgewicht halten. Wenn auch mit dieser Aussage praktisch zunächst offenbar nicht viel gewonnen ist, so scheint mir doch für die Bildung eines klaren Begriffes über das, was man in der Mechanik absolute Bewegung nennt, damit eine sehr geeignete Unterlage gegeben zu sein. Es ist zum mindesten die Aussicht auf einen Weg eröffnet, der nach Auffindung des Wirkungsgesetzes der von den Geschwindigkeiten abhängigen Kräfte zur Bestimmung des Inertialsystems führen würde. Mit anderen Worten: der absolute Raum, von dem das Trägheitsgesetz spricht, wird konstruierbar, ohne dass man dabei die Vorstellung zu opfern braucht, dass im letzten Grunde alle Bewegungen nur relative sind".

experimental value. However, Föppl carried out an experiment in which he considered two reference frames: one is the reference frame of the fixed stars and the other is the reference frame centred on a gyroscope (Föppl, 1904a). Since the gyroscope tends to maintain its axis of rotation oriented towards a fixed direction and is unaffected by the Earth's rotation, when we measure the diurnal rotation of the Earth from a system centred on a gyroscope, it is as if we measure such a rotation from a system theoretically posed in the absolute space. The velocity of the Earth's rotation with respect to the fixed stars is deducible from astronomical observations. If the whole mass of the Earth and its rotation had an inertial effect, this effect would be detectable on the rotation itself of the Earth. This means that the amount of rotation measured in the reference frame of the fixed stars and in the reference frame of the gyroscope should be different. No difference was detected. Once again, it seemed that inertia were an inherent property of any single body.

Wenzel Hofmann is another important author in this context because he first tried to sketch a mathematical theory of relational inertia. He offered his general definition of inertia and claimed that the validity of such definition had to be verified experimentally:

> *Every body is subject to the law of conservation of its relative state of motion or rest with respect to all the other bodies in space; its actual behaviour is then the resultant of all the individual influences.*
>
> However, the extent to which this resultant depends on the mutual separations, the sizes of the various bodies, and their mutual disposition can hardly be found by speculation but only on the basis of experience (perhaps through experiments made specifically for this purpose).[51]

Hofmann stated that the foundation of a relationalist theory of inertia is an enterprise that goes far beyond the capabilities of a single man but some considerations can be offered. Be given two masses M and m, with $M > m$. According to the usual expression for the kinetic energy (or *vis viva*, as Hofmann wrote) $L = \frac{1}{2}mv^2$, the mass M should generate a kinetic energy greater than m. Through a brilliant argument, Hofmann stated that this is not true: suppose that the two masses collide and that an elastic spring posed at the impact point consumes and register the whole kinetic energy. If a reference frame is posed in M, then the entire energy stored in the spring is equivalent to the work by m on the system M. The converse is true if we consider m to be the origin of our reference frame and M in motion. In this case the same energy indicated by the stress of the spring is a measure of the work of M on the system m. Thence, Hofmann wrote:

> *From this we draw an important conclusion: If two mass systems M and m are in relative motion, the vis viva of M relative to m is equal to the vis viva of m relative to M.* [...] From all the foregoing, we can conclude that all inertial phenomena are to be traced to the mutual relationships of the masses to each other, so that the effects of inertia that are achieved are independent of the mass that is chosen as the reference system.[52]

---

[51] Barbour and Pfister (eds.) (1995, p. 129. Italics in the text). We could not check Hofmann's original work, of which a partial English translation is given by Barbour-Pfister, ibid., pp. 128–133.

[52] Ibid., p. 131. Italics in the text.

The first step taken by Hofmann was to change the definition of kinetic energy L, which should not be considered only as a function of a single mass. In the case of two masses M and m one might think of a formula as:

$$L = kMmv^2$$

where k is a constant of proportionality. However, on second thought, it is clear that this formula is too simplistic and unrealistic because it is extremely plausible that the distance r between two bodies also plays a role in the determination of the inertial effect one exerts on the other, so that a more realistic formula will be

$$L = kMmf(r)v^2$$

being $f(r)$ a certain function of the distance between the two masses.

A few months before the publication of Einstein's general theory of relativity Hans Reissner (Reissner, 1914 and Reissner, 1915) added a true mathematical treatment of the same ideas elaborated by Hofmann, although it is not known whether Reissner read Hofmann's work. His treatment is well structured, but is still far from being a complete theory of relationalist inertia. Barbour and Cook in the mid-1970s rediscovered Reissner's ideas and developed them (Barbour, 1975; Cook, 1976).

We summarize the whole issue in these final remarks:

(1) With regard to Newton's concepts of absolute time and space, and of inertia, it is necessary to reflect on the lessons we can draw from the studies developed at the turn of the nineteenth and the twentieth centuries. It is undeniable that these works put the accent on unsatisfactory aspects of Newton's theory. It is true that Newton's absolute concepts are not reducible to observables and are, hence, compromised with something 'metaphysical'. But it is also true that: (a) the definition of inertial reference frame also has a metaphysical feature, namely the concept of a body on which no force acts; (b) if one cannot determine precisely how other masses influence the inertia of a given body and if one cannot offer experiments to confirm new theories, these theories have scarce value in the context of physics. Therefore, the simplifying approach of Newton can be justified, taking into account the huge bulk of results he obtained. The scientists we have examined posed serious problems, but their answers were, at least in great part, unsatisfactory, which is a strong—we cannot say decisive—indication of the enormous difficulty one meets if Newton's theory is to be maintained and only some of its features are changed. Einstein overcame a great part of these difficulties because, to use a Kuhnian expression, he changed the whole paradigm.

(2) The scientists whose doctrines we have analysed are not the only ones who worked on these problems at the turn of the nineteenth and twentieth centuries. Other very important authors such as Stallo, J. Thomson, W. Thomson, Tait, Macgregor, Petzold, Poincaré, Enriques—to mention just a few—might be added to the list. However, for the purposes of this book the authors we have

considered are certainly the most significant. The investigation of their work is sufficient to offer the reader an idea of the general scientific climate we intended to describe in order to throw light on Newton's theory of absolute space and inertia.

(3) The general theory of relativity and the mathematical methods used to develop it are a fundamental watershed for the issue of absolute space and time. Indeed, first of all, thanks to further improvement of the mathematical techniques of differential geometry used in the general theory of relativity, it is possible to give a clear mathematical formulation of the various absolutist or relational-relativistic approaches to Newtonian mechanics. Namely, it is possible to conceive a four-dimensional Newtonian space–time in which, for example, the geometry of theories admitting an absolute space is clearly distinguished from the geometry of those theories which only resort to the notion of inertial reference frame or which rely on a relationalist approach. This is part of the outstanding work carried out by authors such as Sklar, Friedmann and Earman, among others. However, the means they use were not available to Newton and, in great part, to the authors who worked on a relationalist Newtonian theory before Einstein's general relativity. Thus, we do not consider it necessary to discuss their argumentations for a historical-conceptual examination of Newton's ideas, even though, such argumentations significantly enriched the debate on this topic.

(4) A similar consideration can be made for the theories of relationalist inertia developed from the 1950s onwards: they are of great interest, but they resort to means that were unavailable to Newton and, in substance, unavailable also to the physicists who worked at the turn of the nineteenth and twentieth centuries. It is precisely because the latter tried to reform Newtonian mechanics from within that their conceptions are useful to understand the structure of Newton's physics.

7.1.3 Newton's Experimental Philosophy and the Deduction of the Law of Universal Gravitation

Since gravity is the fundamental force that determines the structure of Newton's universe, it is appropriate to illustrate the way he deduced the law of universal gravitation. This is a clear example of the way the precepts of experimental philosophy are intertwined with mathematical deductions. Therefore, before illustrating such a deduction we will analyse the *Regulae Philosophandi*, where Newton expounded such precepts.

At the beginning of the third book, Newton explains the four "regulae philosophandi".[53] They summarize his ideas concerning the logic of scientific discovery. We are interested only in the first three rules:

---

[53] On the *Regulae Philosophandi* see Cohen (1969), Biener (2018), Butts (1993), Di Fate (2011), Ducheyne (2012), Janiak (2008), Mamiani (2001), McGuire (1995), Okruhlik (1989), Shapiro (2004), Smith (2002), Snobelen (2005), Spencer (2004).

> Rule I: We are to admit no more causes of natural things, than such as are both true and sufficient to explain their appearances. (Newton, 1729 II, p. 202)
>
> Rule II: Therefore to the same natural effects we must, as far as possible, assign the same causes. (ibid., p. 202)
>
> Rule III: The qualities of bodies, which admit neither intension nor remission of degrees, and which are found to belong to all bodies within the reach of our experiments, are to be esteemed the universal qualities of all bodies whatsoever. (ibid., p. 203)

The application of this third rule is specified in these terms:

> Lastly, if it universally appears, by experiments and astronomical observations, that all bodies about the Earth gravitate towards the Earth, and that in proportion to the quantity of matter which they severally contain; that the Moon likewise, according to the quantity of its matter, gravitates towards the Earth; that, on the other hand, our Sea gravitates towards the Moon; and that the Planets mutually one towards the another; and the Comets in like manner towards the Sun, we must, in consequence of this rule, universally allow that all bodies whatsoever are endow'd with a principle of mutual gravitation. For the argument from the appearances concludes with more force for the universal gravitation of all bodies, than for their impenetrability; of which among those in the celestial regions, we have no experiments, nor any manner of observation. (ibid., p. 204)

Thence, if all the phenomena we observe show that gravity acts as a universal attraction, there is no reason to think that exceptions exist, especially if we consider that we can verify the validity of the universal gravity attraction in those regions of the universe where it is impossible to verify if other properties of the bodies, such as impenetrability, hold.

The *Regulae* express Newton's experimental philosophy, but they also include a further epistemological thesis, namely, to use a Machian expression, the belief in the economy of scientific thought. It is likely that, behind this epistemic conviction, Newton harbored a presupposition concerning the relation between science and nature: science is not only a system of conventions useful to save phenomena, to offer previsions. Rather, it catches the essence of nature. Therefore, if you can explain a series of phenomena by means of n causes, those phenomena are determined in nature by such n causes at most, perhaps by less than n causes (if science has not yet reached a stage in which the numbers of causes might be reduced), but never by more than n causes.

The second part of the third book is entitled *Phenomena* and it is composed of propositions in which Newton refers to observational evidences and experimental data. He describes six phenomena:

> Phenomenon I: That the circumjovial planets, by radii drawn to Jupiter's centre, describe areas proportional to the times of description, and that their periodic times, the fixed Stars being at rest, are in sesquiduplicate [3/2] proportion of their distances from its centre.[54]
>
> Phenomenon III: That the five primary Planets, Mercury, Venus, Mars, Jupiter and Saturn, with their several orbits, encompass the Sun. (ibid., p. 209)

[54] Newton (1729 II, p. 206). Phenomenon II is the same as phenomenon I, but it is referred to Saturn's satellites (see ibid., p. 208).

> Phenomenon IV: That the fixed Stars being at rest, the periodic times of the five primary Planets, and (whether of the Sun about the Earth, or) of the Earth about the Sun, are in sesquiduplicate proportion of their mean distances from the Sun. (ibid., p. 210)
>
> Phenomenon V: Then the primary Planets, by radii drawn to the Earth, describe areas no wise proportional to the times; But that the areas, which they describe by radii drawn to the Sun, are proportional to the times of description. (ibid., p. 211)
>
> Phenomenon VI: That the Moon by a radius drawn to the Earth's centre, describes an area proportional to the time of description. (ibid., p. 212)

Namely: the observational evidences referred to by Newton show that the area law and the third Kepler law hold for the systems Jupiter-its satellites, Saturn-its satellites as well as for the solar system, supposed the Sun is in the centre of the system. The area law also holds for the system Earth-Moon.

The following section entitled *Propositiones* comprises 33 propositions, but for our purpose, it is enough to focus only on one of these propositions: the fourth one. In the first three propositions, that rely upon those of the first book as well as on the Phenomena, Newton easily proves that: (a) the force which makes Jupiter's satellites rotate around the Planet acts inversely to the square of the distances Jupiter-satellite (Proposition I); (b) the same law holds for the force which makes the planets rotate around the Sun (Proposition II) and for the system Earth-Moon (Proposition III).

In Proposition IV, which is fundamental, Newton proves that the force causing the movement of the Moon around the Earth is terrestrial gravity. This is the first formulation of the law of universal gravitation. Proposition IV reads as follows:

> That the Moon gravitates towards the Earth, and by the force of gravity is continually drawn off from a rectilinear motion, and retained in its orbit. (ibid., p. 215)

Newton's reasoning can be summarised like this: according to the data drawn by several astronomers the mean distance Earth-Moon at the syzigies can be considered approximately equal to 60 terrestrial radiuses, the period of revolution of the Moon around the Earth being 27 days 7 h and 43 min. From this data it is easy to calculate the centripetal force of the Moon in its orbit. Newton knew that the circumference of the Earth was 123,249,600 Parisian feet.[55] Newton then developed a brilliant resoning: he supposed the Moon to be still and, therefore, that it would fall towards the Earth with the same centripetal force (F) that, combined with the inertial tendency, determines the Moon's movement in its orbit. On the basis of previous propositions,[56] it was possible to calculate that the Moon's falling towards the Earth would describe $15\frac{1}{12}$ Parisian feet in the first minute of its descent. Now, since the force increases in inverse proportion of the square distance and since the distance Moon-Earth is equivalent to 60 terrestrial diameters, in the last minute of its descent the Moon will describe a space equivalent to $60 \times 60 \times 15\frac{1}{12}$ Parisian feet, and in the last second

[55] A Parisian foot is equivalent to 32.483 cm. Therefore, if you develop the calculations on the basis of Newton's data, you get a value for the Earth's mean radius which almost coincides with 6,371 km, namely its correct length.

[56] The reference is to the Proposition XXXVI of the first book and to the Corollary 9 of the proposition IV, first book.

of its motion approximately $15\frac{1}{12}$ Parisian feet.[57] But, through experiments with pendulums beating the second (the instrument had been devised by Huygens), it was possible to calculate that a body in free fall under the action of gravity, when the time of descent is a second, traverses exactly the same space as the one traveled by the Moon descending with the same centripetal force F. Thence, according to the first and second of the *regulae philosophandi*, the force causing the Moon's orbit around the Earth and terrestrial gravity turns out to be the same force.

In Proposition V Newton stated:

> That the circumjovial Planets gravitate towards Jupiter; the circumsaturnal towards Saturn; the circumsolar towards the Sun; and by the forces of their gravity are drawn off from the rectilinear motions, and retained in curvilinear orbits (Newton, 1729 II, p. 219)

This deduction is obtained by means of Rule II: since the empirical data show that the orbits of Jupiter's and Saturn's satellites around their planets as well as the orbits of the planets around the Sun follow the same law as the Moon's orbit around the Earth and since the orbit of the Moon is due to the Earth's gravity, then, according to Rule II, Jupiter's and Saturn's gravity are responsible for the rotation of their satellites and the Sun's gravity determines the rotation of the planets around the Sun.

In Proposition VI, Newton proved that the intensity of the gravitational attractions also depends on the masses of the attracting bodies.

Therefore, in the set of propositions IV–VI of the *Principia*'s third book the law of universal gravitation is demonstrated. This is the most important result of the magnificent Newtonian synthesis. This synthesis offers a unitary picture of nature and science: insofar as the actions among distant bodies are concerned, gravity is the dominant force. It determines the structure of Newton's universe. Newton provided the dynamical support to Kepler's kinematical achievements. Hence, unlike Kepler, Newton deduces kinematics from dynamics through mathematics. Inertia and gravity force are the unifying physical magnitudes that determine the kinematic of the universe. Kepler's attempt to find a system of physical causality was fulfilled by Newton, who relied upon different presuppositions. The first and most important of these presuppositions is the full understanding of the concepts of mass and inertia. Together with his extraordinary mathematical ability and physical intuition, these are the concepts that distinguish Newton from the other scientists of his time. The relation between inertia and gravity is encapsulated as a final remark in Rule III (see above, p. 213, n. 15):

> Not that I affirm gravity to be essential to bodies: by their *vis insita* I mean nothing but their *vis inertiae*. This is immutable. Their gravity is diminished as they recede from the Earth. (ibid., pp. 204–205)

In conclusion: the very core of Newton's *philosophia naturalis* is represented by his mathematical presuppositions and deductions, joined to experimental philosophy. His epistemological convictions, explained in the *regulae*, are also significant because they exhibit the new procedures of Newton's *philosophia naturalis* and justify the

---

[57] Actually it is a little bit more, as Newton clarified, but this detail is not relevant to explain the basis of Newton's reasoning.

extension of the law of gravitation to Jupiter's and Saturn's satellites and to the planets. Since it was impossible for Newton to develop experiments on the surface of Jupiter, Saturn and the Sun, he could not have deduced the law of universal gravitation without referring to the *regulae*.

7.2 Newton's Idea of the Universe

What new view of the physical universe and what new concept of nature arose from Newton's *Principia*? To answer these questions we analyse the most famous philosophical text in Newton's writings, the *General Scholium*, which was added as a conclusion to the second edition of the *Principia* in 1713. This text can be subdivided in three parts: in the first the Cartesian theory of vortices, and, therefore, Descartes' cosmology are criticized; in the second Newton's theological views are set out; in the third and last part Newton offers some remarks on gravity. While the first part is a direct consequence of Newton's scientific theory, the theological considerations of the second part are logically neither entailed nor presupposed by his scientific acquisitions; rather, they represent Newton's philosophical interpretation of his own science, and testify to his multifaceted intellectual personality, which is attested also by the many manuscripts he left unpublished on subjects like alchemical investigations, theology, history of religions, and the interpretation of Biblical prophecies. A distinction must be drawn between the actual results of Newtonian science (and the concepts implicated by them) and the philosophical and religious ideas which Newton associated to his scientific research. The validity of this distinction is confirmed by the developments of Newtonian mechanics in the following centuries; already at the end of the eighteenth century, the French astronomer and mathematician Pierre-Simon Laplace in his *Exposition du système du monde* (1796) developed Newton's celestial mechanics leaving aside as unnecessary any theological assumption or implication. Unfortunately, this distinction has not always been preserved in a number of studies on Newton during the last fifty years (see Lotti, 2006).

The first part of the *General Scholium* summarizes the whole array of arguments which Newton had displayed in his work to show that the Cartesian "hypothesis of the vortices" could not explain celestial phenomena (Newton, 1729 II, pp. 387–88). Then, shifting from scientific criticism to a theological conclusion, Newton stated that "this most beautiful System of the Sun, Planets, and Comets, could only proceed from the counsel and dominion of an intelligent and powerful being" (ibid). Newton believed that talking about God starting from phenomena was the proper task of natural philosophy, and he offered his contribution to the argument from design that was so popular in the British tradition of natural philosophy: "Blind metaphysical necessity, which is certainly the same always and every where, could produce no variety of things. All that diversity of natural things which we find, suited to different times and places, could arise from nothing but the ideas and will of a Being necessarily existing" (Newton, 1729, p. 391).

His theology emphasizes God's dominion over the whole universe (our solar system and all the similar planetary systems around fixed stars):

> This Being governs all things, not as the soul of the world, but as Lord over all: And on account of his dominion he is wont to be called *Lord God pantokràtor* or *Universal Ruler*. For *God* is a relative word, and has a respect to servants; and *Deity* is the dominion of God, not over his own body, as those imagine who fancy God to be the soul of the world, but over servants. The supreme God is a Being eternal, infinite, absolutely perfect; but a being, however perfect, without dominion, cannot be said to be Lord God.[58]

Newton's emphatical insistence on the attribute of God's omnipotence and on his active dominion over the world is a typical feature of Puritan theology that he opposed to the pagan reduction of God to an immanent World Soul. However, Newton is here treading on a thin line. Since in his own view time and space are God's attributes, Deity is present everywhere and in every moment:

> He is omnipresent, not *virtually* only, but also *substantially*; for virtue cannot subsist without substance. In him are all things contained and moved; yet neither affects the other: God suffers nothing from the motion of bodies; bodies find no resistance from the omnipresence of God.[59]

How God can exert his all-pervading dominion over the universe preserving his transcendence is a traditional theological difficulty on which Newton did not expand, preferring to invoke the human lack of knowledge of God's substance and of God's modes of existence and action.

If we examine the *Scholium* according to the standard of seventeenth century metaphysics, one cannot avoid the impression of being faced with an amateurish mess. The relationship of God and the universe is described as the personal dominion of the supreme master on his servants, because "the supreme God [...] however perfect, without dominion, cannot be said to be Lord". But then, Newton insists on the substantial omnipresence of God which he conceives as if the universe is contained in the substance of God himself. Finally, realizing the difficulty of clarifying rationally the relationship between God and the universe, Newton takes refuge in the ignorance of God's nature and his action: the manner of his existence "is utterly unknown to us"; equally, "we have no idea of the manner by which the all-wise God perceives and understands all things". Newton claims that God is incorporeal, and that we have knowledge of his attributes, but immediately adds, echoing Lockean agnosticism, that "what the real substance of any thing is we know not". We know

---

[58] Newton (1729, II, p. 389. Italics in the text). Cf. Newton (1726, p. 528): "Hic omnia regit non ut anima mundi, sed ut universorum dominus. Et propter dominium suum, dominus deus Παντοκράτωρ dici solet. Nam deus est vox relativa et ad servos refertur: et deitas est dominatio dei, non in corpus proprium, uti sentiunt quibus deus est anima mundi, sed in servos. Deus summus est ens aeternum, infinitum, absolute perfectum: sed ens utcunque perfectum sine dominio non est dominus deus".

[59] Newton (1729, II, p. 390). Cf. Newton (1726, p. 528–529): "Omnipraesens est non per *virtutem* solam, sed etiam per *substantiam:* nam virtus sine substantia subsistere non potest. In ipso continentur et moventur universa, sed sine mutua passione. Deus nihil patitur ex corporum motibus: illa nullam sentiunt resistentiam ex omnipraesentia dei". Italics in the text.

only how bodies appear to our sense, but "their inward substances are not to be known either by our senses, or by any reflex act of our minds". If this is the limit of our knowledge with regard to natural bodies, it's not suprising that we don't "have any idea of the substance of God". All our discourse about God is allegorical, "for all our notions of God are taken from the ways of mankind by a certain similitude". It sounds as if Newton's theology, for his own admission, is shrouded in an insurpassable ignorance. And yet he confidently concludes that our discourse about God starting from phenomena certainly belongs to "natural philosophy". The *General Scholium* cumulates different views, with no effort to reflect on their consistency. It is an interesting document of Newton's theological ideas, but as a metaphysical text is nothing more than a conceptual quagmire. It deserved D'Holbach's scathing criticism, who defined it an "unintelligible rigmarole" and exclaimed with good reason:

> But, oh man! so great and so powerful, when you were a geometrician; so little and so weak, when you became a theologian; that is to say, when you reasoned upon that which can neither be calculated nor submitted to experience; how could you think of speaking to us of a being who is, by your own confession, to you just what a picture is to a blind man?[60]

The *General Scholium* ends with some remarks on gravity. Newton summarizes the properties of gravity, but admits that he had not been able to discover its cause and that he does not intend to feign hypothesis about it (an implicit critique of the Cartesian conjectural physics); ungrounded hypotheses, which are not deduced from phenomena, should be excluded from experimental philosophy, whatever their nature, "whether metaphysical or physical, whether of occult qualities or mechanical". To search for the cause of gravity remained his main concern, and he envisaged a number of investigations to that effect, concerning chemistry, biological processes, electrical and magnetic phenomena (see, on this topic, not only the end of the *General Scholium*, but also numerous Queries of *Opticks* and some of his alchemical manuscripts). Newton was extremely cautious on making statements on the essence of gravity and limited himself to asserting its properties which he had elucidated in the *Principia.* He claimed that he had discovered by induction from phenomena the laws of motion and of gravitation: "And to us it is enough, that gravity does really exist, and act according to the laws which we have explained, and abundantly serves to account for all the motions of the celestial bodies, and of our sea".[61]

To answer our opening question we may say that, as a result of Newtonian science, for the first time the mechanist approach advocated by Descartes was regarded as inadequate to explain physical phenomena. To be more precise: Newton proved that the Cartesian vortices cannot be stable and, therefore, cannot make up the basic structures of the universe.[62] In particular, the Cartesian theory could not explain

---

60 D'Holbach (1835, p. 229a).

61 Newton (1729, II, p. 392). Cf. Newton (1726, p. 530): "Et satis est quod gravitas revera existat, et agat secundum leges a nobis expositas, et ad corporum caelestium et maris nostri motus omnes sufficiat". For a comment on *General Scholium* and related bibliography see F. Giudice, in Newton (2018, pp. 186–194).

62 Though Newton's proof of the instability of vortices is not completely convincing (see, i.e., Aiton 1972), it is sufficient to falsify the vortex theory as conceived by Descartes.

the motions of the bodies of the solar system. As a consequence, he concluded, philosophically and theologically, that the order of the solar system could not have evolved from a pristine chaos (as Descartes theorized) but had been established from the beginning by the will and the intelligence of God.

This conclusion, however, must be qualified. Newton's criticism of the Cartesian theory of vortices does not imply that he replaced the mechanist physics, based on the transmission of motion by contact, with a physics which admitted the action at a distance through gravitational attraction. It is true that Newtonian physics conceived gravitation as a force which acts instantaneously among bodies, with no apparent mediator, and that, therefore, it was interpreted as implying the notion of action at a distance. However, Newton himself was aware that this was a problematic notion which remained open to investigation in so far as the nature of gravitational force remained unknown. A passage from his letter to Richard Bentley, of 25 February1693, shows his view on this point:

> It is inconceivable, that inanimate brute Matter should, without the Mediation of something else, which is not material, operate upon, and affect other Matter without mutual Contact, as it must be, if Gravitation in the Sense of Epicurus, be essential and inherent in it. And this is one Reason why I desired you would not ascribe innate Gravity to me. That Gravity should be innate, inherent and essential to Matter, so that one Body may act upon another at a Distance through a Vacuum, without the Mediation of anything else, by and through which their Action and Force may be conveyed from one to another, is to me so great an Absurdity, that I believe no Man who has in philosophical Matters a competent Faculty of thinking, can ever fall into it. Gravity must be caused by an Agent acting constantly according to certain Laws; but whether this Agent be material or immaterial, I have left to the Consideration of my Readers. (Newton, 1756, pp. 25–26)

Newton rejected the ancient atomists' view that gravity is inherent to matter because that view would imply the notion that matter is 'self-activating', while he firmly believed that matter is passive and requires external causation. Because of his religious convictions he excluded that gravity could be regarded as a force immanent in matter. It is a subject of discussion among Newtonian scholars if Newton entertained the opinion that the true and direct cause of gravity was God and that gravity was the immaterial instrument by which God exerted his dominion over the cosmos (see Ducheyne, 2012, p. 41 ff.). This can be an interesting topic if we desire to reconstruct historically Newton's ideas, but we should not forget that he never stated this view in his public writings and that the theological foundation of gravity is obviously not required by the Newtonian dynamics.

Newton's view of absolute space and time is another topic in which theology and science are connected by Newton's philosophical interpretation, even if, as outlined in the previous section, no necessary link to theology is required by the conceptual structure of his mechanics, within which the concepts of absolute space and time play a significant role (see Section 7.1.2.). In the *General Scholium* we find how these fundamental presuppositions of Newton's physics are linked to his theological ideas. Here Newton conceives absolute space and time as the divine attributes of omnipresence and eternity:

> [God] is not Eternity and Infinity, but Eternal and Infinite; he is not Duration and Space, but he endures and is present. He endures for ever, and is every where present; and by existing always and every where, he constitutes Duration and Space. Since every particle of Space is *always*, and every indivisible moment of Duration is *every where*, certainly the Maker and Lord of all things cannot be *never* and *no where*.[63]

Again we face the theological difficulty of preserving God's transcendence while God's attributes are interpreted as the framework of physical universe. With respect to the issue of gravity, the difference here is that Newton never asserted the divine nature of gravity while absolute space and time are explicitly defined as divine attributes. However, even in this case it is obvious that the physical role played by the absolute spatio-temporal frame does not require logically any theological foundation: one could easily assign this same frame to a universe conceived in purely naturalistic or materialistic terms. Whatever Newton believed about the possible relation of the essence of space to the essence of God,[64] there is no doubt that his theory of absolute space could be maintained in its physical meaning independently of its metaphysical interpretation.

Newton brought to completion the mathematization of physics that characterized the scientific revolution since its beginnings. In his work phenomena are described and derived from mechanical principles by means of mathematical propositions. Newton's rational mechanics is a synthesis of geometry and physics that surpassed by far the mathematization of nature and of its laws developed by Copernicus, Kepler and Galileo, as well as Descartes' identity of geometry and physics that was mostly verbal and had not produced significant results. Newton's approach is much more sagacious and effective as he combined mathematics with an experimental philosophy based on inductive method (see his Four Rules of Reasoning in Philosophy at the start of *Principia* Book 3). In his methodology and scientific practice a productive conjunction of geometry and experience obtained. This was the true core of Newton's scientific achievements and the key to his success, what Cohen characterized as the "Newtonian style".[65] In early modern natural philosophy the mathematization of physics was linked to theological assertions, in the persuasion that God had created the world "mensura numero et pondere" as it is written in the *Book of Wisdom* 11, 21.[66] This is historically significant, but is logically irrelevant in the sense that the Newtonian science stands intact once you remove this theological frame. The refusal to disentangle the epistemic results of natural science from their historical context entails a sceptical relativism that is associated with the most trivial form of historicism. History of science itself shows unequivocally how the contents of Newtonian

---

[63] Newton (1729, II, p. 390). Cf. Newton (1726, p. 528): "Non est aeternitas et infinitas, sed aeternus et infinitus; non est duratio et spatium, sed durat et adest. Durat semper, et adest ubique, et existendo semper et ubique durationem et spatium constituit. Cum unaquaeque spatii particula sit *semper*, et unumquodque durationis indivisibile momentum *ubique*, certe rerum omnium fabricator ac dominus non erit *nunquam, nusquam*". (Italics in the text).

[64] On this issue see Stein (1970, p. 271f), who argued for the thesis that for Newton space was "a *structure*, or "relational system," which can be conceived of independently of anything else".

[65] Cohen, (1980, p. 15f).

[66] *Biblia Sacra* (1994, p. 106).

science persisted beyond the theological frame of ideas that accompanied them in Newton's mind and in Newton's time.

Newton's world is a dynamical one in which the key role is played by the concept of force raised to an unprecedented scientific status by mathematical determination. As he wrote in his Preface to the first edition of the *Principia*, the content of his natural philosophy was the study of "natural powers":

> [...] we consider chiefly those things which relate to gravity, levity, elastic force, the resistance of fluids, and the like forces, whether attractive or impulsive; and therefore we offer this work as mathematical principles of philosophy. For all the difficulty of philosophy seems to consist in this, from the phenomena of motions to investigate the forces of nature, and then from these forces to demonstrate the other phenomena.[67]

Newton's achievement is even more impressive if we think that in his investigation of natural forces he was capable even of envisaging the future developments of physics. We read again in his Preface these far-sighted words:

> For I am induced by many reasons to suspect that they may all depend upon certain forces by which the particles of bodies, by some causes hitherto unknown, are either mutually impelled towards each other, and cohere in regular figures, or are repelled and recede from each other; which forces being unknown, Philosophers have hitherto attempted the search of Nature in vain. But I hope the principles here laid down will afford some light either to that, or some truer, method of Philosophy.[68]

Newton is aware that a deeper understanding of nature required shifting the focus of investigation from the study of the motion of macroscopic bodies to the microphysics of particles in order to discover the forces binding the elementary components of matter. He reflects on the difference between gravitation, that is only attractive, and magnetism and other phenomena where attraction is not the only disposition of matter and repulsion must also be taken into account. Forty years later, Immanuel Kant in his *Universal Natural History and Theory of the Heavens* (1755) will include repulsion in his cosmological conception, explaining with the interplay of attraction and repulsion the formation of the solar system and of the universe in general.

Newton's view of the universe was shaped by his scientific and theological ideas that he tried to hold together, though he never allowed speculation to alter his scientific theories. One of the most significant examples of the intertwining of these two aspects in Newton's cosmology concerns the issue of the world's decay. One of the arguments advanced by Newton against Cartesian evolutionary cosmogenesis was that the solar

[67] Newton (1729, I, p. ix f). Cf. Newton (1726), *Praefatio ad Lectorem*, p. xiv: "[...] maxime tractamus, quae ad gravitatem, levitatem, vim elasticam, resistentiam fluidorum et ejusmodi vires seu attractivas seu impulsivas spectant: Et ea propter, hac nostra tanquam philosophiae principia mathematica proponimus. Omnis enim philosophiae difficultas in eo versari videtur, ut a phaenomenis motuum investigemus vires naturae, deinde ab his viribus demonstremus phanomena reliqua".

[68] Newton (1729, I, p. x). Cf. Newton (1726), *Praefatio ad Lectorem*, p. xiv "Nam multa me movent, ut nonnihil suspicer ea omnia ex viribus quibusdam pendere posse, quibus corporum particulae per causas nondum cognitas vel in se mutuo impelluntur et secundum figuras regulares cohaerent, vel ab invicem fugantur et recedunt: quibus viribus ignotis, philosophi hactenus naturam frustra tentarunt. Spero autem quod vel huic philosophandi modo, vel veriori alicui, principia hic posita lucem aliquam praebebunt".

system appears to be so well contrived that it shows the counsel and design of the divine Mind. God created the "hard and solid particles" of matter and set them in order: "And if he did so, it's unphilosophical to seek for any other Origin of the World, or to pretend that it might arise out of a Chaos by the mere Laws of Nature".[69] This is a purely philosophical interpretation of astronomical data concerning the orbits of the comets and of the planets: Newton is convinced that "blind Fate" could never create a system in which the excentric orbits of the comets, which move "in all manner of positions", would combine harmoniously with the orbits of the planets, which are concentric and are set, more or less, on the same plane. This argument might have been directed against classical atomism, but it was not pertinent to Descartes, since in the Cartesian view the Laws of Nature are set by God and the evolution of matter from chaos to cosmos is not triggered by "blind Fate". Kant was to side with Descartes against Newton on this point, asserting that "there is a God precisely because nature can proceed even in chaos in no other way than regularly and orderly".[70] More significant than the different philosophical interpretations of how the ordered system of the world came to be formed (through evolution for Descartes and Kant, and by direct creation for Newton), is the fact that Newton himself admitted that the system of the world was subject to irregularities: even if its order had to be explained as resulting from God's direct creation without evolution from chaos to cosmos, it was an imperfect order which needed correction from time to time. Newton wrote that the irregularities were the effect of "the mutual Actions of Comets and Planets upon one another, and which will be apt to increase, till this System wants a Reformation".[71] The idea that gravitational systems are unstable is coupled with another opinion that Newton expressed in the same Query: "Motion is much more apt to be lost than got, and is always upon the Decay".[72] Newton does not accept the Cartesian idea of the conservation of the quantity of motion on a universal scale and this denial depends on Newton's idea that motions are caused by forces acting in natural phenomena:

> Seeing therefore the variety of Motion which we find in the World is always decreasing, there is a necessity of conserving and recruiting it by active Principles, such as are the cause of Gravity, by which Planets and Comets keep their Motions in their Orbs, and Bodies acquire great Motion in falling; and the cause of Fermentation, by which the Heart and Blood of Animals are kept in perpetual Motion and Heat; the inward Parts of the Earth are constantly warm'd, and in some places grow very hot; Bodies burn and shine, Mountains take Fire, the Caverns of the Earth are blown up, and the Sun continues violently hot and lucid, and warms all things by his Light. For we meet with very little Motion in the World, besides what is owing to these active Principles.[73]

These are the typical philosophical reflections that Newton made public in the Queries of the *Optics*: they outline new lines of investigation and show how he

[69] Newton (1721, p. 378). The quotation is from Query 31, that had been published as Query 23 in the 1706 Latin edition of the *Optícks*: cfr. Newton (1706, p. 345).

[70] Kant (1981, p. 86).

[71] Newton (1721, p. 378).

[72] Ibid., p. 373.

[73] Ibid., p. 375.

pursued an explanatory theory for the nature of forces and of the origin of motion. Newton excluded, as we have seen, that forces might be conceived as intrinsic to the material structure of bodies. He adopted the traditional metaphysical view that matter is inert and that forces are exerted in nature by God either directly or indirectly. This granted the providential action of God, that Newton embraced fearing the danger of atheistic naturalism.[74] The notion of God's active intervention in the universe exposed Newton to the criticism of Leibniz, who was much more skilled than him in philosophical debates and more attentive to the rational coherence of his metaphysical conceptions. Leibniz did not hesitate to ridicule Newton's views on the issue of the relationship between God and the universe:

> Sir Isaac Newton and his followers also have a very odd opinion concerning the work of God. According to their doctrine, God Almighty needs to wind up his watch from time to time, otherwise it would cease to move. He did not, it seems, have sufficient foresight to make it a perpetual motion. No, the machine of God's making is so imperfect, according to these gentlemen, that he is obliged to clean it now and then by an extraordinary concourse, and even to mend it, as a clockmaker mends his work; he must consequently be so much the more unskillful a workman as he is more often obliged to mend his work and to set it right. According to my opinion, the same force and vigor always remains in the world and only passes from one part of matter to another in agreement with the laws of nature and the beautiful pre-established order.[75]

Clarke, who defended Newton's ideas, replied with an apologetic counterattack, pointing out the dangerous materialism lurking behind the idea of a perfect natural mechanism that does not need God's providential intervention:

> The notion of the world's being a great machine, going on without the interposition of God as a clock continues to go without the assistance of a clockmaker, is the notion of materialism and fate, and tends (under pretence of making God a *supramundane intelligence)* to exclude providence and God's government in reality out of the world. And by the same reason that a philosopher can represent all things going on from the beginning of the creation without any government or interposition of providence, a skeptic will easily argue still farther backwards and suppose that things have from eternity gone on (as they now do) without any true creation or original author at all, but only what such arguers call all-wise and eternal nature.[76]

This exchange of accusations shows the apologetic attitude of both parties, each of which points to the weakest spot of the other in a theological competition for orthodoxy. This was a common posture in metaphysical debates of the age. What is more significant is how the two philosophical positions are related to modern science. Clearly, Leibniz's rationalism is more in tune with modern science than the Biblical providentialism to which Newton remained faithful. Leibniz states here a sort of principle of conservation, while Newton refrains from it as he wants to underline the continuous dependence of nature on God. Newton's metaphysics falls short of his own science and is less in line with modern scientific attitudes than was Leibniz's metaphysical rationalism.

---

[74] See Kubrin (1967, pp. 325–330).

[75] Leibniz and Clarke (2000, p. 4).

[76] Ibid., p. 6.

References

Aiton, E. J. (1972). *The vortex theory of planetary motions*. New York: American Elsevier Publishing Company.

Arthur, R. (1994). Space and relativity in Newton and Leibniz. *The British Journal for the Philosophy of Science, 45*(1), 219–240.

Assis, A.K.T. (2003). The relationship between Mach's principle and the principle of physical proportions. In M. Sachs & A.R. Roy (Eds.), *Mach's principle and the origin of inertia* (pp. 37–44). Apeiron

Assis, A. K. T. (1989). On Mach's principle. *Foundations of Physics Letters, 2*, 301–318.

Assis, A. K. T. (1999). Arguments in favour of action at a distance. In A. E. Chubykalo, V. Pope, & R. Smirnov-Rueda (Eds.), *Instantaneous action at a distance—Pro and contra* (pp. 45–56). Commack: Nova Science Publishers.

Assis, A. K. T., & Graneau, P. (1995). The reality of Newtonian forces of inertia. *Hadronic Journal, 18*, 271–289.

Barbour, J. B., & Pfister, H. (eds.) (1995). *Mach's principle from Newton's Bucket to quantum gravity* (pp. 159–171). Boston-Basel-Berlin: Birkhäuser.

Barbour, J. B. (1975). Forceless Machian dynamics. *Nuovo Cimento, 26B*, 16–22.

Barbour, J. B. (1982). Relational concepts of space and time. *British Journal for the Philosophy of Science, 33*, 251–274.

Barbour, J. B. (2001). *The discovery of mechanics*. Oxford: Oxford University Press.

Barbour, J. B., & Bertotti, B. (1982). Mach's principle and the structure of dynamical theories. *Proceedings of the Royal Society of London, A, 382*, 295–306.

Belkind, O. (2007). Newton's conceptual argument for absolute space. *International Studies in the Philosophy of Science, 21*(3), 271–293.

Belkind, O. (2012). *Physical systems. Conceptual pathways between flat space-time and matter*. Dordrecht-Heidelberg-London-New York: Springer.

Berman, M. S. (2008). On the Machian origin of inertia. *Astrophysics and Space Science, 318*, 269–272.

Bertoloni Meli, D. (1993). *Equivalence and priority: Newton versus Leibniz*. Oxford: The Clarendon Press.

Biblia Sacra iuxta vulgatam versionem. (1994). Rec. R. Weber, editionem quartam emendatam praep. R. Gryson. Stuttgart; Deutsche Bibelgesellschaft.

Biener Z. (2018). Newton's *Regulae Philosophandi*. In C. Smeenk, E. Schliesser (Eds.), *Oxford Handbook for Isaac Newton's*. Oxford: Oxford University Press.

Boudri, J. C. (2002). *What was mechanical about mechanics: The concept of force between metaphysics and mechanics from Newton to Lagrange*. Dordrecht-Boston-London: Kluwer Academic Publishers.

Bussotti, P. (2017c). Il ruolo dello spazio assoluto nella fisica di Newton. In A. Neri (Ed.), *Là dove il pensiero incontra l'esperienza* (Vol. I, pp. 111–162). ETS.

Butts, R. E. (1993). Whewell on Newton's rules of philosophizing. *Historical Pragmatics: Philosophical Essays, 155*, 293–312.

Cassini, A. (2005). Newton and Leibniz on Non-substantival space. *THEORIA. An International Journal for Theory, History and Foundations of Science, 20*(1), 25–43.

Chandrasekhar, S. (1995). *Newton's Principia for the common reader*. Oxford: Clarendon Press.

Cohen, B. I. (1969). Hypotheses in Newton's Philosophy. In I. Bernard Cohen-Marx W. Wartofsky (eds.), *Boston Studies in the Philosophy of Science. Proceedings of the Boston Colloquium for the Philosophy of Science 1966/1968*, 5: 304–326.

Cohen, B. I. (1980). *The Newtonian revolution. With illustrations of the transformations of scientific ideas*. Cambridge: Cambridge University Press.

Cook, R. J. (1976). Is gravitation a results of Mach's inertial interaction? *Il Nuovo Cimento B, 35*(1), 25–34.

Davidson, W. (1957). General relativity and Mach's principle. *Monthly Notices of the Royal Astronomical Society, 117*(2), 212–224.
De Gandt, F. (1995). *Force and geometry in Newton's* Principia. Princeton: Princeton University Press.
Di Fate, V. J. (2011). Is Newton a 'radical Empiricist' about method? *Studies in History and Philosophy of Science*, Part A4 2(1), 28–36.
DiSalle, R. (1991). Conventionalism and the origins of the inertial frame concept. *Chicago Journal, 2*, 139–147.
DiSalle, R. (1994). On dynamics, indiscernibility, and spacetime ontology. *British Journal for the Philosophy of Science, 45*, 265–287.
DiSalle, R. (1995). Spacetime theory as physical geometry. *Erkenntnis, 42*(1995), 317–337.
DiSalle, R. (2002). Newton's philosophical analysis of space and time. In I. B. Cohen & G. E. Smith (Eds.), *The Cambridge companion to Newton* (pp. 33–56). Cambridge: Cambridge University Press.
DiSalle, R. (2006). *Understanding Space-Time: The Philosophical Development of Physics from Newton to Einstein.* Cambridge: Cambridge University Press.
DiSalle, R. (2020). Absolute space and Newton's theory of relativity. *Studies in History and Philosophy of Modern Physics, 71*, 232–244.
Ducheyne, S. (2012). *The main business of natural philosophy. Isaac Newton's natural-philosophical methodology.* Dordrecht: Springer.
Einstein, A., & Infeld, L. (1938). *The evolution of physics. From early concepts to relativity and quanta.* Cambridge: Cambridge University Press.
Ellis, B. D. (1962). Newton's concept of motive force. *Journal of the History of Ideas, 23*(2), 273–278.
Earman, J. (1989). *World enough and space-time. Absolute versus relational theories of space and time.* Cambridge, Massachusetts-London: The MIT Press.
Föppl, A. (1904a). Über einen Kreiselversuch zur Messung der Umdrehungsgeschwindigkeit der Erde. *Akademie der Wissenschaften, München, Mathematisch-Wissenschaftliche Klasse, Sitzungsberichte,* Sitzung vom 6. Februar, 1904, 5–28.
Föppl, A. (1904b). Über absolute und relative Bewegung. *Akademie der Wissenschaften, München, Mathematisch-Wissenschaftliche Klasse, Sitzungsberichte*, Sitzung vom 5. November 1904, 383–395.
Föppl, A. (1894). *Einführung in die Maxwell'sche Theorie der Elektricität*. Leipzig: Teubner.
Frank, P. (1909). Gibt es eine absolute Bewegung? Vortrag, gehalten am 4. Dezember 1909. *Wissenschaftliche Beilage, 1910*, 1–10.
Frank, P., & Rothe, H. (1911). Über die Transformation der Raumzeitkoordinaten von ruhenden auf bewegte Systeme. *Annalen Der Physik, 34*, 825–855.
Freudenthal, G. (1986). *Atom and individual in the age of Newton. On the genesis of mechanistic world view.* Dordrecht: Reidel Publishing Company.
Friedlaender, B., & Friedlaender, I. (1896). *Absolute oder relative Bewegung?* Berlin: Leonhard Simion.
Friedman, M. (1983). *Foundations of space-time theories. Relativistic physics and philosophy of science.* Princeton: Princeton University Press.
Giné, J. (2009). On the origin of inertia: The modified Newtonian dynamics theory. *Chaos, Solutions and Fractals, 41*(4), 1651–1660.
Giulini, G. (2002). Das Problem Der Trägheit. *Philosophia Naturalis, 39*, 343–374.
Gould, J. A. (1962). The existence of absolute space. *The Ohio Journal of Science, 62*(2), 101–104.
Graneau, P., & Graneau, N. (2003). Machian inertia and the Isotropic universe. *General Relativity and Gravitation, 35*, 751–770.
Grünbaum, A. (1963). *Philosophical problems of space and time.* Boston: Alfred Knopf.
Guicciardini, N. (1999). *Reading the* Principia. *The debate on Newton's mathematical methods for natural philosophy from 1687 to 1736*. Cambridge: Cambridge University Press.

Heimann, P. M., & McGuire, J. E. (1971). *Newtonian forces and Lockean powers: Concepts of matter in eighteenth-century thought*. Philadelphia: University of Philadelphia Press.
Henry, J. (2011). Gravity and *De Gravitatione*: The development of Newton's ideas on action at a distance. *Studies in History and Philosophy of Science Part A, 42*(1), 11–27.
Hill, B. (2003). Newton's *De Gravitatione et Aequipondio Fluidorum* and the Lockean four-dimensionalism. *British Journal for the History of Philosophy, 11*(2), 309–321.
Hofmann, W. (1904). *Kritische Beleuchtung der beiden Grundbegriffe der Mechanik: Bewegung und Trägheit und daraus gezogene Folgerungen betreffs der Achsendrehung der Erde und des Foucault'schen Pendelversuchs*. Wien und Leipzig: M. Kuppitsch.
Hugett, N. (2008). Why the parts of absolute space are immobile. *The British Journal for the Philosophy of Science, 59*(3), 391–407.
Janiak, A. (2008). *Newton as philosopher*. New York: Cambridge University Press.
Kant, I. (1981). *Universal natural history and theory of the heavens*. Translation with introduction and notes by S. L. Jaky. Edinburgh: Scottish Academic Press.
Kochiras, H. (2012). Newton on matter and space. http://philsci-archive.pitt.edu/9176/1/Kochiras.NewtonOnMatter&SpaceInDeGrav(Excerpt_for_7th_Fellows_Conf).pdf
Kubrin, D. (1967). Newton and the cyclical cosmos: Providence and the mechanical philosophy. *Journal of the History of Ideas, 28*, 325–346.
Lange, L. (1885a). Über das Beharrungsgesetz. *Berichte über Verhandlungen der Königlich Sächsischen Gesellschaft der Wissenschaften* (pp. 333–351). Mathematisch-physikalische Klasse.
Lange, L. (1885b). Über die wissenschaftliche Fassung des Galileischen Beharrungsgesetzes. *Philosophische Studien, 2*, 266–297.
Lange, L. (1885c). Nochmals Über Das Beharrungsgesetz. *Philosophische Studien, 2*, 539–545.
Lange, L. (1886). *Die geschichtliche Entwicklung des Bewegungsbegriffs und ihr voraussichtliches Endergebnis. Ein Beitrag zur historischen Kritik der mechanischen Prinzipien*. Leipzig: W. Engelmann.
Lange, L. (1902). Das Inertialsystem vor dem Forum der Naturforschung. *Philosophische Studien, 20*, 1–71.
Lange, L. (2014). On the law of inertia. Translation (by H. Pfister) of Über das Beharrungsgesetz. *The European Physical Journal H, 39*(2), 251–262.
Laymon, R. (1978). Newton's Bucket experiment. *Journal of the History of Philosophy, 16*(4), 399–413.
Leibniz, G. W., & Clarke, S. (2000). *Correspondence*. Ed. with an introduction by R. Ariew. Cambridge: Hackett Publishing Company.
Lotti, B. (2006). Filosofia naturale e teologia nello *Scholium Generale* di Newton. In C. Giuntini & B. Lotti (a cura di), *Scienza e teologia fra Seicento e Ottocento. Studi in memoria di Maurizio Mamiani* (pp. 57–80). Leo S. Olschki.
Lucas, J. R. (1984). *Space, time, and causality*. Oxford: Oxford University Press.
Mach, E. (1919). *The science of mechanics: A critical and historical account of its development*. Chicago-London: The open court publishing Co. English translation of the fourth edition of Mach's *Die Mechanik in ihrer Entwickelung historisch-kritisch dargestellt* (1st edn., 1883).
Mamiani, M. (2001). To twist the meaning: Newton's *regulae philosophandi* revisited. In J. Z. Buchwald & I. B. Cohen (eds.), *Isaac Newton's natural philosophy* (pp. 3–14). Cambridge, Mass: MIT Press.
McGuire, J. E. (1978a). Existence, actuality and necessity: Newton on space and time. *Annals of Science, 35*, 463–508.
McGuire, J. E. (1978b). Newton on place, time and God: An unpublished source. *The British Journal for the History of Science, 11*(38), 114–129.
McGuire, J. E. (1995). *Tradition and innovation: Newton's metaphysics of nature*. Boston: Kluwer Academic Publishers.
McLaughlin, T. (2008). Nature and inertia. *The Review of Metaphysics, 62*, 251–284.
Nerlich, G. (1976). *The shape of space*. Cambridge University Press.
Neumann, C. (1870). *Über die Principien der Galilei-Newtonschen Theorien*. Leipzig: Teubner.

Newburgh, R. (2007). Inertial forces, absolute space, and Mach's principle: The genesis of relativity. *American Journal of Physics, 75*, 427–430.
Newton, I. (1666). *De Gravitatione et Aequipondio Fluidorum*. http://www.newtonproject.ox.ac.uk/catalogue/record/THEM00093
Newton, I. (1706). *Optice: Sive de Reflexionibus, Refractionibus, Inflectionibus et Coloribus Lucis Libri Tres*. Latine reddidit Samuel Clarke. Londini: Impensis Sam. Smith & Benj. Walford.
Newton, I. (1721). *Opticks: Or, a treatise of the reflections, refractions, inflections and colours of light,* The third ed. corrected. London: W. and J. Innys.
Newton, I. (1726). *Philosophiae Naturalis Principia Mathematica* (3rd edn). London: Innys.
Newton, I. (1729). *The Mathematical Principles of Natural Philosophy*. (A. Motte, Trans.). 2nd Latin edn (1713). London: B. Motte.
Newton, I. (1756). *Four letters from sir Isaac Newton to doctor Bentley. Containing some arguments in proof of a Deity*. London: Print. for R. and J. Dodsley.
Newton, I. (1962). *On the gravity and equilibrium of fluids*. In A. Rupert & H.-M. B. Hall (Eds.), *Unpublished scientific papers of Isaac Newton* (pp. 121–156). Cambridge: Cambridge University Press.
Newton, I. (1999). The Principia. *Mathematical principles of natural philosophy*. A new translation by I. B. Cohen, A. Whitman and assisted by J. Budenz. Preceded by a guide to Newton's *Principia* by I. B. Cohen. Berkeley-Los Angeles-London: University of California Press.
Newton, I. (2018). *Principi matematici della filosofia naturale*. A cura di F. Giudice. Torino: Einaudi.
Norton, J. D. (1995). Mach's principle before Einstein. In J. B. Barbour & H. Pfister (Eds.), *Mach's principle from Newton's Bucket to quantum gravity* (pp. 9–55). Boston-Basel-Berlin: Birkhäuser.
Okruhlik, K. (1989). The foundation of all philosophy: Newton's third rule. In J. R. Brown & J. Mittelstrass (Eds.), *An intimate relation: Studies in the History and Philosophy of Science* (pp. 97–114). Boston: Kluwer Academic Publishers.
Pfister, H. (2014). Ludwig Lange and the law of inertia. *The European Physical Journal H, 39*, 245–250.
Poincaré, H. (1902). *La science et l'hypothèse*. Paris: Flammarion.
Putz, V. (2019). A theory of inertia based on Mach's Principle. *Universe, 5*(188), 1–8.
Raine, D. J. (1981). Mach's principle and space-time structure. *Reports on Progress in Physics, 44*, 1151–1195.
Reichenbach, H. (1957). *The philosophy of space and time*. New York: Dover. (Original German edition 1927).
Reichenbach, H. (1960). *The theory of relativity and a priori knowledge*. Berkeley: University of California Press. (Original German edition 1920).
Reissner, H. (1914). Über die Relativität der Beschleunigungen in der Mechanik. *Physikalishe Zeitschrift, 15*, 371–375.
Reissner, H. (1915). Über eine Möglichkeit die Gravitation als unmittelbare Folge der Relativität der Trägheit abzuleiten. *Physikalishe Zeitschrift, 16*, 179–185.
Russell, B. (1903). *The principles of mathematics*. Cambridge: Cambridge University Press.
Rynasiewicz, R. (1995). By their properties, causes and effects: Newton's scholium on time, space, place and motion. *Studies in History and Philosophy of Science*. First part *26*(1), 133–153. Second part *26*(2), 295–321.
Rynasiewicz, R. (2004). Newton's views on space, time, and motion. In E. N. Zalta (Ed.), *Stanford Encyclopedia of Philosophy*. https://plato.stanford.edu/entries/newton-stm/
Sachs, M., & A.M. Roy, (Eds.). (2003) *Mach's principle and the origin of inertia*. Montreal: Apeiron.
Sciama, D. W. (1953). On the origin of inertia. *Monthly Notices of the Royal Astronomical Society, 113*, 34–42.
Sciama, D. W. (1957). Inertia. *Scientific American, 196*(2), 99–109.
Sciama, D. W. (1959). *The unity of the universe*. London: Faber and Faber.
Shapiro, A. (2004). Newton's "Experimental philosophy". *Early Science and Medicine, 9*(3), 185–217.

Sklar, L. (1974). *Space, time and spacetime.* Berkeley-Los Angeles-London: University of California Press.

Slowik, E. (2002). *Cartesian spacetime. Descartes' physics and the relational theory of space and motion.* Springer.

Smith, G.E. (2002). The methodology of the *Principia.* In I. B. Cohen & G. E. Smith (Eds.), *The Cambridge companion to Newton* (pp. 138–173). Cambridge: Cambridge University Press.

Snobelen, S. D. (2005). The true frame of nature: Isaac Newton, heresy and the reformation of natural philosophy. In J. Brooke & I. Maclean (Eds.), *Heterodoxy in early modern science and religion* (pp. 223–262). Oxford: Oxford University Press.

Sonego, S. (2020). *General relativity.* Lecture notes of the course of mathematical physics held at the Department of Mathematics, Informatics, Physics. University of Udine.

Spencer, Q. (2004). Do Newton's rules of reasoning guarantee truth... must they? *Studies in History and Philosophy of Science Part A 35*(4), 759–782.

Stein, H. (1967). Newtonian space-time. *Texas Quarterly, 10*(3), 174–200.

Stein, H. (1970). On the notion of field in Newton, Maxwell and beyond. In R.H. Stuewer (Ed.), *Historical and philosophical perspectives of science*, pp. 264–310. Minneapolis: University of Minneapolis Press.

Streintz, H. (1883). *Die physikalischen Grundlagen der Mechanik.* Leipzig: Teubner.

Torretti, R. (1999). *The philosophy of physics.* Cambridge: Cambridge University Press.

Toulmin, S. (1959). Criticism in the history of science: Newton on absolute space, time, and motion. I. *the Philosophical Review, 68*(1), 1–29.

Westfall, R. S. (1971a). *Force in Newton's physics. The science of dynamics in the seventeenth century.* London-New York: McDonald-Elsevier.

Westfall, R. S. (1971b). *The construction of modern science: Mechanism and mechanic.* New York: Wiley.

Westfall, R. S. (1981). *Never at rest: A biography of Isaac Newton.* Cambridge: Cambridge University Press.

Will, C. M. (1993). *Theory and experiment in gravitational physics* (Revised). Cambridge: Cambridge University Press.

Zylbersztajn, A. (1994). Newton's absolute space, Mach's principle and the possible reality of fictitious forces. *European Journal of Physics, 15*(1), 1–8.

Chapter 8
Leibniz: The Philosopher-Scientist

In the context of an interdisciplinary approach aimed at pointing out the interconnections between science and philosophy in the early modern centuries, no author is probably more interesting than Leibniz. He was a man of great learning, with a wide variety of intellectual interests, and was profoundly involved in mathematics, science and philosophy from his youth. His relevance in the history of mathematics and of logic is well known: in mathematics he invented the infinitesimal calculus and in logic he devised the *characteristica universalis* and the *calculus ratiocinator* with the purpose of contriving a formal system of reasoning. In philosophy he proposed 'monadology', an idealistic, panorganicist metaphysics in which he tried to reinterpret the scholastic notions of individual substance and final causes in conjunction with modern science. He revised Cartesian mechanicism introducing the idea of living force (*vis viva*) which, though probably conceived in a physical context, played a role also in his metaphysics, since monads are active individual substances of incorporeal nature. Monadology was a view of reality centred around the concepts of infinity and continuity, two key ideas in Leibniz's thought with logical, mathematical and theological implications. This ambitious program could not be fulfilled satisfactorily in a systematic way and was illustrated in a variety of texts, letters and sketches, many of them he left unpublished.

The cosmological perspective is intrinsic to Leibniz's monadology: every monad mirrors the universe in its peculiar way and differs from other monads for infinitesimal degrees in a continuum series (therefore constituting an uninterrupted ontological chain).[1] The core tenet of monadology is that the same universal reality is reflected and multiplied in the infinity of monads:

> Now this mutual connection or accommodation of all created things to each other and of each to all the rest causes each simple substance to have relations which express all the others and consequently to be a perpetual living mirror of the universe. (*Monadology* § 56: Leibniz, 1969, p. 648)

---

[1] See on this topic the classical study Lovejoy (1936, 1964).

P. Bussotti and B. Lotti, *Cosmology in the Early Modern Age: A Web of Ideas*,
Logic, Epistemology, and the Unity of Science 56,
https://doi.org/10.1007/978-3-031-12195-1_8

> Just as the same city viewed from different sides appears to be different and to be, as it were, multiplied in perspectives, so the infinite multitude of simple substances, which seem to be so many different universes, are nevertheless only the perspectives of a single universe according to the different points of view of each monad. (ibid. § 57)

This is, however, a *metaphysical* cosmology and we need to see how Leibniz's *physical* cosmology fits into this speculative picture. Unlike the other authors we examined so far, Leibniz's metaphysics influenced profoundly how he conceived science. In Sect. 8.1 we'll trace a historical-conceptual path concerning the relations of Leibniz' s science and philosophy, focusing on his most significant scientific ideas and describing, in particular, his planetary model. In Sect. 8.2, we will comment on those aspects of Leibniz's metaphysics more closely connected to his scientific theories.

8.1 Leibniz's Cosmological Ideas

In this section we will address three aspects of Leibniz's physics and metaphysics that are crucial for his cosmology: (1) Leibniz's mechanistic conception of physics and his ideas on gravity which are connected to such a mechanistic background; (2) the concept of force, through which Leibniz reintroduced the final causes in physics; (3) the structure of the solar system he described in *Tentamen de motuum coelestium causis*.

8.1.1 Mechanism and Gravity

Leibniz wrote his first scientific essay in 1671 when he was 25. The title of his booklet is *Hypothesis physica nova* (*A new physical hypothesis*). Leibniz mathematical knowledge was, at that time, very poor, but he knew Cartesian physics and philosophy very well and he was brimming with ideas. In the *Hypothesis physica nova* Leibniz presented a mechanistic conception of cosmology and cosmogony.[2] He based it on four presuppositions: (1) Vortices of ethereal matter exist. The ether of a vortex permeates all the objects that lie within the vortex itself. Some objects contain a high percentage of ether, whereas others a lower percentage; (2) all celestial bodies rotate around their axes; (3) All bodies are elastic. Their elasticity depends on the ether they contain, for the ether moves within the cavities of the bodies behaving as a mobile elastic pillow; (4) each body is composed of discrete entities that Leibniz

[2] On the *Hypothesis physica nova* see: Bassler (2002), Beeley (1999), Bernstein (1980), Bussotti (2015, pp. 73–81), Copek (1966), Duchesneau (1989, 1994a), Iltis (1973), McDonough (2007), revised 2019; Wilson (1990), Woolhouse (2000).

called *bullae* (bubbles). Bubbles are not atoms because they are infinitely divisible.[3] They are made up of particles, which, in their turn, are divisible. The bubbles can mutually join to form other bubbles. The universe, considered as a whole, is composed both of particles and bubbles. Skies are made up of particles, not of bubbles, while bodies are composed of bubbles, not of single particles, in the sense that you cannot find any isolated particle in a body. If all the particles existing at the beginning of the universe would remain at rest, no solid body could exist. On the Earth, as the result of specific mechanisms,[4] the particles of ether and of a supposed primitive terrestrial matter join to form new unities, the bubbles. Solid bodies are made up of bubbles in which the presence of Earth's particles is preponderant, whereas elements like water are made up of bubbles in which the ethereal particles are preponderant. The bubbles are set one next to the other without empty spaces among them (Leibniz, like Descartes, denies the existence of void). A bubble has to be imagined as a sort of small object (not necessarily of spherical form) with a shell, which separates the bubble from the rest of the world. Within the shell there are particles. Leibniz refused Descartes' distinction of three kinds of particles. In his view, each particle can differ from another particle as regard to its form, density, magnitude, etc., but the particles are not subject to so specific differences as to be divided into three kinds. In this early work, Leibniz addressed a plurality of topics: the origin of the solar system, the nature of bubbles, the composition of the Earth, gravity, reflexion and refraction (both addressed as collision-problems), specific weight, atmospheric pressure, electricity and magnetism. One of the main issues of Leibniz's entire scientific production is the question of gravity, which is possibly the most important issue for the physics and cosmology of the time.

Leibniz referred first to the ideas of Hobbes and Torricelli, and implicitly of Descartes: the Sun rotates and, in doing so, produces the rotation of its vortex. This rotation attracts the Earth towards the Sun since the lightest particles of the vortex tend to go to the external layers of the vortex and, therefore, they push the Earth inwards. A consequence of this same mechanism would be that the Earth should attract stones and other bodies, as Descartes had theorized (in fact, the whole argument is erroneous because in a rotating vortex, the denser bodies move towards the external layers, not towards the internal ones).[5] Leibniz accepted this idea only with regard to the attraction exerted by the Sun on the Earth, but denied that gravity on the Earth might be explained like this. He actually refused this sort of "vorticistic universal gravitation": he was convinced that the Cartesian vortex theory, with the improvements and specifications added by Hobbes and Torricelli, might explain the curved motions of the planets around the Sun, but he also believed that this theory

[3] Leibniz mentioned Anaxagoras' homeomeries as objects that share this property with his bubbles, though for many other significant aspects he considered homeomeries and bubbles very different objects (Leibniz, 1671, §§ 44–45, pp. 41–42).

[4] These mechanisms depend on the motions of the ether's particles in proximity of the Earth and on the difference of temperature in different points of the atmosphere and of the Earth's surface: see Leibniz (1671, §§ 10–11, p. 23).

[5] For an explanation of this phenomenon, see Bussotti and Lotti (2018, p. 110), in particular note 37.

could not explain, without any further hypothesis, the rectilinear motions. More particularly, a theory of gravity should explain the rectilinear motions with which bodies fall on the Earth. In fact, the explanation of gravity offered by Descartes in his *Principia* is highly problematic (see Bussotti & Lotti, 2018, pp. 88–89). Therefore, Leibniz introduced a further conjectural phenomenon: the Sun emits particles whose motion is not uniform either in direction or in speed. Most of these particles return on the Sun surface after having described a certain path; if it were not so, the Sun would be quickly exhausted. However, some of these particles exit from the Sun's range of action. Thus, the ether of the planets is perturbed by the particles emitted by the Sun. According to Leibniz, at the beginning of universe history, the terrestrial globe was made up of homogeneous matter. Since the particles emitted by the Sun penetrate the Earth with different angles and speeds, the particles of the terrestrial ether, hit by the solar particles, in their turn, penetrate the terrestrial globe with different angles and speeds. This action, joined to a rotation caused by the "wind" of the solar particles, contributed to create bubbles made up of ether and primordial Earth's matter. The bubbles compose the objects; the bubbles in which the percentage of ether is low are hard, dense and heavy (i.e. stones are composed of these bubbles), whereas the bubbles made up of a high percentage of ether are light (water and, even more, air). It follows that gravity depends on the composition of a body, that is on the nature of its bubbles. It has nothing to do with the attraction exerted by a body on another body. To use modern terms, it seems that, in this early work, Leibniz is ascribing gravity to the different densities of bodies.

In the course of the years Leibniz developed a series of ideas concerning gravity. He rethought the bubbles-theory and eventually discarded it, but his explanations of gravity remained within a mechanistic framework.

Leibniz's early cosmology deserves some comments:

(1) In this early work, Leibniz did not claim to have offered absolutely precise explanations of the phenomena he dealt with. It is not by chance that the first term of the title is *Hypothesis*. Nonetheless, he was convinced that such explanations might be achieved only within the frame he had outlined.

(2) The influence of Descartes' *Principia* is conspicuous. Just like Descartes, Leibniz tried to explain all physical phenomena by a modified vortex theory and adopted a mechanistic point of view. Leibniz was also well aware of the development of vortex theory after Descartes. He actually referred to the ideas of Torricelli, Hobbes, Huygens. Furthermore, he was also deeply interested in new scientific fields of research. Indeed, Leibniz mentioned Boyle and his chemistry more than once. One has the impression of a work full of passion and intuitions, but unsatisfactory under many respects. If Descartes's physics is a romance, then Leibniz's *Hypothesis* is no less fictional.

(3) The mathematical treatment is almost completely missing.[6] In this perspective, the *Hypothesis* does not represent a progress if compared to Descartes's *Principia*. For example, the conjecture of particles emitted by the Sun to explain

[6] The lack of mathematical treatment connotes almost all Leibniz's works on gravity: see Leibniz (1677, 1689), Leibniz (1690), Bussotti (2015), Chap. 5, pp. 71–114.

gravity might have been a good idea. However, if no mathematical treatment is developed so to obtain predictions that can be verified or falsified by observations or experiments, it is impossible to test the fruitfulness of this conjecture from a scientific point of view. In this respect, it is significant—as we have seen in Sect. 6.1—that Huygens tried to offer a quantitative theory of gravity based on vortex theory, while Leibniz never tackled the subject mathematically in his long sequel of reflections. Probably this depends on the fact that at that time Leibniz had still a partially qualitative and intuitive view of science. Before attempting a mathematical treatment, he wanted to grasp an intuitive idea of how gravity worked. He could not accept gravity as a given phenomenon. On the contrary, Newton assumed gravity as given—without dealing with its nature and origin—and developed a mathematical treatment relying on his well-known axioms or laws of motion. This is a crucial difference: Newton realised that science should only focus on phenomena (though very general as gravity), while Leibniz still had the idea of a complete system in which no phenomenon can be studied if its origin is not explained. Thus, we may argue that Leibniz had a metaphysical approach to science.

(4) A theory denying the void cannot explain density in a satisfactory way. If one believes in the existence of the void, density can be explained by claiming that a body is the more dense the more it contains matter compared to void. But how to explain density without the void? Descartes stated that the particles of the first element are very quick, small and tenuous. Leibniz stated that bubbles have different densities. However, it seems extremely problematic to guess what tenuous or dense means without referring to the ratio between void and matter.

(5) Finally, the most significant aspects of *Hypothesis Physica nova* are two: (a) Leibniz realised that an intermediate level between the particles and the bodies has to exist if one wants to explain the composition of matter. The Cartesian analysis based only on particles is too unrefined. To use an analogy, if the atoms are compared to the particles, then bubbles have to be compared to molecules. Leibniz's theory of matter is certainly naïve and unsatisfactory, but the problems faced by Leibniz and the way he tried to solve them are not trivial; (b) he realised that Descartes's theory of gravity is incomplete, if not fallacious, and grasped that gravity would have been the testbed for the new physics. This explains why gravity was one of the most recurrent subjects in Leibniz's speculation.[7] Even though Leibniz modified substantially his conceptions over the years, the *Hypothesis* is a significant work since it provides the basis for his subsequent ideas on gravity.

---

[7] In the course of the years Leibniz developed several hypotheses on the origin of gravity: see Leibniz (1977), in LSB, II, 1, pp. 441–466; Leibniz (1689, 1860, 1962), VI, pp. 144–161; Leibniz (1690?, 1860, 1962, VI, pp. 161–187); Leibniz (1690, 1860, 1962, VI, pp. 193–20); Leibniz (1706, 1860, 1962), Pars I et Pars II plus Beilage, VI, pp. 254–280. See also Bussotti (2015), Chap. 5, pp. 71–114. Despite their remarkable differences, all of these hypotheses offered mechanical explanations of gravity. Therefore, the conceptions expounded in the *Hypothesis Physica Nova* are indicative of how Leibniz addressed the problems of gravity along his entire scientific career.

8.1.2 *Beyond Mechanism: The Concept of* Vis

Starting at least from 1686 Leibniz did not think anymore that mechanism could explain the real foundations of physics. In that year Leibniz published a brief essay in the *Acta Eruditorum* that became quite renowned: *Brevis Demonstratio Erroris memorabilis Cartesii [...] (A Brief Demonstration of a Notable Error of Descartes [...]*, see Leibniz, 1686a) whose content is also a part of the celebrated *Discours de metaphysique* (Leibniz, 1686b). In his *Principles of Philosophy* Descartes claimed that in the interactions the quantity of motion is conserved, that is, to use modern language the quantity $p = mv$. However, he considered this quantity as a scalar rather than a vector. Following Descartes's presuppositions, Leibniz proved that the scalar quantity which is conserved is different from what Descartes had thought. This quantity *vis viva*. The *vis viva* is the double of what we call kinetic energy, that is the *vis viva* of a body of mass m and speed v is mv^2. In modern terms, it represents the capability of a body to do work. Leibniz thought of *vis viva* as a quantity that is conserved and that, on a metaphysical level, indicates the creatures' activity. *Vis viva* is connected to an order of causality which is superior to mechanical causes. This order is that of final causes. The principle of *vis viva* conservation is interpreted by Leibniz as the expression of a final cause because the activity of the creatures (of which the *vis viva* represents the intensity) is a significant aspect of the divine project whose final scope is the conservation of the best possible world. If God had preserved the best possible world only through miracles, that is through continuous direct interventions, this would have been evidence of his lack of power, as if he were not able to establish a regular order. On the contrary, principles of conservation are evidence of a steady natural order established by God at the moment of creation. Therefore, Leibniz believed that he had reintroduced the final causes in physics. They had been abhorred by mechanistic philosophers and scientists because they were interpreted as a residue of Aristotelism and were inserted within an anthropocentric-teleological vision of the world that they did not accept. However, Leibniz tried to justify the scientific necessity of final causes and their fundamental metaphysical role. Let us sketch this development following the way in which Leibniz specified his concept of *vis*.[8]

In the *Essay de dynamique sur les loix du mouvement [...] (Essay of dynamics on the Laws of Motion [...]*, see Leibniz 1692?), a manuscript probably datable around 1691–1692, Leibniz introduced the term *progrès* to indicate the vector we call quantity of motion and proved that this quantity is conserved in the interactions, whereas Descartes' scalar quantity is not. Nonetheless, he claimed that a more fundamental scalar quantity has to exist. He called this quantity "*absolute force*". Its intensity is determinable by the effects that it produces. Leibniz stated that the force of a body

[8] On various aspects of Leibniz's dynamics and its relation with metaphysics see Allen (1984), Bernstein (1981), Carlin (2006), Dumitrescu (2011), Gabbey (1971), Gale (1973, 1984), Garber (2008), Garber (2009), Gueroult (1934), Iltis (1971), Knobloch (1995), Lindsay (1975), Lodge (1997), Lodge (2001), Miller (1982), Papineau (1977), Parkinson (1974), Rozemond (2009), Stevenson (1997).

can be determined as the product of its mass multiplied by the height that the body can reach, given its motion. Leibniz seems to think of the quantity mgh, namely what nowadays we call potential energy (see Bussotti, 2017b, pp. 110–114): a body of mass m which falls from the height h in a gravitational field with acceleration g, will be able to re-ascend exactly to the height h. Leibniz offers the example of the pendulum, whose bob will reach exactly the same height from which it started, excluding the friction. This quantity indicates the force of the body when it is in equilibrium. However, when the equilibrium is broken, the effects of the body are not measured anymore by mgh but by mv^2, the *vis viva*. Thence, Leibniz seems to offer two different definitions of the absolute force of a body, one in terms of *vis viva* and one in terms of the product mgh. However, if one considers only the free fall of bodies, a body starting from the height h arrives at the ground with a speed $\sqrt{2gh}$ (which results from Galileo's law of falling bodies), so that the quantity mv^2 is equal to $2mgh$, and not to mgh as Leibniz erroneously stated. Apart from this error, what Leibniz claimed is correct in the specific case of the free fall of bodies. It seems that he guessed something profound with regard to the principle of energy conservation: one quantity, that is mgh, plays the role of potential energy while another, namely the *vis viva*, that of kinetic energy. However, he never defined the quantity $mgh + mv^2$ as the one which is conserved. To clarify: when a body is at rest at a certain altitude h, Leibniz defined its force to be mgh; when the body begins to move, its force is defined mv^2. Leibniz, though, did not consider the sum of the two quantities which is the equivalent of our mechanical energy. What is truly conserved is this sum ($mgh + mv^2$), and not the two quantities considered separately. When a body is at rest at height h, its potential energy is mgh, while the kinetic energy is 0. When the body reaches the soil, its potential energy is 0 (if, as in most problems concerning the motions on Earth, we pose the zero of the potential at the soil's level) and its kinetic energy is mv^2 (actually, $1/2mv^2$).[9] Therefore, as to the initial and final positions of a body, what Leibniz claimed is correct. However, his analysis does not take into account the intermediate positions of a body: if, the body being in those positions, one defines, as Leibniz seems to do, its "force" as *vis viva*, it is false that such a *vis viva* is conserved. As a matter of fact, it is necessary to add the potential energy of the body in any single position. What is conserved in any single position of the body is the sum of its *vis viva* and its potential energy. Leibniz is paving the way to the physics of the conservation-principles, but his concepts are still embryonic and not fully developed. They are intuitions rather than structured scientific concepts.

Specimen Dynamicum (Leibniz, 1695) is an essential work to understand the relation between physics and metaphysics. Leibniz distinguished between *active*

---

[9] We offer here a brief explanation why kinetic energy is today considered equal to $1/2mv^2$. Kinetic energy K of a particle moving at the speed v is defined as the work of an external force by the velocity of the particle to increase the speed from 0 to v. The work is the scalar product $F{\cdot}s$ (actually $F{\cdot}ds$ considering the infinitesimal form), where F is the force and s the displacement of the particle. Therefore, the work is the difference of two kinetic energies, the initial and final one. In an accelerated motion, from Galileo's law one has $s = 1/2at^2$, where a is the acceleration caused by F and t is the time during which the motion takes place. Therefore, one has that the work $F{\cdot}s = 1/2ma^2\ t^2$, since $F = ma$. Because $v = at$, one has $F{\cdot}s = 1/2mv^2$.

force and *passive force*. These concepts are certainly related to those of *vis viva* and *mgh* but cannot be identified with them.

The *primitive active force* indicates the action-capability inherent in any corporeal substance. It is compared by Leibniz to Aristotle's first entelechy and corresponds to the soul or substantial form.

The *derivative active force* is the force considered as a physical magnitude. Leibniz claimed that it is, i.e., the force involved in the bodies' collision.

The *primitive passive force* is the mere resistance of a body, considered as resistance to be moved or to any further action potentially causing a modification in the state of the body. It is a typical feature of what Aristotle called the *first matter*.

The *derivative passive force* is a typical property of the *second matter*, namely of the bodies insofar as they are a *synolon* of form and matter.

With regard to these four kinds of forces, Leibniz wrote:

> *Active force* (which might not inappropriately be called power [*virtus*], as some do) is twofold, that is, either *primitive*, which is inherent in every corporeal substance *per se* (since I believe that it is contrary to the nature of things that a body be altogether at rest), or *derivative*, which, resulting from a limitation of primitive force through the collision of bodies with one another, for example, is found in different degrees. Indeed, primitive force (which is nothing but the first entelechy) corresponds to the *soul or substantial form*. But, for that reason, it pertains only to general causes, which are insufficient to explain the phenomena. And so we agree with those who deny that we should appeal to forms when treating the individual and specific causes of sensible things. [...]
>
> Similarly, passive force is also twofold, either primitive or derivative. And indeed, the *primitive force of being acted upon* [*vis primitiva patiendi*] or of *resisting* constitutes that which is called *primary matter* in the schools, if correctly interpreted. This force is that by virtue of which it happens that a body cannot be penetrated by another body, but presents an obstacle to it, and at the same time is endowed with a certain laziness, so to speak, that is, an opposition to motion, nor, further, does it allow itself to be put into motion without somewhat diminishing the force of the body acting on it. As a result, the *derivative force of being acted upon* later shows itself to different degrees in *secondary matter*.[10]

These passages are significant because of the connection they show between Leibniz's physics and metaphysics. Active force is an inherent property of bodies. Under this respect it is a metaphysical, an ontological concept. It indicates the activity of substances. However, since activity implies movement, primitive active

[10] Leibniz (1989a, pp. 119–120). Cf. Leibniz (1860, 1962), VI, pp. 236–237: "Duplex autem est Vis Activa (quam cum nonnullis non male Virtutem appelles), nempe ut primitiva, quae in omni substantia corporea per se inest (cum corpus omnimode quiescens a rerum natura abhorrere arbitrer), aut derivativa, quae primitivae velut limitatione, per corporum inter se conflictus resultans, varie exercetur. Et primitiva quidem (quae nihil aliud est, quam ἐντελέχεια ἡ πρώτη) animae vel formae substantiali respondet, sed vel ideo non nisi ad generales causas pertinet, quae phenomenis explicandis sufficere non possunt. Itaque illis assentimur, qui formas in rerum sensibilium causis propriis specialibusque tradendis adhibendas negant. [...]. Similiter vis quoque passiva duplex est, vel primitiva vel derivativa. Et quidem vis primitiva patiendi seu resistendi id ipsum constituit, quod materia prima, si recte interpreteris, in Scholis appellatur, qua scilicet fit, ut corpus a corpore non penetretur, sed eidem obstaculum faciat, et simul ingnavia quadam, ut sic dicam, id est ad motum repugnatione sit praeditum, neque adeo nisi fracta nonnihil vi agentis impelli se patiatur. Unde postea vis derivativa patiendi varie in materia secunda sese ostendit".

force also has a physical meaning, as we will see. Primitive active force depends on the soul or on the form of the substance. Insofar as it is the first entelechy of the substance, the primitive active force is the principle of the monad's internal actions, according to Leibniz's conception of the monad as self-sufficient. If, from the metaphysical level we descend to the phenomenal level, it is necessary to explain how bodies change their kinetic state in consequence of collisions. In order to achieve this explanation, the inherent force is not sufficient. It is necessary to introduce a relational magnitude, called by Leibniz derivative active force. It indicates the potentiality of a body to change the kinetic state of another body. It is remarkable that Leibniz considered only the collisions as actions which can modify the kinetic state of bodies. Clearly the Cartesian view still exerted an influence on him, though his picture is different from the Cartesian one. Leibniz introduced some elements that are an unequivocal reworking of Aristotle's final causes and that, certainly, Descartes would not have approved.

Passive force indicates the *inertia* of a body. Here inertia has to be interpreted as follows: the primitive passive force indicates the tendency, the potentiality of a body to maintain its state of rest, in Leibniz's terms *ignavia* (laziness), or *ad motum repugnatio* (opposition to motion). It is an inherent property of bodies; but, while the active force is due to the soul of the substance, the passive force is due to its matter. The derivative passive force is the actual resistance opposed by a body to the movement when it collides with other bodies.

One of the main interpretative problems is to understand if this concept of inertia, which is different from Newton's, is, nonetheless, compatible with Newton's. As we will see, Leibniz spoke of inertial state as a state of rest. On the other hand, while developing his physics and his planetary theory he considered that uniform rectilinear motion tends to be maintained. How is it possible to reconcile these views? The answer is not easy, because Leibniz is not explicit on this point; however, reasonable conclusions can be drawn by examining what he wrote starting from the 1690s, the decade in which the *Specimen Dynamicum* was composed. In a letter to Sturm, datable between the end of June and the beginning of July 1697, Leibniz wrote:

> [. . .] and hence, as the resistance constitutes the general matter of a body, so the *nisus* constitutes the peculiar form of each body or its first activity (as I would translate *entelecheian ten proten*, according to my meaning), since the division itself of the extended [matter] into parts, and hence the form, arises from such *nisus*. And the resistance does not set up only bodies' impenetrability, but also the natural inertia, which is commonly less acknowledged. The natural inertia was called in this manner by Kepler. It is the property for which matter is not indifferent to motion and rest (as many think), but rather opposes a new motion in proportion to its mass (*molis*), in the same way we see that a loaded ship is transported more slowly by the same wind.[11]

[11] Our translation. Cf. LSB, II, 3, pp. 339–340: "[...] atque adeo ut resistentia generalem corporis materiam, ita nisus peculiarem cujusque corporis formam vel primam activitatem (ut ἐντελέχειαν τὴν πρώτην ex meo sensu interpreter) constituit; cum ipsa etiam extensi in partes divisio atque adeo figura, ex ipso oriatur. Et resistentia non tantum facit corporum impenetrabilitatem sed et aliud minus vulgo expensum nempe inertiam naturalem, a Keplero sic appellatam, qua fit ut materia non sit ad motum quietemque (ut plurimi arbitrantur) indifferens sed potius novo motui proportione molis suae repugnet, quemadmodum videmus navem magis oneratam, eodem vento tardius ferri".

This letter confirms that matter is the passive aspect of each body. Inertia is the manifestation of the passivity of matter. Form is the active aspect and entelechy is its expression. In the quoted passage, Leibniz explains that matter is not indifferent to motion and rest as many thinkers erroneously thought. Who are these thinkers? Our idea is that Leibniz included Newton among them. Indeed, according to Newton—to use Leibniz's language—matter is indifferent to rest or uniform rectilinear motion.

It seems thus legitimate to conclude that, according to Leibniz, at a metaphysical level, Newton's principle of inertia is false. Matter tends to rest. Rest is a metaphysical state which is different from uniform rectilinear motion. Leibniz's idea that, if a body is in motion, its inertia opposes a new motion, cannot be interpreted as if he thought that inertia pertains also to non-accelerated motions. Rather his conception seems to be the following:

(1) matter has the tendency to rest, due to its inertia;
(2) once a body is in motion, inertia continues to oppose motion, but entelechy tends to make the body move;
(3) if a new action is not exerted, the state of motion does not change. That is: there is an equilibrium between the derivative active principle of the bodies, and the derivative passive force, namely inertia in Leibniz's, not in Newton's view. This equilibrium is maintained until new actions do not change the kinetic state of a body.

According to this reading it can be asserted that Leibniz accepted Newton's inertia principle at a physical level, but rejected it at the metaphysical one. Our idea is confirmed by other passages from Leibniz's letters.

In a letter to Denis Papin on 28 February 1699, we read:

> Now, for me, rest is nothing but a privation. It follows that the mass in itself resists motion and this is what I call, with Kepler, inertia. But when the body is in movement and resists rest, then I think there is a force or entelechy, which makes the body to continuously tend to movement. Thence, it follows that mass resists continuously entelechy.[12]

Less than one month later, on 24 March 1699, Leibniz wrote to De Volder:

The editors of the *Akademie Ausgabe* of Leibniz's works dated this letter between the end of June and before the fifth of July 1697. For an analysis of Leibniz's concept of inertia, see Bernstein (1981), Gabbey (1971), Ghins (1990), Giorgio (2011), Giulini (2002), Lariviere (1987), Look (2011), Ranea (1986), Woolhouse (2000). On the concept of inertia in Leibniz see also Arthur (1998), Bertoloni Meli (1993), Bouquiaux (2008), Crockett (2008), Duchesneau (1994a), Garber (1994, 2009), Jauernig (2008), Papineau (1977), Puryear (2012), Roberts (2003), Suisky (2009) (in particular Chap. 2). Several ideas of this section are drawn from Bussotti (2015), Chap. 3.

[12] Our translation. Cf. LSB III, 8, p. 69: "Or selon moy le repos n'estant autre chose qu'une simple privation; il s'ensuit que c'est donc la masse en elle même qui resiste au mouvement, et c'est ce que j'appelle avec Kepler, inertie. Mais quand le corps est en mouvement, et resiste au repos, alors je tiens qu'il a une force ou entelechie, qui le fait tendre à continuer le mouvement. D'où il s'ensuit que la masse resiste continuellement à l'entelechie, et ainsi qu'il y a action et reaction dans le corps même".

> Since matter in itself therefore resists motion by a general passive force of resistance but it is set in motion by a special force of action, or entelechy, it follows that inertia also constantly resists the entelechy or motive force during its motion.[13]

As Leibniz clarified in the second part of the *Specimen Dynamicum*,[14] a state of absolute rest can never be realized because entelechy always opposes to rest and any substance always has its entelechy.

A problem remains: Leibniz, as we will see in Sect. 8.2, had a relativistic and relationalist conception of space and time, whereas, from what we have read in these quotations, he distinguished rest and motion as two completely different states. But, if space and time as absolute entities do not exist, how is it possible to conceive, at least in theory, the distinction between an absolute state of rest and an absolute state of motion? Is not this contradictory?

Some scholars point out that Leibniz believed in absolute motion (see Section 8.2) and that this idea is not inconsistent with his conception of space and time.[15] Their arguments are convincing. The situation might be summarised like this: space only indicates the relations of mutual positions among bodies; time only indicates the order of succession of events, and neither time nor space have an existence separate from bodies. However, quantity of matter, entelechy, or passive forces (matter) and active forces (form) exist in themselves and are inherent to bodies. Impacts between bodies also exist. In the impact between two bodies *A* and *B* the derivative active force of *A* (the one which is considered to hit the other, but the situation is perfectly reversable) produces a modification of the kinetic state of *B*, which reacts in function of his derivative passive force. Thence, in the instant of the impact, the kinetic state is determined by the derivative active force of *A* and by the derivative passive force of *B*. After the impact, the derivative force of *B* determines what can be called the absolute kinetic state of the body: the derivative passive force tends to stop the body, whereas the derivative active force tends to produce a motion. There is an equilibrium between these two tendencies and the body continues in its state of motion until a further impact does not change it.

Therefore, active and passive forces are metaphysical properties of the substance because active force is an indication of substance's activity insofar as any substance is spiritual, endowed with a soul; passive force is tied to the material aspect of the bodies. However, Leibniz tried to deduce dynamical consequences from them, which is a clear indication that the whole of his dynamics relies on the metaphysical background he conceived in the course of his philosophical and scientific speculation. The fundamental problem for Leibniz is how to transcribe these metaphysical conceptions

---

[13] Leibniz (1989b, p. 517). Cf. LSB, II, 3, p. 547: "Cum igitur materia motui per se repugnet vi generali passiva resistentiae; at vi speciali actionis seu entelechiae in motum feratur; sequitur ut etiam inertia durante motu Entelechiae seu vi motrici perpetuo resistat". The correspondence Leibniz–De Volder has been edited and translated by P. Lodge, see Leibniz (2013).

[14] Leibniz (1860, 1962, VI, p. 252).

[15] On this subject, see Bertoloni Meli (1993, pp. 76–78), Garber (2009, pp. 173–189), Jauernig (2008, pp. 19, 33), note 16, Puryear (2012), Roberts (2003).

applied to dynamics into a mathematical form concretely usable within physics. This is a difficult and not completely realizable task.

Let us see whether there is a connection between the concepts of active and passive force and those of *vis viva* and *vis mortua*. *Vis viva* is the essential concept of Leibniz's dynamics (as we said, its mathematical expression is mv^2), while the notion of *vis mortua* is by far more obscure. A reasonable interpretation is that, in Leibniz's intentions, *vis mortua* is connected to two slightly different ideas, which, however, Leibniz does not distinguish adequately: that of potential motion and that of incipient or infinitesimal motion of an object. Leibniz considered the centrifugal force as a *vis mortua* because a rotating object has the tendency to recede along the tangent, but, in fact, it does not recede. Gravity is a *vis mortua* too because a body has always the potentiality to fall, but, in fact, it falls only in some circumstances, i.e. when it is free to fall. In this respect the *vis mortua* is connected to the potential motion of a body. Leibniz claimed that the *vis viva* is generated by an infinity of continuous impressions of *vis mortua*. Therefore, the *vis mortua* would be equal to something like mass by an infinitesimal velocity, expressible by mdv. Integrating such infinitesimal velocities one obtains the *vis viva*.

With regard to the relation between dead and living forces Leibniz wrote:

> From this it follows that *force* is also twofold. One force is elementary, which I also call *dead force,* since motion *[motus]* does not yet exist in it, but only a solicitation to motion *[motus]*, as with the ball in the tube, or a stone in a sling while it is still being held in by a rope. The other force is ordinary force, joined with actual motion, which I call *living force*. An example of dead force is centrifugal force itself, and also the force of heaviness *[vis gravitatis]* or centripetal force, and the force by which a stretched elastic body begins to restore itself. But when we are dealing with impact, which arises from a heavy body which has already been falling for some time, or from a bow that has already been restoring its shape for some time, or from a similar cause, the force in question is living force, which arises from an infinity of continual impressions of dead force.[16]

Thus, centrifugal force and gravity are dead forces insofar as they are connected to potential motion. Centrifugal force is related to the tendency to recede along the tangent (Newtonian inertia), but when the body rotates such a tendency is not actualised. A body in equilibrium in a gravitational field (like a ball suspended at a vertical thread) only has the potentiality to move, because the movement actually takes place only when the equilibrium is broken. Therefore, it seems that the *vis mortua* holds some similarities to our concept of potential energy. A different situation is given when a stretched elastic body begins to restore itself to its original shape: in that case

[16] Leibniz (1989a, pp. 121–122). Cf. Leibniz (1860, 1962, VI, p. 238): "Hinc Vis quoque duplex: alia elementaris, quam et mortuam appello, quia in ea nondum existit motus, sed tantum solicitatio ad motum, qualis est globi in tubo, aut lapidis in funda, etiam dum adhuc vinculo tenetur; alia vero vis ordinaria est, cum motu actuali conjuncta, quam voco vivam. Et vis mortuae quidem exemplum est ipsa vis centrifuga, itemque vis gravitatis seu centripeta, vis etiam qua Elastrum tensum se resistuere incipit. Sed in percussione, quae nascitur a gravi jam aliquamdiu cadente, aut ab arcu se aliquamdiu restituente, aut a simili causa vis est viva, ex infinitis vis mortuae impressionibus continuatis nata". For interesting insights on the concept of *vis mortua* in Leibniz, see: Boudri (2002), in particular pp. 91–98; Capecchi (2012), in particular pp. 197–199; Garber (2008); Iltis (1971); Pasini (2016) ; Roberts (2003).

the dead force is connected to the incipient, infinitesimal motion, what nowadays we call an act of motion. The two meanings of dead force are connected because the equilibrium is broken in the first instant of motion of a body, so that one can consider the infinitesimal quantity dv, which is correlated to the dead force.

A further observation concerns the way in which a living force can be deduced by a dead force. As we have noted previously, according to what Leibniz claimed in the mentioned passage of the *Specimen Dynamicum*, the living force should be calculated as $\int m dv$ ("infinity of continual impressions of dead force"), but this integral is equal to mv, not to mv^2, that is to the quantity of motion, not to *vis viva*. In this intricate series of concepts, Leibniz explained in a letter to De Volder that, at least on some circumstances, the living force is the integral of the dead force by velocity, not simply the integral of the dead force. In this case the expression $\frac{1}{2}mv^2$ for the *vis viva* is obtained:

> Consequently, in the case of a heavy body receiving an increase of speed equal and infinitely small at every moment of its fall, the dead and the living force can be calculated at the same time. The speed increases uniformly with time but the absolute force as the square of the time, that is, as the effect. So according to the geometric analogy or our analysis, the solicitations [dead forces] are as dx, the speed is as x, the forces [living force] as xx, or $\int x dx$.[17]

It seems rather strange that Leibniz spoke of an integral with respect to dv (or dx to use his notation in the letter to De Volder), namely the differential of velocity, but explained neither how the integral of the dead force with respect to time or space should be interpreted nor what quantity integrated with respect to time or space produces the living force. This was and is very unusual in physics because the basic variable of integration is time or, on some circumstances, space. Despite this anomalous kind of deduction, *vis viva* is a clearly determined physical magnitude, while *vis mortua* is not a well defined concept because it comprehends different physical magnitudes. Johann Bernoulli had the merit to offer a better clarification of the notion of *vis mortua*.[18]

What are the relations between living force-dead force and active–passive forces? The *vis viva* can be identified with the *derivative active force*. To be more precise: the *vis viva* is one of the physical manifestation of entelechy. The *passive derivative force* has some similarities to Newton's inertia, but we have just seen that Newton's concept of inertia is not a primitive notion of Leibniz's dynamics. Leibniz interprets Newtonian inertia as the equilibrium between active and passive forces. It thus seems quite difficult to identify a precise physical-mathematical magnitude that can be considered as a derivative passive force.

Finally: it seems that the active and passive forces are referred by Leibniz to modification and change of state in general, not only to movement and kinetic state of a

---

[17] Capecchi (2012, p. 197). Cf. Leibniz (1879, 1978, II, p. 156): "Eodem modo etiam fit, ut gravi descendente, si fingatur ei quovis momento nova aequalisque dari celeritatis accessio infinite parva, vis mortuae simul et vivae aestimatio observetur, nempe ut celeritas quidem aequabiliter crescat secundum tempora, sed vis ipsa absoluta secundum spatia seu temporum quadrata, id est secundum effectus. Ut ita secundum analogiam Geometriae seu analysis nostrae solicitationes sint ut dx, celeritates ut x, vires ut xx seu ut $\int x dx$".

[18] See, e.g. Capecchi (2012, pp. 199–201); Heimann (1977).

body. In the *Specimen Dynamicum* he spoke almost exclusively of movement, but his forces are potentially referable to other changes of state, too. The *primitive active force* is the main metaphysical concept expressed in the *Specimen Dynamicum*. This force is compared to entelechy, which is a finalistic Aristotelian concept. It can be interpreted as an indication of the aim, of the *telos* that any substance has in the Leibniz's harmonic universe. It can also be regarded as the archetype of the *derivative active force*. Considered only in its physical effects and from our human point of view, the manifestation of the *primitive active force* is the *derivative active force*. But the *derivative active force* is the *vis viva* and the principle of *vis viva* conservation is at the basis of Leibniz's physical conception. Given this relation between the *derivative active force* and the *primitive active force*—which is interpreted as a finalistic concept—, we may understand why Leibniz considered the principle of *vis viva* conservation as a *finalistic principle*. Leibniz identified three ontological levels to which three different epistemological approaches *correspond*: (a) the phenomena considered only on their mechanical aspects. It is correct to deal with physics through a mechanistic approach, if one considers only the physical world in itself; (b) mechanical science has a foundation in dynamics, based on the principles of conservation (the most important of which is the principle of *vis viva* conservation); (c) physical principles do not grasp the final and most profound level, which is that of metaphysical principles and of ontological finalism. This is based on the concept of monad as a complete notion and on the principle of pre-established harmony. To those who might object against the meaningless idea of a monad "without windows" (i.e. without any causal interaction with the rest of the world), Leibniz might have answered that the notion of monad describes the real state of affairs at the metaphysical level. However, since men are not God and cannot have an intuitive vision of such ultimate foundation, they have to deal with the world as it appears to them. They can deduce the metaphysical truth only by reasoning and cannot grasp it by intuition.

The relations between science and metaphysics cannot be more profound than in Leibniz. As we shall discuss in Sect. 8.2, the strong connection between physics and metaphysics was Leibniz's *desideratum*: as a matter of fact, he tried to ground physics, and above all dynamics, on some metaphysical principles, while, at the same time, he did not admit the recourse to spiritual notions for explaining physical phenomena. However, the relation between the two fields is anything but clear and unproblematic.

8.1.3 Leibniz's Model of the Solar System

We have analysed the fundamental concepts introduced by Leibniz to illustrate his view of the physical world and the relations between his physics and metaphysics. An important step to understand Leibniz's cosmological conception is his model of the solar system, which contains kinematic, dynamical and structural elements. Since

1689 Leibniz wrote several essays on this topic.[19] His general idea was to adopt the vortex theory, and to enrich it with many mathematical elements that were missing in Descartes's theory and that Huygens did not supply, at least insofar as the kinematics and dynamics of the solar system were concerned. At all appearances, Leibniz's model seems conceived to show that the same results obtained by Newton could be achieved within a different physical theory. Leibniz claimed to have elaborated his model before reading Newton's *Principia*, but this is doubtful.[20] Even though there are many interesting elements in Leibniz's cosmology, several of his ideas appear to have been constructed ad hoc and are difficult to be unified in a satisfactory dynamical perspective. We will not enter into all the mathematical details of his model because our aim is to develop a conceptual analysis; we will thus refer only to the necessary items.[21]

The general structure of the model. To explain the kinematics of the solar system Leibniz resorted to two ideas: the 'harmonic circulation' and the 'paracentric motion'.

(1) Each planet rotates in a vortex around the Sun. It is not completely clear if the Sun is at the centre of the vortex. The vortex is moved by a particular kind of circulation that Leibniz called *Circulatio harmonica*:

> I call a *circulation* a *harmonic* one if the velocities of circulation in some body are inversely proportional to the radii or distances from the centre of circulation, or (what is the same) if the velocities of circulation round the centre decrease proportionally as the distances from the centre increase; or, most briefly, if the velocities of circulation increase proportionally to the closeness.[22]

If only the *circulatio harmonica* existed, the planetary path would be a circle: a concentric one if the Sun were in the geometrical centre of the orbit; an eccentric one, if the Sun were not in the geometrical centre. In fact, the orbit is an ellipse. Thus, Leibniz introduced a second motion for each planet, independent of the motion of the vortex:

(2) Leibniz called this motion *motus paracentricus*. He imagined that a planet, while moving in a layer of the vortex, moves, simultaneously, on a segment of a straight line. If this rectilinear motion is directed to the Sun, the planet approaches our star; if it is directed in the opposite sense, it moves away from the Sun. Giving appropriate values to the paracentric motion, it is possible to

---

[19] Leibniz wrote the following papers on the planetary theory: Leibniz (1689, 1860, 1962); Leibniz (1690?, 1860, 1962); Leibniz (1706, 1860, 1962) (he published only the first writing). Several letters or parts of letters are also dedicated to this subject.

[20] With regard to the possible influence of Newton's *Principia* on Leibniz's planetary theory, see Bertoloni Meli (1993).

[21] For a complete treatment of all the mathematical aspects of Leibniz's planetary theory, see Bussotti (2015). See also Bussotti and Pisano (2017).

[22] Bertoloni Meli (1993, pp. 129–130). Cf. Leibniz (1689, [1860] [1962], VI, pp. 149–150): "Circulationem voco Harmonicam, si velocitates circulandi, quae sunt in aliquo corpore, sint radiis seu distantiis a centro circulationis reciproce proportionales, vel (quod idem) si ea proportione decrescant velocitates circulandi circa centrum, in qua crescunt distantiae a centro, vel brevissime, si crescant velocitates circulandi proportione viciniarum".

determine the observed ellipticity of the orbit, both in case the Sun is supposed to be in the centre or eccentric in the harmonic vortex.

Until the fundamental studies by Aiton in the 1960s, Leibniz's *velocitas circulandi* was interpreted as the speed of the planet around the Sun. If this were the case, Leibniz would have made a very serious mistake.[23] Instead, Aiton convincingly proved that Leibniz's conception is correct: the *velocitas circulandi* is not the module of velocity. It is the transverse velocity, and this magnitude satisfies Leibniz's law: it decreases as the distance from the Sun. The *motus paracentricus* is, instead, the radial motion. Thus, according to Aiton, Leibniz adopted an interesting physical perspective. He examined motion from the point of view of the rotating planet using what we call now polar coordinates. Thence, he divided the velocity in the *transverse* and *radial component*. There is no mistake in Leibniz's theory.

Furthermore, if a planet moves of *circulatio harmonica*, it is trivial to prove that it satisfies Kepler's area law. One may thus suspect that Leibniz invented ad hoc the harmonic vortex precisely because he knew that in a *circulatio harmonica* the area law is satisfied.

Some specifications. The *paracentric motion* depends on two opposite tendencies (Leibniz, 1689, 1860, 1962, § 9, p. 152): (a) *Impressio excussoria circulationis*, translated by Bertoloni Meli as "outward impression of circulation"; (b) *Attractio solaris* (the Sun's attraction) that Leibniz considers a specific kind of solicitation of gravity (see below). Leibniz does not use the term 'attraction' in the sense of action at a distance.

The *impressio excussoria circulationis* is the centrifugal force due to the velocity of the planet moving of *circulatio harmonica*. The *impressio excussoria* exists in each point of the trajectory. Leibniz gave both a geometrical representation and an analytical expression for the instantaneous outward *impressio excussoria circulationis* (Leibniz, 1689, 1860, 1962, § 10, p. 152). He named this quantity *conatus centrifugus* or *conatus excussorius circulationis* (ibidem).

If the motion is circular and uniform, the *conatus* is as V^2, where V is the transverse velocity. If two or more circles are considered in which the movement is uniform, then the *conatus* are as V^2/R, where R is the radius. From this Leibniz deduced another expression which is fundamental in his reasoning: if a body moves with a harmonic circulation, the *conatus centrifugus* is inversely proportional to the radius vector. This happens because of the inverse proportion between transverse velocity and radius vector in the *circulatio harmonica* and because of the relation $c = V^2/R$, where c is the centrifugal conate. Leibniz considered, then, a fixed elementary area, swept by the radius-vector in an infinitesimal time dt (in accordance with the area law), which Leibniz indicated by ϑa and assumed it equal to the double of the elementary triangle $M_2M_3\Theta$, namely equal to $D_2M_3 \cdot \Theta M_2$ (see Fig. 8.1).

The expressions $\Theta\ M_n$ can be indicated by $r =$ radius, because the difference between $\Theta\ M_i$ and ΘM_{i-1} is infinitesimal, which can be neglected in this calculation.

[23] See Aiton (1960), (1962), (1964), (1965), (1972). In the first book of his *Principia*, Section II, Proposition IV, Corollary VI, Newton proved that if the third Kepler law holds, then the velocities are as the square roots of the distance Sun-planet, and not as such a distance.

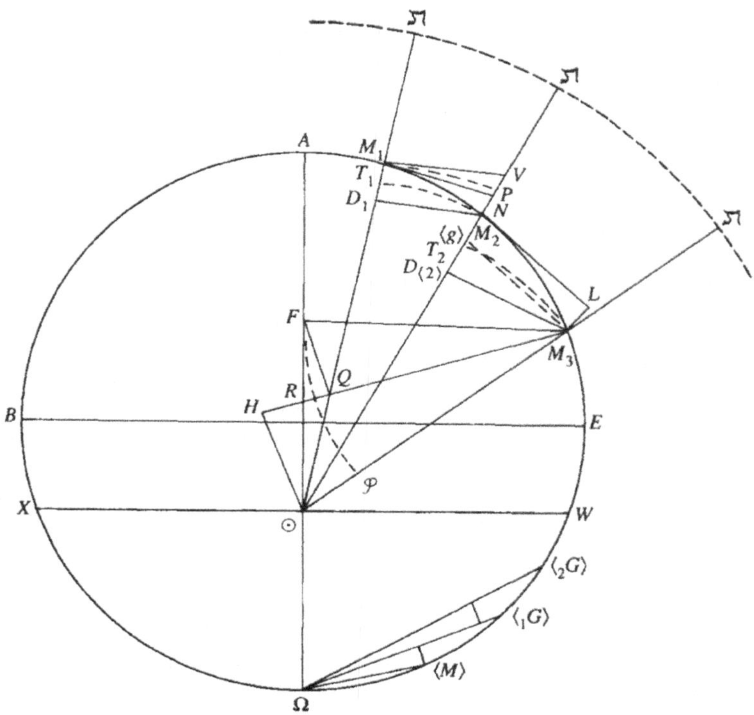

Fig. 8.1 The figure used by Leibniz to explain his planetary theory, with slight modifications to make it more intelligible. *Source* Bertoloni Meli (1993, p. 129)

Therefore, $D_2M_3 = \vartheta a/r$ and the centrifugal conate $D_2T_2 = (D_2M_3)^2/2\Theta M_3$. Hence, we get:

$$D_2T_2 = \frac{\vartheta^2 a^2}{2r^3}$$

This is the analytical expression of the *conatus centrifugus* obtained by Leibniz. This result is consistent with the interpretation that he is using, though not explicitly, a non-inertial reference frame in polar coordinates. Leibniz's expression for *conatus centrifugus* is correct .[24]

Thus, from the point of view of the planet, the acceleration along the radius is given by two components:

---

[24] An expression for the centrifugal force is $\frac{L^2}{mr^3}$, where L is the angular momentum. The centrifugal conate is expressed by Leibniz as $D_2T_2 = \frac{\vartheta^2 a^2}{2r^3}$. Since ϑa represents an infinitesimal area, it can be indicated by dA. In modern terms the relation between the infinitesimal area swept by the radius vector and the angular momentum is $L = 2m\frac{dA}{dt}$. If one does not take into account the constant factor m (the mass) and considers—as Leibniz did—a unitary infinitesimal time, then the relation becomes $L = 2dA$. Therefore, if we exclude a constant factor 8, Leibniz's result is perfectly correct.

(a) one outwards, which is the *conatus centrifugus* due to the harmonic circulation;
(b) the other one is due to Sun's gravity, the specific *solicitatio gravitatis* acting in the specific case of the planetary motion.

Leibniz was able to find a mathematical expression for the *solicitatio gravitatis* and determined the instantaneous acceleration of the planet as the difference between the *solicitatio gravitatis* and the *conatus centrifugus*.

All the concepts introduced by Leibniz serve to study the motion of a body rotating in an ellipse in which the Sun is in one of the two foci and in which the dynamical elements are the *conatus centrifugus* and the *solicitatio gravitatis*. Through a long and complicated reasoning of infinitesimal geometry Leibniz was able to prove that:

> If a body is moved in an ellipse, the velocity of circulation around a focus is at the paracentric velocity (that is the velocity with which the body descends towards the focus or moves away from the focus) as the minor or transverse axis is to the square root of the difference between the square of the focal distance and the square of the difference of the mobile's distances from the two foci.[25]

From this proposition several important corollaries follow, which describe the properties of the motion in an elliptical orbit where the centre of the force is in one of the foci.

The first corollary claims: in an ellipse, given a point P, the ratio between the paracentric (radial) velocity and the velocity of circulation (transverse velocity) is proportional to the ordinate. This means that the ratio between the velocity with which the planet approaches to or recedes from the Sun is to the velocity of circulation as the distance of the planet from the apses-line.

Two other significant corollaries state that:

(1) in an ellipse, given the mobile point P, the ratio between the velocity in the orbit and the velocity of circulation is as the mean proportional between the distances of P from the foci.
(2) The velocities, with which a point M_3 (see Fig. 8.1) changes its distance from the minor axis BE, are as the velocities with which it changes its distance from the focus Θ.

By far the most interesting property proved by Leibniz is that, in the specific case of the elliptical motion in which the Sun is in one of the two foci, the inverse square laws holds, that is the solicitation of gravity is inversely as the square of the distance Sun-planet. Namely, the so called direct problem of central forces is solved for this kind of motion.

Through the outlined theory, Leibniz was able to describe the planetary motion in accordance with the first two laws of Kepler.

Commentaries. We have presented the most important technical features of Leibniz's

---

[25] Our translation. Cf. Leibniz (1690?, 1860, 1962) VI, p. 174: "Si quid moveatur in Ellipsi, velocitas circulandi circa focum est ad velocitatem paracentricam, nempe descendendi ad focum vel a foco recedendi, ut axis minor seu transversus est ad latus differentiae inter potestatem distantiae focorum inter se et potestatem differentiae distantiarum mobilis a focis".

planetary model. We will now attempt to offer a correct historical and conceptual interpretation of Leibniz's work.

(a) Let us start from Leibniz's point of view. He considered his theory as an alternative to Newton's: he had reconstructed the planetary orbits so that Kepler's first two laws were satisfied and had determined both the kinematics and dynamics of the solar system. Nothing was missing for a satisfactory explanation of the planetary movements. Throughout his scientific career, Leibniz also tried to explain the origin of gravity through several reworkings of the vortex theory, so that, from his point of view, his theory was preferable to Newton's since there is no hypothesis on the origin of gravity in the *Principia*.

(b) Our interpretation diverges from Leibniz's view. Leibniz's planetary theory is a magnificent example of mathematical ability and creativity in physics, but it is not even comparable with Newton's results. Leibniz did not develop a general theory of forces, from which the motion of the planets around the Sun derives as a mere application; rather, his ingenious concepts seem forged ad hoc to explain the specific phenomenon of planetary motion, but are not inserted within a complete theory of dynamics. In this respect, there are some similarities with Kepler's dynamical notions: for any characteristic of the planetary movement (mean motion, ellipticity, latitude, etc.) an ad hoc force was postulated. Since Leibniz wrote the *Tentamen* eighty years after Kepler's *Astronomia Nova*, physics had progressed enormously and Leibniz knew that the correct law is the inverse square law. Therefore, his theory matched the data but was, nonetheless, an ad hoc invention.

In a letter to Leibniz dated 11 July 1992, Huygens pointed out many difficulties concerning the concept of *circulatio harmonica*:

> It is sure that the gravities (pesanteurs) of the planets are in inverse double reason as their distances from the sun, which, together with the centrifugal virtue (vertu), provides Kepler's eccentrical ellipses. But I was never able to understand, from your explanation given in the *Acta* of Leipzig [the published version of the *Tentamen*], how you deduce the same ellipses, by replacing your harmonic circulation and maintaining the same proportions of gravities. I do not see how you find the place for a kind of Descartes's deferent-vortex, which you want to maintain, since the mentioned proportion of gravity, joined with the centrifugal force, produce—by themselves—Keplerian ellipses, according to the proof given by Mr. Newton. You have long promised me to clarify this difficulty.[26]

Huygens claims that the postulation of the harmonic vortex is completely useless because the inverse square law and the planetary ellipses derive immediately from

[26] Our translation. Cf. LSB, III, 5, p. 337: "Il est certain que les pesanteurs des Planetes estant posees en raison double reciproque de leurs distances du soleil, cela, avec la vertu Centrifuge, donne les Eccentriques Elliptiques de Kepler. Mais comment en substituant vostre Circulation Harmonique, et retenant la mesme proportion des pesanteurs, vous en deduisez les mesmes Ellipses, c'est ce que je n'ay jamais pu comprendre par vostre explication qui est aux *Acta* de Leipsich; ne voiant pas comment vous trouvez place à quelque espece de Tourbillon deferant de des Cartes, que vous voulez conserver; puisque la dite proportion de pesanteur, avec la force Centrifuge produisent elles seules les Ellipses Keplerienes selon la demonstration de Mr Newton. Vous m'aviez promis il y a longtemps d'eclaireir cette difficulté".

Newton's demonstrations. Therefore, why introduce a vortex that rotates in such a particular manner as to agree with the area law?

Aiton makes a similar remark when he writes:

> Since the harmonic vortex played no part in the motion of a planet in its orbit, this vortex may be left out of account in the analysis of Leibniz's theory. (Aiton, 1964, p. 112)

And again:

> What he [Leibniz] still failed to see clearly was that the harmonic circulation of the planet followed from the attraction, so that his resolution of the orbital motion into transverse and radial components, which gave a correct mathematical representation, had a sufficient physical foundation in the attraction without the addition of the harmonic vortex. (Aiton, 1972, p. 136)

It is true that from a physical-mathematical point of view the harmonic vortex is useless. Nonetheless, within Leibniz's conception, it plays the significant role of allowing him to prove the area law without resorting to the immediate action at a distance of a centripetal force. Thence, Leibniz thought to have proved the area law through the action of a really existing object, the harmonic vortex, eliminating any action at a distance. However, Huygens's objection is fully justified because admitting a harmonic circulation means to presuppose, not to prove, the area law.

(c) A further historiographic question concerns the originality of Leibniz's distinction between transverse and radial velocity in a curvilinear motion. In Kepler's *Epitome* we find a passage in which the idea that transverse velocity decreases as the distance from the Sun is already expressed.[27] Kepler's language is rather convoluted and the topic is touched in a cursory way so that an excellent knowledge of his works is required in order to identify and appreciate this passage. Leibniz knew Kepler's works well and admitted that some of his conceptions were inspired by Kepler's. Was that passage of *Epitome* the source of Leibniz's *velocitas circulandi*? The question is legitimate, but the answer can only be conjectural. What is certain is that Leibniz formulated his notion with a clarity that was lacking in Kepler's passage.

(d) Let us come to the most important issue: Aiton claims that Leibniz analysed the planetary motion from the point of view of a non-inertial reference frame. This is true from a purely mathematical standpoint. In a non-inertial frame centrifugal forces appear. From a physical perspective, the question arises as to whether Leibniz understood that the centrifugal forces are apparent forces which appear only in relation to the reference system we use, while the true force is, in our case, gravity. Newton was sure that Leibniz did not grasp this crucial issue. While objecting to Leibniz's planetary theory, Newton wrote:

> Propositions 20th (sic) and 25th are false, because they show a centrifugal force which is less than planet's gravity towards the sun. Therefore, they are false. The motion of a planet in its

[27] See KGW, VII, p. 377, lines 27–36. Bussotti 2015, pp. 121–123 analyzes of this passage by Kepler with reference to its possible influence on Leibniz's concept of *velocitas circulandi*. The sixth chapter of Bussotti 2015 deals with Kepler's influence on Leibniz's planetary theory.

> orbit does not depend on the excess of gravity upon centrifugal force (as Leibniz believes), but the orbit is incurved only by gravity's action, to which the centrifugal force (as reaction or resistance) is always equal and opposed, according to the third law posed by Newton.[28]

Newton thought that in the case of planetary motion only a real force is acting: gravity. The centrifugal force is only an apparent one. Instead, according to Newton, Leibniz believed in the existence of two real forces: gravity and centrifugal force. The core of Newton's criticism seems to be that Leibniz thought the centrifugal forces to be real forces, not apparent ones. In fact, nothing in Leibniz's writings can lead us to think that he believed the centrifugal forces to be fictitious. Our idea is that Leibniz regarded his theory as the real theory of the planetary movement, not merely as a hypothesis. Therefore, it can be asserted that he was working in a non-inertial reference frame, but from his own point of view that frame was simply the only one that allows to grasp the real laws of nature. Among these, there is also the law of centrifugal force. For Leibniz, it is Newton's treatment which does not catch the true nature of things.

(e) David Gregory summarized Leibniz's planetary theory in the proposition LXXVII of his work *Astronomiae physicae et geometricae elementa* (Gregory, 1702, 99–101). In the proposition LXXVIII (pp. 101–104) he commented critically on Leibniz's theory. Let us examine his most significant remarks together with Leibniz's answers:

 (i) The paths of comets along the Zodiac are quite oblique, sometimes they form right angles with the Zodiac, sometimes they move in the opposite direction to the planets. Their motion satisfies the area law. Therefore, if we accept Leibniz's reasoning, each comet should be afloat in a vortex. This vortex would interact with the planetary vortices. This situation entails one problem: one should admit that the cometary vortices and the planetary vortices do not interact at all, otherwise the mutual interaction would modify the motion of the vortices and, hence, would destroy the *circulatio harmonica*. Leibniz replied that in nature there are vortices which do not mutually interfere such as when little stones are thrown into the water and form circular vortices. They do not interfere with each other, although all the vortices belong to the same medium, that is water. This also occurs when sound-waves permeate the air: they do not disturb each other. However, Leibniz never addressed the problem of the vortex stability that had been raised by Newton in the first edition of the *Principia*.

---

[28] Our translation. Cf. Newton 1712?, p. 313: "Propositio vigesima (sic) prima et Propositio vigesima quinta, minorem exhibent vim centrifugam quam gravitatem Planetae in Solem ideoq: falsae sunt. Motus Planetae in orbe non pendet ab excessu gravitatis supra vim centrifugam (ut credit Leibnitius) sed Orbis incurvatur a gravitatis actione sola, cui vis centrifuga (ut reactio vel resistentia) semper est equalis et contraria per motus Legem tertiam a Newtono positam" (Newton is here speaking in third person. Edleston dates this text to 1712). It is not exactly true that the centrifugal force is related to the centripetal by the third law of Newton, but this aspect is not relevant in our context.

(ii) Gregory objected that a planet moving in its vortex would meet a resistance, so that it would be impossible for the planet to maintain its regular motion. Leibniz's reply to this objection seems to be consequent upon this reasoning: (I) his model is mathematically equivalent to the Newtonian one; (II) Newton's model is based on the assumption that space is void and forces act at a distance; (III) since these two assumptions are wrong, while the mathematical model is correct, the physical counterpart of the model must be different from the Newtonian one; (IV) therefore, a physical counterpart to Newton's action at a distance and void space has to exist in which the planets move without suffering any friction, exactly as in Newton's void space. Leibniz supposed that his harmonic vortices were the only reasonable alternative to Newton's ideas: a planet which moves in a harmonic vortex does not suffer any friction caused by the vortex's particles. This solution avoids both the problem of the action at a distance, since the planet moves being afloat in the vortex, and the problem of the void space since the vortex is frictionless. Leibniz wrote:

> The new and marvellous property of the Harmonic Circulation, already shown at the end of § 12, helps us. This property has the feature that the bodies carried in a resistant medium, which, however, circulates harmonically, can move absolutely freely, as if they were moved in a non-resistant medium.[29]

A body in a vortex with harmonic circulation moves as if it were afloat, without experiencing any resistance. Furthermore, given the equivalence between the motion in a void space and that in a harmonic vortex, if the harmonic circulation were interrupted for any reason, it would be quickly restored. Leibniz did not offer any direct physical proof of the harmonic vortices' existence. His reasoning is constructed as a logical deduction based on the assumptions that no action at a distance and no void space can exist.

Leibniz added a further reason in favour of his theory. He observed that the planets rotate in an orbit with a slight inclination on the Sun equator, and that they rotate around their axes in the same sense of rotation as the Sun around its axis. The satellites too rotate in a plane with a slight inclination on the planetary orbit, and they rotate around their axes coherently with the axis-movement of the planets. Gravity in itself cannot explain this situation, while a series of vortices moving in the same direction can. The possible objection to this assumption is that the multiplication of objects is not an explanation. Leibniz explained the observed data by postulating the existence of a series of vortices which fulfil a priori the necessary conditions; but the reason why each vortex has the inclination adequate to determine the planetary inclination is as mysterious as the planetary inclinations themselves.

(iii) Gregory claimed that the planets moving in harmonic vortices do not satisfy Kepler's third law. Here Gregory's objection relies on the interpretation that

---

[29] Our translation. Cf. Leibniz (1706, 1860, 1962, VI, p. 269): "Hic ergo succurrit nobis nova et pulcherrima Circulationis Harmonicae jam sub finem § 12 ostensa proprietas qua efficitur, ut quae feruntur in medio resistente [...] sed tamen harmonice circulante, moveri possint liberrime, non minus quam si moverentur in medio resistendi non capace".

Leibniz's velocity of circulation is the module of the vector velocity and not the transverse velocity. Thus, if Aiton is correct, this objection misunderstands Leibniz's planetary theory. Moreover, Leibniz seized the opportunity to include Kepler's third law in his system. He assumed that all planetary vortices had the same *vis viva* and that the matter of the vortices is limited to the tiny ring between aphelion and perihelion. Relying upon these hypotheses, these statements follow:

(1) the matter m of the vortex in which the planets rotate is as the circumference because the ring between aphelion and perihelion is so thin that can be approximated by a circumference. Such matter is, therefore, as the radius or the distance from the Sun;
(2) since *vis viva* is equal to mv^2, the distances from the Sun are inversely proportional to the squares of the velocities;
(3) if the motion is circular and uniform (which is a reasonable approximation for the planetary motion), the circumferences are as the periodical time by the velocity with which the body moves; therefore, the distances from the Sun have this same proportion;
(4) the velocities are, thus, directly as the distances and inversely as the periodical times;
(5) the squares of the velocities are, therefore, as the squares of the distances directly and the squares of the periodical times inversely;
(6) but, according to (2) the squares of the velocities are inversely proportional to the distances from the Sun;
(7) thence, from (5) and (6) it follows that the squares of the periodical times are as the cubes of the distances from the Sun.

In this way Leibniz was able to accomodate even Kepler's third law into his system.

How to evaluate Leibniz's planetary theory? From an historical point of view it is extremely interesting. It is a clear example of the various elements which converge in Leibniz's science: there is the mathematical level; there is the underlying conviction that the cause of every action must be determined and that the cause of gravity cannot be the action at a distance among bodies, but that gravity depends on some properties of the vortices; there is the idea that a vortex causes not only the attraction of the Sun towards the planet, but also the transverse component of the planet's velocity, namely the *velocitas circulandi*. In addition to the harmonic vortex, the *vis viva* also plays a role because Leibniz deduced Kepler's third law from the, admittedly, bizarre idea that all planetary orbits have the same *vis viva*. Finally, though the metaphysical conception of the pre-established harmony is not directly mentioned by Leibniz in his planetary theory, it represents the philosophical background of the theory itself. It is not a coincidence that he chose the term *harmonica* to indicate the *circulatio* of the planets around the Sun. The *circulatio harmonica* and the fact that all the orbits have the same *vis viva* are features of the harmonic world created by God. The ad hoc hypotheses that Leibniz introduced to explain the planetary motions are arbitrary from a scientific point of view, but they appeared understandable to him and especially necessary in a harmonic universe. Thus, from a scientific perspective, Leibniz's planetary theory is a wonderful piece of mathematical ability, but its physical content cannot be considered of outstanding value: the theory nonetheless presents several interesting elements such as the use of non-inertial reference frames. To understand Leibniz's way of thinking his planetary theory is extremely valuable.

8.2 The Attempt to Unify Mathematics, Metaphysics and Science in Leibniz's Cosmology

Leibniz intended to join metaphysics, mathematics and physics in a rational nexus which admitted epistemic distinctions but with no significant gap in the realm of ontology. For this attitude, he represents the perfect specimen of the intertwining of science and philosophy that we have documented in early modern cosmologies. He aimed at a synthesis which should keep together Christian theology and modern science, mechanicism and teleology, freedom and rational determination, contingency and necessity, individuality and infinity. Obviously, his ambitious intent, though pursued with a sophisticated intelligence and a wide learning, could have hardly succeeded, and the difficulties of his philosophy were the inevitable outcome of this conciliatory project.

A fundamental principle of Leibniz's philosophy is the so called principle of sufficient reason which means simply that nothing is without a reason (nihil est sine ratione) and that there is no effect without a cause. This quintessentially rationalistic principle—endowed with an all-encompassing epistemic value—has a theological version of cosmological relevance: creation is explained as the result of God's choice to create the best possible world among infinite possible worlds. In God's mind there are infinite concatenations of events, each of which is a possible world; among them, God chooses to create the best possible world, which is the world with the fullest presence of beings. Assuming the traditional equation in western metaphysics between being and good, God is determined from his benevolent nature to create the best possible world; he does not enjoy the freedom either not to create any world or to create a world different from the existing one, which would have been inevitably worse. In a writing of 1697, *De rerum originatione radicali* ("On the Ultimate Origination of Things"), Leibniz explained this theological view by means of physical analogies:

> We can now understand in a wonderful way how a kind of divine mathematics or metaphysical mechanism is used in the origin of things and how the determination of the maximum takes place. So the right angle is the determined one of all angles in geometry, and so liquids placed in a different medium compose themselves in the most spacious figure, a sphere. But best of all is the example in ordinary mechanics itself; when many heavy bodies pull upon each other, the resulting motion is such that the maximum possible total descent is secured. For just as all possibilities tend with equal right to existence in proportion to their reality, so all heavy objects tend to descend with equal right in proportion to their weight. And just

> as, in the latter case, that motion is produced which involves the greatest possible descent of these weights, so in the former a world is produced in which a maximum production of possible things takes place.[30]

This passage illustrates both Leibniz's reappropriation of finalism and his predilection for the harmonious correspondence between physics and metaphysics, but his analogical reasoning conceals several problems. What is the exact meaning of the image of all possible worlds striving to come to existence with equal rights? Do they have a sort of ontological impulse in themselves independently of the divine choice? In truth, they cannot claim 'equal right' to existence, otherwise everything which is possible would exist necessarily; only the world involving the maximum of being has a superior right to existence and therefore exists. But why is it so? How can Leibniz distinguish the infinite subsets of the set of possible worlds in God's mind in such a way to prove that they are incompatible with each other and that only one of them can exist? There is no logical criterion that justifies this assumption, since each of the possible worlds is distinct from the other and consistent in itself; Leibniz himself recognized in *De veritatibus primis* (1686–1689) that "it is as yet unknown to men, whence arises the incompossibility of diverse terms or how it can happen that diverse essences are opposed to each other, since all the purely positive terms appear to be compatible with each other".[31] The superficial and somewhat obscure analogies with the mechanics of fluids and falling bodies cannot shed any light on these difficulties.

From the principle of sufficient reason stems another famous Leibnizian principle of equal cosmological significance: that of the identity of the indiscernibles. It is formulated in *Monadology* § 9 in these terms:

> Each monad must be different from every other. For in nature there are never two beings which are perfectly like one another, and between which it would not be possible to find an internal difference, that is, a difference founded on an intrinsic denomination [i.e. on an intrinsic quality].[32]

---

[30] Leibniz (1989b, p. 488). Cf. Leibniz (1697, (1875–1890), VII (1890), p. 304): "Ex his jam mirifice intelligitur, quomodo in ipsa originatione rerum Mathesis quaedam Divina seu Mechanismus Metaphysicus exerceatur, et maximi determinatio habet locum. Uti ex omnibus angulis determinatus est rectus in Geometria, et uti liquores in heterogeneis positi sese in capacissimam figuram nempe sphæricam componunt, sed potissimum uti in ipsa mechanica communi pluribus corporibus gravibus inter se luctantibus talis demum oritur motus, per quem fit maximus descensus in summa. Sicut enim omnia possibilia pari jure ad existendum tendunt pro ratione realitatis, ita omnia pondera pari jure ad descendendum tendunt pro ratione gravitatis, et ut hic prodit motus, quo continetur quam maximus gravium descensus, ita illic prodit mundus, per quem maxima fit possibilium productio".

[31] Transl. by B. Russell (1900, p. 20, note 1). Cf. Leibniz (1686–1689, [1875–1890], VII [1890], p. 195): "Illud tamen adhuc hominibus ignotum est, unde oriatur incompossibilitas diversorum, seu qui fieri possit ut diversae essentiae invicem pugnent, cum omnes termini pure positive videantur esse compatibiles inter se". Cf. Mathieu (1986, pp. 59–65); Mogens Laerke (2016, p. 143).

[32] Leibniz 1965, p. 149. Cf. Leibniz 1714, 2001, p. 12: "Il faut même, que chaque Monade soit différente de chaque autre. Car il n'y a jamais dans la nature deux Êtres, qui soient parfaitement l'un comme l'autre et où il ne soit possible de trouver une différence interne, ou fondée sur une dénomination intrinsèque". See also Leibniz (1896, bk. II chap. xxvii § 1, p. 238): "It is always necessary that besides the difference of time and place there be an internal *principle of distinction*,

This principle defines the individual nature of the monads, those spiritual unities which for Leibniz constitute the entire reality. The identity of indiscernibles derives from the principle of sufficient reason on the assumption that no reason might explain the existence of two monads entirely identical in their properties and different only in number, that is different only for their extrinsic position in time and space. God could not create two identical monads because he would not have any reason to do so. Since monads are metaphysical 'atoms', in the sense that they are elementary centres of force without material extension, their differences cannot be merely quantitative, but must be qualitative, pertaining to their interior spiritual essence: these differentiating properties concern the capacity of each of them to represent the whole universe in an infinite variety of points of view, like mirrors reflecting each other and the whole from a single individual perspective.

Though the identity of indiscernibles is a metaphysical principle rooted in the panpsychist character of Leibniz's monadology, he illustrates it with examples drawn from the occurrences of daily life and from the scientific experience available at that time: telling the anecdote of Princess Sophia who, testing the validity of the principle, could not find two identical leaves in her garden, Leibniz comments that even "two drops of water, or milk, viewed with a microscope, will appear distinguishable from each other".[33] This principle served to disprove atomism, he added. Thus, it can be useful to discuss it beyond Leibniz's metaphysics and to query whether it is compatible with our physical knowledge. Being a statement concerning the ultimate components of reality, no objection drawn from the external appearance of objects is relevant, as we understand from Leibniz's remark on the microscopic observation of two drops of water. Therefore, the principle may be tested only if we consider the microphysics of particles. In this respect, the identity of indiscernibles cannot be held if we consider the epistemic status of elementary particles: how can one electron, photon or quark be distinguishable from another by means of intrinsic properties? In quanta physics two particles cannot even be distinguished on the basis of their extrinsic properties such as their position and speed. Moreover, the complementarity between wave and particle makes the individual identity of ultimate components of a physical field undeterminable. Louis De Broglie wrote that "the individuality of

---

and, though there are many things of the same kind, it is nevertheless true that none of them are ever perfectly alike: thus although time and place (i.e. external relation) serve us in distinguishing things which we do not easily distinguish by themselves, the things do not cease to be distinguishable in themselves. The essence of *identity* and *diversity* consists, then, not in time and place, although it is true that the diversity of things is accompanied by that of time or of place because they bring with them different impressions of the thing". Cf. Leibniz (1704, (1875–1890), 1882, 1978, V, p. 213): "Il faut tousjours qu'outre la difference du temps et du lieu, il y ait un principe interne de distinction, et quoyqu'il y ait plusieurs choses de même espece, il est pourtant vray qu'il n' y en a jamais de parfaitement semblables: ainsi quoyque le temps et le lieu (c'est à dire le rapport au dehors) nous servent à distinguer les choses, que nous ne distinguons pas bien par elles mêmes, les choses ne laissent pas d'estre distinguables en soy. Le precis de l'identité et de la diversité ne consiste donc pas dans le temps et dans le lieu, quoyqu'il soit vray, que la diversité des choses est accompagnée de celle du temps ou du lieu, parce qu'ils amenent avec eux des impressions differentes sur la chose".

[33] Leibniz and Clarke (2000), Leibniz's Fourth Letter § 4, p. 22.

identical particles cannot be 'followed' when the possible realms of presence of these particles merge one with the other in space".[34]

The principle of the identity of indiscernibles establishes the irreducible individuality of every monad, but it is coupled with another fundamental principle of Leibniz's ontology, i.e. the principle of continuity. We read a statement of this principle in a letter of 1702 by Leibniz to the French mathematician Pierre Varignon, with whom he discussed the topic of the infinitesimal calculus:

> The law of continuity demands that *when the essential determinations of a being approach those of another so that likewise accordingly all the properties of the first must gradually approach those of the last*, it is necessary that all the orders of natural beings form only one chain, in which the different classes, like so many links, connect so closely the one to the other, that it is impossible for the senses and the imagination to fix the precise point where any one begins or ends.[35]

Each monad is characterized by its individual difference, but there is a continuous transition from one monad to the other, so that each monad differs from another through infinitesimal variations. The principle of continuity has its origin in Leibniz's mathematics, it is applied to the theory of motion in physics, and has also a metaphysical version, as we can read in *The Monadology* (1714): "I also take it as agreed that every created being is subject to change, and therefore the created monad also, and further that this change is continuous in each one".[36]

Leibniz's refusal to ground individual identity on merely quantitative spatio-temporal parameters is linked also to his view of space and time as derivative and relative dimensions, in stark contrast to the Newtonian conception of absolute space and time. The main metaphysical objection advanced by Leibniz in his correspondence with the Newtonian Samuel Clarke is that if space is "a real absolute Being", then it is infinite and eternal; this entails that space is either an attribute of God (his immensity) or God himself. But this conclusion is theologically unacceptable because space is divisible in parts which is incompatible with the idea of God.[37] A further objection is that the idea of absolute space violates the principle of sufficient reason: space is uniform in all its points that, considered in themselves, do not differ from each other; so, if space were absolute—which means that it existed independently of the created things—God wouldn't have had any reason to locate bodies at

[34] De Broglie (1937, pp. 272 f). See also p. 256: "Le développement des statistiques quantiques a conduit à renoncer complètement à la possibilité de distinguer deux corpuscules de même nature engagés dans un même système et à regarder comme identiques et indiscernables deux états d'un système ne différant l'un de l'autre que par la permutation de deux corpuscules de même nature".

[35] Leibniz (1896, p. 713). Cf. Leibniz (1992, Bd. IV, p. 264). In the *New Essays* Leibniz defines the law of continuity "one of my great maxims, and one of the most verified". It corresponds to the traditional motto that "natura non facit saltus". In physics its meaning is that the transition from rest to motion and from one state of motion to the following occurs through infinitesimal intermediate degrees. In metaphysics it implies that there are infinite degrees of perceptions which are the representative essence of monads. The law of continuity entails the existence of actual infinite. (Leibniz, 1896, pp. 50–51). See: Duchesneau (1994b), Jorgensen (2009).

[36] Leibniz (1989b), p. 643.

[37] Leibniz and Clarke (2000), Leibniz's Third Letter § 3, p. 14.

the creation in a region of space rather than in another (ibid., § 5, p. 15). Therefore, the creation of the world in an absolute space would violate the principle of sufficient reason. The same objection applies to time. If time were absolute, there should not be any reason to explain why God had created things at a determinate instant of time rather than at another instant (ibid., § 6, p. 15). For Leibniz the correct view is that space and time are purely relative to existing things: space is the order of coexistence of things, while time is their order of succession. Points and instants do not exist by themselves apart from the things whose relations of order we perceive in the forms of space and time.

Clarke had an easy reply with regard to the violation of the principle of sufficient reason: he observed that even in Leibniz's view of space and time as relative orders, there would not be any intelligible reason other than the freedom of God's will to explain why things have been created in a particular order instead than in another one.[38] More generally, Clarke refused the absoluteness of the principle of sufficient reason when applied to God's will and his creative actions:

> It is very true that nothing is without a sufficient reason why it is, and why it is thus rather than otherwise. And, therefore, where there is no cause, there can be no effect. But this sufficient reason is often times no other than the mere will of God. There can be no other reason but the mere will of God, for instance, why this particular system of matter should be created in one particular place, and that in another particular place, when (all place being absolutely indifferent to all matter) it would have been exactly the same thing vice versa, supposing the two systems (or the particles) of matter to be alike.[39]

Clarke's reply highlights the fundamental opposition between Leibniz's rationalistic and Newton's voluntaristic approach to theology. The two views are irreconcilable and so in their correspondence Leibniz's and Clarke's arguments and counterarguments are based on incompatible assumptions. For instance, if we examine Clarke's reply on the creation of bodies, it is obvious that Leibniz would not accept the idea that God could choose between "two systems of Matter" which are alike; only one system of Matter (the most perfect) can be created and its structure involves its particular spatio-temporal order.

Clarke was in a more difficult position when he tried to answer Leibniz's theological objection on the relations between God and absolute space. Struggling to defend the shaky theology formulated by Newton in the General Scholium, Clarke asserted that absolute space is not an entity, but only a property of the divine essence and that space is not actually divisible in parts: "for infinite space is *one,* absolutely and essentially *indivisible*, and to suppose it *parted* is a contradiction in terms, because there must be space in the *partition itself*, which is to suppose it *parted* and yet *not parted* at the same time".[40] Thus, Clarke concluded, the term 'parts' may be referred to space only in a 'figurative' way. Leibniz found Clarke's argument incorrect: "that space has no parts because its parts are not separable and cannot be removed from one another by being plucked out" is irrelevant: parts may exist even if they are not

[38] Ibid., Clarke's Third Reply, § 2, p. 18.

[39] Ibid., Clarke's Second Reply, § 1, p. 11.

[40] Ibid., Clarke's Third Reply, § 3, p. 19.

separable and the partition of space is obtained "either by the bodies that are in it or by lines and surfaces that may be drawn and described in it".[41]

The relation of God to the universe is a major topic of the correspondence with Clarke. Leibniz labelled Newton's theological ideas as "ridiculous" as he attacked the Newtonian thesis that God had to intervene from time to time to regulate the mechanics of the world, which would naturally tend to defective alterations due to cumulative anomalies produced by gravity attraction.[42] To this picture of God as a clockmaker who is forced to repair the imperfect machine he has built, Leibniz opposed his idea of a perfect rational order pre-established by the divine Intellect:

> According to my opinion, the same force and vigor always remains in the world and only passes from one part of matter to another in agreement with the laws of nature and the beautiful pre-established order. And I hold that when God works miracles, he does not do it in order to supply the needs of nature, but those of grace. Whoever thinks otherwise, must necessarily have a very mean notion of the wisdom and power of God.[43]

Clarke replied that Leibniz's view of the Deity as an Intelligence above the world (*intelligentia supramundana*) severed the providential link between God and his creatures and turned out to be in agreement with materialistic determinism in conceiving the universe like a self-moving machine which does not require any divine presence.[44] This exchange of accusations reveals how thorny was the issue of the relation of God to the universe for authors such as Newton and Leibniz (and Descartes before them) who wanted to keep together their scientific knowledge with their metaphysical and theological opinions. A satisfactory balance between transcendence and immanence was impossible to achieve: Newton's God seemed to merge with the world through the divine properties of absolute space and time, but at the same time was proclaimed to be a Lord *Pantokrator* whose free will rules the creation, in which he intervenes directly from time to time with his 'miraculous' repair of the decaying machine; Leibniz's God was a God of pure reason, the architect of the pre-established rational harmony among all monads, but precisely because his essence and actions obeyed to the logical principle of identity and to the ontological principle of sufficient reason, God and the Universe were bound together in the infinite continuity of a deterministic a priori rationality.[45]

One of the most significant aspects of Leibniz's cosmological views which deserves further discussion is his notion of relative space and time especially because, after Mach and Einstein, it has been appraised as being in accordance with our current

---

[41] Ibid., Leibniz's Fifth Letter, § 51, p. 48f.

[42] In the Latin edition of *Opticks* (1706) Newton held that the quantity of motion in the system of the world is bound to decrease due to the growing irregularities stemming from the mutual attractions among the planets. Newton entertained this thesis for philosophical reasons: "He was led to this concept of decay, at least in part, by his growing realization that a mechanical universe in which the amount of motion remained the same could be used by atheists to argue against the existence of a Deity" (Kubrin, 1967, p. 334).

[43] Leibniz and Clarke (2000), Leibniz's First Letter § 4, p. 4.

[44] Ibid., Clarke's First Reply § 4, p. 6.

[45] On the different role and incidence of Newton's and Leibniz's theology and metaphysics within their physics, see our Conclusion.

scientific theories. He considers space and time as relations and consequently as ideal beings (*entia rationis*), as rational constructions: if relations are mistakenly transformed in real things, entities are multiplied without necessity contrary to the Occam principle.[46] Does this mean that for Leibniz reality is spaceless and timeless? No, because he distinguishes duration and extension, which belong to things, from time and space which are only the abstract frame in which we order the mutual relations of things which have their own duration and extension:

> It appears that the author [sc. Clarke] confounds immensity, or the extension of things, with the space according to which that extension is taken. Infinite space is not the immensity of God; finite space is not the extension of bodies, as time is not their duration. Things keep their extension, but they do not always keep their space. Everything has its own extension, its own duration, but it does not have its own time and does not keep its own space.[47]

There is, then, an objective, ontological counterpart of our spatio-temporal relational order, though, unlike Newton, extension and duration do not have any autonomous existence (any 'absolute reality') but inhere to the existence of things. Having defined the metaphysical grounding of space and time, Leibniz in the correspondence with Clarke explains how the notion of space arises from a process of rational abstraction. Space, as we said, is the order of coexistence of things; this order consists in their distance. The idea of distance, by means of which we determine the spatial 'situation' of things, presupposes the idea of place:

> When it happens that one of those coexistent things changes its relation to a multitude of others which do not change their relation among themselves, and that another thing, newly come, acquires the same relation to the others as the former had, we then say that it is come into the place of the former; and this change we call a motion in that body in which is the immediate cause of the change (ibid., § 47, p. 46).

In this way we form the notion of 'place' (*locus*) that is connected to the idea of local motion. We call 'space' that general idea which comprehends all places. To achieve the idea of space is sufficient to consider the relations and the rules of change (i.e. the laws of motion) among real things, "without needing to fancy any absolute reality *out of* the things whose situation we consider" (ibid.). Place, however, should not be confused with the actual 'relation of situation' of a body towards other bodies: while the place (being an abstraction) remains the same, the 'relation of situation' between a fixed body *C* and two bodies (*A* and *B*) which occupy the same place in two different times is not the same, because the individual properties of the bodies *A* and *B* are different. In sum, space is an ideal entity conceived by mind to describe the real relations of existing things, but we must take care to understand that the two levels do not coincide. In Aristotelian terms, space is not a substance, because only existing things are substances; space is not even an accident, because an accident inheres always to a single being and cannot belong to a plurality of them at the same time. Space is a "mere ideal thing" used to express "the truth of relations" among real beings (ibid., p. 47).

---

[46] Manuscript annotation by Leibniz to his Fourth Paper. Leibniz and Clarke (1957, pp. 100–101). Cf. Leibniz (1967, vol I, p. 322, note 9).

[47] Leibniz and Clarke (2000), Leibniz's Fifth Letter § 46, p. 45. Cf. Vailati (1997), p. 112.

This sample of analysis, which has been summarized here in the most concise way, illustrates how Leibniz considered together metaphysical, mathematical and physical aspects, in an effort both to connect and to distinguish them. As we can expect, this difficult attempt to harmonize the various levels of consideration left several problems undiscussed and unsolved. For instance, in what sense space and time are only ideal frames to describe relations of things if things themselves have both duration and extension? ‘Things’ are called bodies in physics, but in metaphysics they must be conceived as monads, i.e. isolated (and yet coordinated by God’s plan of creation) spiritual unities: how can we interpret extension and duration as true ontological properties of incorporeal entities? Moreover, at a metaphysical level, what is the relation of the duration and extension of creatures to the immensity and eternity of God and how do the latter differ from the former?[48]

To complete our examination of space and time we will now discuss the commonly held view that Leibniz’s relational and ideal doctrine of space and time is closer to contemporary science than the absolutist Newtonian conception. In his Fourth Paper Leibniz states that “if there were no creatures, space and time would be only in the ideas of God”.[49] This implies that the science of space and time is not merely a human construction, but concerns a set of eternal truths, an ideal order, immutably conceived by the mind of God. In an annotation to the correspondence with Clarke, Leibniz wrote that since time and space (like numbers and other ideal things) belong to essences and not to existences, God cannot alter the order of instants and of places.[50]

Here we see that Leibniz’s and Newton’s concepts of space and time have similarities and differences: for Leibniz the ideal order of the chronological and spatial relations is certainly not relative to human observers, but is fixed forever, exactly like the mathematical order of the Newtonian absolute space and time. Leibniz’s space is the set of all possible relations among objects with no restriction to the positions of existing objects; in this respect, it is no different from Newton’s absolute space that, being independent of the existence of physical bodies, pre-exists their creation and has the capacity of hosting whatever possible relations among bodies God might choose to create. The true difference lies in the fact that for Leibniz time and space indicate relations among bodies and don’t have an objectual existence. They have the same logical status as mathematical relations. For instance, if one considers the relation “to be bigger”, it is absolutely true that 5 is bigger than 3, independently of the fact that any human being knows such truth. This does not mean that the relation “to be bigger” is an object which exists independently of the numbers to which it is applied. Similarly, for Leibniz it is absolutely objective that two events have a

---

[48] Ibid., § 106, p. 61: “The immensity and eternity of God are things more transcendent than the duration and extension of creatures, not only with respect to the greatness, but also to the nature of the things. Those divine attributes do not imply the supposition of things extrinsic to God, such as are actual places and times. These truths have been sufficiently acknowledged by theologians and philosophers”. We see here how Leibniz fell back in traditional theological modes contenting himself with unexplained assertions.

[49] Ibid., Leibniz’s Fourth Letter, § 41, p. 27.

[50] Leibniz and Clarke (1957, pp. 100–101). Cf. Leibniz (1967, vol. I, p. 323, note 11).

certain distance *d* or that they are separated by an interval of time *t*. However, this does not imply that the sets of all distances or of all time intervals pre-exist physical events like two entities subsisting by themselves, which was Newton's conception. Leibniz's view of the relational nature of time and space is coherent with his idea that the void does not exist: if no void space exists, certainly a space *à la* Newton cannot have any physical reality. Space and time for Leibniz have only an ideal meaning.

As for actual physical space, the Newtonian view was defended by Clarke with a precise reference to Newton's bucket experiment and to the explanation of centrifugal forces within an absolute spatial frame. Clarke wrote in his Fourth Reply:

> To this argument no answer has ever been given. It is largely insisted on by Sir Isaac Newton in his *Mathematical Principles* (definition 8) where, from the consideration of the properties, causes, and effects of motion, he shows the difference between real motion, or a body's being carried from one part of space to another, and relative motion, which is merely a change of the order or situation of bodies with respect to each other. This argument is a mathematical one, showing from real effects that there may be real motion where there is none relative, and relative motion where there is none real; it is not to be answered by barely asserting the contrary.[51]

In his reply to Clarke Leibniz did not discuss this point. He conceded that "there is a difference between an absolute true motion of a body and a mere relative change of its situation with respect to another body", but repeated that he had refuted absolute space in virtue of his principle of sufficient reason.[52] Another difference between Newton and Leibniz here comes to the fore: Newton conceived absolute space metaphysically as an attribute of God, but did not resort to God to assert its physical existence; rather, he proposed the bucket experiment. As Sklar remarks:

> Newton's argument [in favour of absolute space] changes the whole atmosphere of the philosophical debate. If he is correct, we cannot resolve the ontological issue of the "reality" of space by applying to our common-sense knowledge of the world the critical apparatus of the philosopher. We must, in addition, take account of particular fundamental features of the best available contemporary science; and as this science develops and changes, we must perhaps change our metaphysics accordingly. (Sklar, 1974, pp. 158–159)

---

[51] Leibniz and Clarke (2000), Clarke's Fourth Reply § 13, p. 31.

[52] Ibid., Leibniz's Fifth Letter, § 53, p. 49. Leibniz's admission of absolute motion derives from his dynamics. In a critical note on art. 25 of Part II of Descartes's *Principles of Philosophy*, Leibniz wrote: "If motion is nothing but the change of contact or of immediate vicinity, it follows that we can never define which thing is moved. For just as the same phenomena may be interpreted by different hypotheses in astronomy, so it will always be possible to attribute the real motion to either one or the other of the two bodies which change their mutual vicinity or position. Hence, since one of them is arbitrarily chosen to be at rest or moving at a given rate in a given line, we may define geometrically what motion or rest is to be ascribed to the other, so as to produce the given phenomena. Hence if there is nothing more in motion than this reciprocal change, it follows that there is no reason in nature to ascribe motion to one thing rather than to others. The consequence of this will be that there is no real motion. Thus, in order to say that something is moving, we will require not only that it change its position with respect to other things but also that there be within itself a cause of change, a force, an action" ("Critical Thoughts on the General Part of the Principles of Descartes", 1692: Leibniz, 1989b, p. 393).

On the contrary, to refute absolute space in his correspondence with Clarke, Leibniz invoked the metaphysical principle of sufficient reason, assuming that even God obeyed it in its creation.[53]

More than any other author examined in our book Leibniz shows how cosmological views are shaped by a combination of interconnected perspectives stemming from theology, metaphysics, mathematics and physics. His monadology is a theory of the universe in which the spiritual activity of each monad and their infinite variety play a central role. Leibniz thought that his physics (centred on dynamics) and his mathematics (the discovery of the calculus which entailed the concepts of infinitesimals and of the continuum) converged with his metaphysical view. His extremely ambitious project, which still attracts those philosophers who love to reconcile the irreconcilable, could not avoid the difficulties arising from this combination. Holding together different perspectives (such as mechanicism in physics and teleology in metaphysics) merely by hinting at different levels of knowledge and consideration is a philosophical expedient that cannot be accepted at face value: the connections among different levels of knowledge or discourse should be clearly determined and motivated; the transitions and shifting from one realm to another should be clearly explained and justified: these are the missing links in Leibniz's philosophy.[54]

Leibniz's rationalism may in some instances appear closer to early-modern science than the Newtonian theology of a personal Biblical God (see above, p. 246f.), but we should not forget that Leibniz's metaphysics was also influenced by theological and apologetic concerns. An example of how he bent his own views for apologetic purposes is given in a passage of his *Essais de Théodicée* (1710). Here he evokes the theory of gravitational attraction (that he himself denied) to show that the substantial presence of God in the Eucharist is a plausible or at least not an absurd doctrine, since it shares a presupposition with a physical theory, namely the Newtonian attraction at distance:

> Thence we see that the dogma of real and substantial participation can be supported […] by a properly understood analogy between *immediate operation* and *presence*. Many philosophers have deemed that, even in the order of Nature, a body may operate from a distance immediately on many remote bodies at the same time. So do they believe, all the more, that nothing can prevent divine Omnipotence from causing one body to be present in many bodies together, since the transition from immediate operation to presence is but slight, the one perhaps depending upon the other. It is true that modern philosophers for some time now have denied the immediate natural operation of one body upon another remote from it, and I confess that I am of their opinion. Meanwhile remote operation has just been revived in England by the admirable Mr. Newton, who maintains that it is the nature of bodies to be attracted and gravitate one towards another, in proportion to the mass of each one, and the rays of attraction it receives. Accordingly the famous Mr. Locke, in his answer to Bishop Stillingfleet, declares that having seen Mr. Newton's book he retracts what he himself said, following the opinion of the moderns, in his *Essay concerning Human Understanding*, to wit, that a body cannot operate immediately upon another except by touching it upon its surface

---

[53] Cf. above p. 279f. On this topic see Bussotti (2017c, pp. 136–143).

[54] Jammer (1957, p. 158), wrote that "it is almost axiomatic for Leibniz's philosophical and scientific reasoning that the physico-mechanical and the theologico-teleological conceptions of the world should not exclude each other".

> and driving it by its motion. He acknowledges that God can put properties into matter which cause it to operate from a distance. Thus the theologians of the Augsburg Confession claim that God may ordain not only that a body may operate immediately on divers bodies remote from one another, but that it even exist in their neighbourhood and be received by them in a way with which distances of place and dimensions of space have nothing to do. Although this effect transcends the forces of Nature, they do not think it possible to show that it surpasses the power of the Author of Nature. For him it is easy to annul the laws that he has given or to dispense with them as seems good to him, in the same way as he was able to make iron float upon water and to stay the operation of fire upon the human body.[55]

We wonder how Leibniz could defend a theological dogma with a physical analogy based on a theory that he himself regarded as false. Ironically, it seems an appropriate strategy to explain the mystery of transubstantiation! To understand the anomaly of the previous passage in the context of Leibniz's thought, we must consider that Leibniz had embraced mechanicism at a very early age, according to his autobiographical letter to Nicolas Remond.[56] He had critized Cartesian mechanicism and introduced the notion of force, but his 'reform of mechanicism' never led him to admit attractive forces acting at a distance between bodies. In his *Specimen Dynamicum*

---

[55] Leibniz (1985, p. 85f). ("Preliminary Dissertation on the Conformity of Faith with Reason", § 19). Cf. Leibniz (1710, [1875–1890], 1885, 1978, VI, pp. 61–62): "On voit par là que le dogme de la participation réelle et substantielle se peut soutenir [...] par une Analogie bien entendue entre l'operation immediate, et la presence. Et comme plusieurs philosophes ont jugé que, même dans l'ordre de la nature, un corps peut operer immediatement en distance sur plusieurs corps éloignés tout à la fois, ils croient, à plus forte raison, que rien ne peut empêcher la toute-puissance divine de faire qu'un corps soit present à plusieurs corps ensemble, n'y ayant pas un grand trajet de l'operation immédiate à la presence, et peut-être l'une dependant de l'autre. Il est vrai que, depuis quelque temps, les philosophes modernes ont rejetté l'operation naturelle immediate d'un corps sur un autre corps éloigné, et j'avoue que je suis de leur sentiment. Cependant l'operation en distance vient d'être rehabilitée en Angleterre par l'excellent M. Newton, qui soutient qu'il est de la nature des corps de s'attirer et de peser les uns sur les autres, à proportion de la masse d'un chacun et des rayons d'attraction qu'il reçoit; sur quoi le celebre M. Locke a déclaré, en repondant à M. l'évêque Stillingfleet, qu'après avoir v le livre de M. Newton, il retracte ce qu'il avait dit lui-même, suivant l'opinion des modernes, dans son Essai sur d'entendement, savoir qu'un corps ne peut operer immediatement sur un autre qu'en le touchant par sa superficie, et en le poussant par son mouvement, et il reconnoit que Dieu peut mettre des proprietés dans la matiere, qui la fassent opérer dans l'éloignement. C'est ainsi que les Theologiens de la Confession d'Augsbourg soutiennent qu'il depend de Dieu, non seulement qu'un corps opere immediatement sur plusieurs autres éloignés entr'eux, mais qu'il existe même auprès d'eux et en soit reçu d'une maniere dans laquelle les intervalles des lieux et les dimensions des espaces n'ayent point de part. Et quoyque cet effet surpasse les forces de la nature, ils ne croyent point qu'on puisse faire voir qu'il surpasse la puissance de l'Auteur de la nature, à qui il est aisé d'abroger les loix qu'il a données ou d'en dispenser comme bon luy semble, de la même maniere qu'il a pû faire nager le fer sur l'eau et suspendre l'operation du feu sur le corps humain". On Leibniz's apologetic interest in the Eucharist and transubstantiation see Fouke (1992, p. 145): "As part of his general defense of Christianity and his efforts at ecclesiastical reunion, the Eucharist remained a philosophical concern for Leibniz from the 1660's to the end of his life. Each stage of his philosophical development and every modification of his system, was accompanied by an explanation of the Eucharist and a demonstration that his metaphysical system is compatible with Transubstantiation".

[56] Leibniz to Remond, 10 January 1714, in Leibniz (1875–1890, vol. 3, p. 606), where Leibniz recalls that at the age of fifteen he had resolved to dismiss the scholastic substantial forms in favor of mechanicism.

(1695), he criticized "the philosophical opinions of certain great mathematicians [...] who admit empty space and seem not to shrink from the theory of attraction". Leibniz is, indeed, certain that "true attraction is not found in nature".[57] In the Preface to his *Nouveaux Essais sur l'Entendement Humain* (first draft completed in 1704), Leibniz commented on Locke's passage to which he refers in the *Essais de Théodicée*: he approves the conception of the moderns (that was also shared initially by Locke) "that one part of matter works immediately upon another only by pushing it by contact"[58]; he then takes notice of Locke's retractation and praises him for his "modest piety", in so far as Locke recognized "that God can do more than we can understand, and that thus there may be inconceivable mysteries in the articles of faith"; however, he concludes that we should not be "obliged to recur to the miracle in the ordinary course of nature and to admit powers and operations absolutely inexplicable".[59] In 1705, advocating his system of pre-established harmony, Leibniz wrote that "it has the advantage of conserving, in its full rigor and generality, the great principle of physics that a body never receives a change in motion except through another body in motion which pushes it: *corpus non moveri nisi impulsum a corpore contiguo et moto*".[60] In his famous correspondence with Clarke (1715–16), only five years after Leibniz wrote the passage quoted from his *Essais de Théodicée* (1710), "the admirable Mr. Newton" and "the famous Mr. Locke" had become two major examples of that dangerous materialism that was infecting British philosophy.[61] There is a strong continuity in Leibniz's thought about the impossibility of action at a distance, which is reasserted even when Leibniz resorts to the Newtonian theory for apologetic purposes.

What is surprising in the quoted passage is the mixing of, not to say confusion between, theological and scientific discourse. Initially, Leibniz presents the analogy between immediate operation (in physics) and presence (in theology). He then rejects the immediate operation by attraction as a physical theory, without examining it: he refers only to Locke's remarks on the Newtonian theory, which he interprets as based purely on the omnipotence of God, without considering Locke's underlying sceptical theory of our ignorance of the essences of substances. Finally, Leibniz concludes by juxtaposing the Newtonian-Lockean doctrine with the *Confessio Augustana*, resorting again to the argument of the supreme omnipotence of God who can suspend the laws of nature. This conclusion amounts to the nullification of his previous considerations, since Leibniz could have said from the very beginning that the substantial presence of Jesus Christ in the sacrament of the holy communion is a supernatural event, and therefore there is no need to find, and no possibility of finding, an analogy

---

[57] Leibniz, (1989b, p. 449).

[58] Leibnitz, (1896, p. 54).

[59] Ibid., p. 55.

[60] Leibniz (1989b, p. 587) ("Considerations on Vital Principles and Plastic Natures, by the Author of the System of Pre-established harmony").

[61] These accusations were formulated in the letter by Leibniz to Caroline, princess of Hanover, that occasioned the dispute between him and Samuel Clarke: see Koyré (1957, p. 235f).

in wrong physical theories. The eucharistic presence is an event that could be classified among the miracles and the articles of faith, but as Leibniz has clarified in the Preface to the *Nouveaux Essais* the ordinary course of nature is a rational sequence of events that has to be understood without any recourse to miracles (such as the mysterious attraction).

The passage we have examined is a clear example of how theological concerns interfered with scientific and philosophical views in early modern thinkers. In this case there is no real attempt at conciliating distinct, not to say antithetical domains: Leibniz did not change his ideas on attraction with a view to harmonizing his physical theories with a revealed supernatural doctrine. The lesson we can draw from this case is much simpler: it is evidence of the weight of apologetic purposes in Leibniz's eclectic philosophy, and it's also a typical example of apologetic procedures in which every kind of argument, even the weakest and contradictory ones, were employed *ad majorem Dei gloriam.*

References

Aiton, E. J. (1960). The celestial mechanics of Leibniz. *Annals of Science, 16*(2), 65–82.

Aiton, E. J. (1960). The celestial mechanics of Leibniz in the light of Newtonian criticism. *Annals of Science, 18*(1), 1–41.

Aiton, E. J. (1964). The celestial mechanics of Leibniz: A new interpretation. *Annals of Science, 20*(2), 111–123.

Aiton, E. J. (1965). An imaginary error in the celestial mechanics of Leibniz. *Annals of Science, 21*(3), 169–173.

Aiton, E. J. (1972). *The vortex theory of planetary motions.* New York: American Elsevier Publishing Company.

Allen, D. (1984). From *vis viva* to primary force in matter. In A. Heinekamp (Ed.), *Leibniz' dynamica* (pp. 55–61). Steiner.

Arthur, R. (1998). Cohesion, division and harmony: Physical aspects of Leibniz's continuum problem (1671–1686). *Perspectives on Science, 6*(1–2), 110–135.

Bassler, O. (2002). Motion and mind in the balance: The transformation of Leibniz's early philosophy. *Studia Leibnitiana, 34*(2), 221–231.

Beeley, P. (1999). Mathematics and nature in Leibniz's early philosophy. In S. Brown (Ed.), *The young Leibniz and his philosophy (1646–76)* (pp. 123–145). Dordrecht-Boston-London: Kluwer Academic Publishers.

Bernstein, H. R. (1980). *Conatus*, Hobbes and the young Leibniz. *Studies in History and Philosophy of Science, 11*, 25–37.

Bernstein, H. R. (1981). Passivity and inertia in Leibniz's dynamics. *Studia Leibnitiana, 13*(1), 97–113.

Bertoloni Meli, D. (1993). *Equivalence and priority: Newton versus Leibniz.* Oxford: The Clarendon Press.

Boudri, J. C. (2002). *What was mechanical about mechanics: The concept of force between metaphysics and mechanics from Newton to Lagrange.* Dordrecht-Boston-London: Kluwer Academic Publishers.

Bouquiaux, L. (2008). Leibniz against the unreasonable Newtonian physics. In M. Dascal (Ed.), *Leibniz: What Kind of Rationalist?* (pp. 99–110). New York: Springer.

Bussotti, P. (2015). *The complex itinerary of Leibniz's planetary theory: Physical convictions, metaphysical principles and Keplerian inspiration.* Basel: Springer-Birkhäuser.

Bussotti, P. (2017b). *Leibniz. Dal calcolo infinitesimale al linguaggio dei computer*, serie "Grandangolo Scienza", "Corriere della Sera". RCS Media Group.

Bussotti, P. (2017c). Il ruolo dello spazio assoluto nella fisica di Newton. In A. Neri (Ed.), *Là dove il pensiero incontra l'esperienza* I (pp. 111–162). Pisa: ETS.

Bussotti, P., & Lotti, B. (2018). The problem of circular motion in René Descartes. *Giornale Critico della Filosofia Italiana*, settima serie vol. XIV, anno XCVII (XCIX), fasc. I, Gennaio-Aprile 2018 (pp. 76–114).

Bussotti, P., & Pisano, R. (2017). Historical and philosophical details on Leibniz's planetary theory as physical-structural model. In R. Pisano, M. Fichant, P. Bussotti, & A. R. E. Oliveira (Eds.), *The dialogue between sciences, philosophy and engineering. New historical and epistemological insights. Homage to Gottfried W. Leibniz 1646–1716* (pp. 49–92). London: College Publications.

Capecchi, D. (2012). *History of virtual works law*. Milan-Heidelberg-New York-Dordrecht-London: Springer.

Capek, M. (1966). Leibniz's thought prior to the year 1670, from atomism to a geometrical kinetism. *Revue Internationale De Philosophie, 20*, 249–256.

Carlin, L. (2006). Leibniz on final causes. *Journal of the History of Philosophy, 44*(2), 217–233.

Crockett, T. (2008). Space and time in Leibniz's early metaphysics. *The Leibniz Review, 18*, 41–79.

De Broglie, L. (1937). *La physique nouvelle et les quanta*. Paris: Flammarion.

Duchesneau, F. (1994b). Leibniz on the principle of continuity. *Revue Internationale de Philosophie*, 48, *188*(2), 141–160.

Duchesneau, F. (1989). Leibniz's *Hypothesis Physica Nova*: A conjuction of models for explaining phenomena in an intimate relation. *Studies in the history and philosophy of science. Boston Studies in the Philosophy of Science, 116*, 153–170.

Duchesneau, F. (1994a). *La dynamique de Leibniz*. Paris: Vrin.

Dumitrescu, M. (2011). L'accord entre les causes efficiente et finale dans la théorie Leibnizienne de l'harmonie préétablie. In H. Breger, J. Herbts, & S. Erdner (Eds.), *Natur und Subjekt, IX internationaler Leibniz Kongress unter der Schirmerrscahft des Bundespräsidenten, Hannover 26. September bis 1. Oktober 2011. Vorträge, Teil 1* (pp. 255–261). Hannover: Hartmann.

Fouke, D. C. (1992). Metaphysics and the eucharist in the early Leibniz. *Studia Leibnitiana*, Bd. 24, H. 2, 145–159.

Gabbey, A. (1971). Force and inertia in seventeenth-century dynamics. *Studies in History and Philosophy of Science, II*, 1–67.

Gale, G. (1973). Leibniz' dynamical metaphysics and the origins of the *vis viva* controversy. *Systematics: The Journal of the Institute for the Comparative Study of History, Philosophy and the Sciences, 11*(3), 184–207.

Gale, G. (1984). Leibniz' force: Where physics and metaphysics collide. In A. Heinekamp (Ed.), *Leibniz' dynamica* (pp. 62–70). Stuttgart: Steiner.

Garber, D. (1994). Leibniz: Physics and philosophy. In L. Jolley (Ed.), *The Cambridge companion to Leibniz* (pp. 270–352). Cambridge: Cambridge University Press.

Garber, D. (2008). Dead force, infinitesimals, and the mathematicization of nature. In U. Goldenbaum & D. Jesseph (Eds.), *Infinitesimal differences: Controversies between Leibniz and his contemporaries* (pp. 281–306). De Gruyter.

Garber, D. (2009). *Leibniz: Body, substance, monad*. Oxford: Oxford University Press.

Ghins, M. (1990). *L'inertie et l'espace-temps absolu de Newton à Einstein: une analyse philosophique*. Bruxelles: Palais des Académies.

Giorgio, E. (2011). Trägheit und *Circulatio Harmonica*: Formen der Subjektivität im "Tentamen" von 1689 und in der "Theodizee". In H. Breger, J. Herbts, & S. Erdner (Eds.), *Natur und Subjekt, IX internationaler Leibniz Kongress unter der Schirmerrscahft des Bundespräsidenten*, Hannover 26. September bis 1. Oktober 2011. Vorträge, Teil 1 (pp. 292–299). Hannover: Hartmann.

Giulini, G. (2002). Das Problem Der Trägheit. *Philosophia Naturalis, 39*, 343–374.

Gregory, D. (1702). *Astronomiae physicae et geometricae elementa*. Oxford: Sheldonian Theatre.

Gueroult, M. (1934). *Leibniz: Dynamique et métaphysique; suivi d'une note sur le principe de la moindre action chez Maupertuis*. Paris: Les Belles-Lettres.

Heimann, P. M. (1977, March). Leibniz and Johann Bernoulli's theory of motion. *Centaurus* (pp. 1–26).

Iltis, C. (1971). Leibniz and the *Vis Viva* controversy. *Isis, 62*(1), 21–35.

Iltis, C. (1973). The decline of Cartesianism in mechanics: The Leibnizian-Cartesian debates. *Isis, 64*(3), 356–373.

Jammer, M. (1957). *Concepts of force. A study in the foundation of dynamics*. Cambridge (Mass.): Harvard University Press.

Jauernig, A. (2008). Leibniz on motion and the equivalence of hypotheses. *The Leibniz Review, 18*, 1–40.

Jorgensen, L. M. (2009). The principle of continuity and Leibniz's theory of consciousness. *Journal of the History of Philosophy, 47*(2), 223–248.

Knobloch, E. (1995). Harmony and cosmos: Mathematics serving a teleological understanding of the world. *Physis, 32*(1), 54–89.

Koyré, A. (1957). *From the closed world to the infinite universe*. Baltimore: John Hopkins Press.

Kubrin, D. (1967). Newton and the cyclical cosmos: Providence and the mechanical philosophy. *Journal of the History of Ideas, 28*, 325–346.

Lariviere, B. (1987). Leibnitian relationalism and the problem of inertia. *Canadian Journal of Philosophy, 17*(2), 437–448.

Leibniz, G. W. (1671). *Hypothesis physica nova*. In Leibniz ([1849–1863], 1962), (Vol. VI, pp. 17–59).

Leibniz, G. W. (1677). *Leibniz an Honoratus Fabri*. In Leibniz ([1849–1863], 1962), (Vol. VI, pp. 81–106).

Leibniz, G. W. (1686a). *Brevis Demonstratio Erroris memorabilis Cartesii et aliorum circa Legem naturalem, secundum, quod volunt a Deo eandem semper quantitatem motus conservari, qua et in re mechanica abutuntur* plus Beilage. In Leibniz ([1849–1863], 1962), (Vol. VI, pp. 117–123).

Leibniz, G. W. (1686b). *Discours de metaphysique*. In Leibniz ([1875–1890], 1978), (Vol. IV, pp. 427–463).

Leibniz, G. W. (1686–1689). *De veritatibus primis*. In Leibniz ([1875–1890], 1978), (Vol. VII, pp. 194–195).

Leibniz, G. W. (1689). *Tentamen de Motuum Coelestium Causis* (Erste Bearbeitung). In Leibniz ([1849–1863], 1962), (Vol. VI, pp. 144–161).

Leibniz, G. W. (1690). *De Causa gravitatis et defensio sententiae Autoris de viris Naturae Legibus contra Cartesianos*. In Leibniz ([1849–1863], 1962), (Vol. VI, pp. 193–201).

Leibniz, G. W. (1690, 1860, 1962). *Tentamen de Motuum Coelestium Causis* (Zweite Bearbeitung). In Leibniz ([1849–1863], 1962) (Vol. VI, pp. 161–187).

Leibniz, G. W. (1695). *Specimen Dynamicum pro admirandis Naturae Legibus circa corporum vires et mutuas actiones detegentis et ad suas causas revocandis*. Pars I et Pars II. In Leibniz ([1849–1863], 1962), (Vol. VI, pp. 234–254).

Leibniz, G. W. (1697). *De rerum originatione radicali*. In Leibniz ([1875–1890], 1890) (Vol. VII, pp. 302–308).

Leibniz, G. W. (1704). *Nouveaux essais sur l'entendement humain par l'auteur du systeme de l'harmonie preestablie*. In Leibniz ([1875–1890], 1882, 1978) (Vol. V).

Leibniz, G. W. (1706, 1860, 1962). *Illustratio Tentaminis de Motuum Coelestium Causis*, Pars I et Pars II plus Beilage. In Leibniz ([1849–1863], 1962), (Vol. VI, pp. 254–280).

Leibniz, G. W. (1840). *Opera philosophica Omnia*. In J. E. Erdmann (Ed.). Sumtibus G. Eichleri.

Leibniz, G. W. ([1849–1863], 1962). *Mathematische Schriften* (7 vols). In C. I. Gerhardt (Ed.) Georg Olms.

Leibniz, G. W. ([1875–1890], 1978). *Philosophische Schriften* (7 vols) In C. I. Gehrardt (Ed.). Georg Olms.

Leibniz, G. W. (1896). *New essays concerning human understanding*. Together with an Appendix consisting of some of his shorter pieces (A. G. Langley, Transl). New York: The Macmillan Company.

Leibniz, G. W. (1903). *Opuscules et fragments inedites de Leibniz extraits des manuscrits de la Biblioteque Royale de Hannover* (ed. L. Couturat). Alcan.
Leibniz, G.W. (1923, 1938, 1950). *Sämtliche Schriften und Briefe* (abbreviated LSB). Darmstad, 1923, Leipzig, 1938, Berlin, 1950: Deutsche Akademie der Wissenschaften zu Berlin.
Leibniz, G. W. (1967). *Scritti filosofici*, a cura di D.O. Bianca. Torino: UTET.
Leibniz, G. W. (1969). *Philosophical papers and letters*, (L. E. Loemker (Ed.) by Transl.). Dordrecht; Reidel.
Leibniz, G. W. (1985). *Theodicy. Essays on the Goodness of God, the freedom of man and the origin of evil*. Ed. with an Introd. by A. Farrer. (E. M. Huggard, Transl.). Chicago and La Salle: Open Court.
Leibniz, G.W. (1989a). *Philosophical essays*. Translated by R. Ariew and D. Garber. Indianapolis & Cambridge: Hackett.
Leibniz, G. W. (1989b). *Philosophical papers and letters*. In L. E. Loemker (Ed.). Dordrecht: Kluwer.
Leibniz, G. W. (2013). *The Leibniz-De Volder correspondence: with selections from the correspondence between Leibniz and Johann Bernoulli*. Translated by P. Lodge. New Haven: Yale University Press.
Leibniz, G. W., & Clarke, S. (1957). *Correspondance Leibniz-Clarke* presentée d'après les manuscrits originaux des Bibliothèques de Hanovre et de Londres, par A. Robinet. Paris: Presses Universitaires de France.
Leibniz, G. W., & Clarke, S. (2000). *Correspondence*. Ed. with an Introduction by R. Ariew. Indianapolis-Cambridge: Hackett Publishing Company.
Lindsay, R. B. (1975). The introduction of *vis viva* and related concepts. In R. B. Lindsay (Ed.), *Energy: Historical development of the concept* (pp. 108–118). Stroudsburg Pa: Dowden, Hutchinson and Ross.
Lodge, P. (1997). Force and the nature of body in "Discourse on metaphysics" 17–18. *Leibniz Society Review, 7*, 116–124.
Lodge, P. (2001). Primitive and derivative forces in Leibnizian bodies. In H. Poser, C. Asmuth, & U. Goldenbaum (Eds.), Nihil sine ratione*: Mensch, Natur und Technik im Wirken von G.W. Leibniz Schirmherrschaft; VII. Internationaler Leibniz-Kongreß* (pp. 720–727). Hannover: Gottfried-Wilhelm-Leibniz-Gesellschaft.
Look, B. C. (2011). Leibniz's theory of causation. In B. C. Look (Ed.), *The continuum companion to Leibniz* (pp. 174–191). London: Continuum Publishing Corporation.
Lovejoy, A. O. (1936, 1964). *The great chain of being. A study of the history of an idea*, The William James Lectures delivered at Harvard University, 1933. Harvard University Press.
Mathieu, V. (1986). *Introduzione a Leibniz*. Roma-Bari: Laterza.
McDonough, J. K. (2007, revised 2019). Leibniz's philosophy of physics. In E. N. Zalta (Ed.), *Stanford Encyclopedia of Philosophy*. https://plato.stanford.edu/entries/leibniz-physics/.
Miller, R. B. (1982). *Force and substance: A study of the interrelation of dynamics and metaphysics in Leibniz*. New Brunswick: Rutgers University Press.
Mogens Laerke, M. (2016). *Compossibility, compatibility, congruity*. In G. Brown, & Y. Chieck (Eds.), *Leibniz on compossibility and possible worlds* (pp. 125–144). Cham (Switzerland): Springer.
Newton, I. (1712?) Epistola cujusdam ad amicum. In J. Edleston (Ed.), *Correspondence of Sir Isaac Newton and Professor Cotes, including letters of other eminent men* (pp. 308–314). Parker (1850).
Papineau, D. (1977). The *vis viva* controversy: Do meanings matter? *Studies in History and Philosophy of Science, 8*(2), 111–142.
Parkinson, J. H. R. (1974). Science and metaphysics in Leibniz's *Specimen Inventorum. Studia Leibnitiana, 6*(1), 1–27.
Pasini, E. (2016). G. W. Leibniz's anti-death perspective: Spontaneity of death and absolute immortality. In C. Tandy (Ed.), *Death and anti-death, volume 14: Four decades after Michael Polanyi, three centuries after G. W. Leibniz* (pp. 169–186). Ann Arbour: Ria University Press.

Puryear, S. (2012). Motion in Leibniz's middle years: A compatibilist approach. In D. Garber, & D. Rutherford (Eds.), *Oxford studies in early modern philosophy* (Vol. VI, pp. 135–170). Oxford: Clarendon Press.

Ranea, A. G. (1986). A priori, inercia y acción motriz en la dinámica de Leibniz. *Revista De Filosofía, 26–27*, 164–169.

Roberts, J. T. (2003). Leibniz on force and absolute motion. *Philosophy of Science, 70*, 553–573.

Rozemond, M. (2009). Leibniz on final causation. In S. Newlands, L.M. Jorgensen (Eds.), *Metaphysics and good: themes from the philosophy of Robert Merryhew Adams*, pp. 272–294. Oxford: Oxford University Press.

Russell, B. (1900). *A critical exposition of the philosophy of Leibniz*. Cambridge: Cambridge University Press.

Sklar, L. (1974). *Space, time and spacetime*. Berkeley-Los Angeles-London: University of California Press.

Stevenson, G. (1997). Miracles, force, and Leibnizian laws of nature. *Studia Leibnitiana, 29*, 167–188.

Suisky, D. (2009). *Euler as Physicist*. Berlin: Springer.

Vailati, E. (1997). *Leibniz and Clarke. A study of their correspondence*. Oxford: Oxford University Press.

Wilson, C. (1990). Atom, minds, and vortices in *De Summa Rerum*: Leibniz vis-à-vis Hobbes and Spinoza. In S. Brown (Ed.), *The young Leibniz and his philosophy* (pp. 223–243). Dordrecht: Kluwer Academic Publishers.

Woolhouse, R. S. (2000). Leibniz's collision rules for inertialess bodies indifferent to motion. *History of Philosophy Quarterly, 17*(2), 143–157.

Conclusion

In the history of science it would be a naive and anachronistic error to assess past theories as if they were only progressive steps towards the formation of the body of scientific knowledge that we possess today. No serious historian of science makes this mistake anymore. More generally, no intellectual historian evaluates past doctrines and views only in so far as they are anticipations of our current standard ideas. It is unquestionable that every scientific theory was and is elaborated to resolve specific problems that are typical of the intellectual and historical context in which researchers operate. This is a truth that applies to the past as well as to the present. A scientific theory can be understood only if we take into account the state-of-the-art knowledge at a specific period of time, the set of available observational data, the existing technology with which to conduct observations and experiments, as well as the social and cultural environment in all its aspects. These are the conditions that affect every scientific enterprise, the actual 'circumstances' in which every scientific investigation is carried out. Besides, in the history of science as well as in other areas of historical research it is impossible to trace a linear, one-sided development. The interrelation and complexity of the elements involved are so great that it would be historically meaningless to select a group of topics, regarding them as anticipatory of contemporary science, and to neglect all remaining aspects, discarding them as the ballast of the past. Sifting gold from sand is not an appropriate metaphor to describe the job of an historian. In the mines of history there are no golden nuggets that can be easily detected and cleaned of the surrounding dirt.

The foregoing considerations are commonly accepted and need not be insisted upon. However, to obtain a correct historiographical perspective we cannot limit ourselves to the awareness that the past cannot be interpreted as a simple preparation for the present. There is another side of the coin that we risk overlooking in our emphasis on historical contextualization. Benedetto Croce used to say that all history is always contemporary history, meaning that historical investigation is driven by the historians' intellectual interests which are necessarily linked to the present in which

P. Bussotti and B. Lotti, *Cosmology in the Early Modern Age: A Web of Ideas*, Logic, Epistemology, and the Unity of Science 56,
https://doi.org/10.1007/978-3-031-12195-1

every historian lives.[1] We may modify the meaning of Croce's thesis to apply it to the history of science, pointing out that our present scientific knowledge is the inevitable and indispensable background from which we are able to reconstruct and evaluate the emerging of modern science. It is simply impossible to understand how problems were addressed and phenomena were explained in early modern science without taking into account our wider and higher understanding of the physical universe. It is not a question of choice, it is an epistemic necessity, as well as an inevitable historical situation. We come later and we know better than the natural philosophers and scientists of the early modern centuries. This is indisputable unless one denies, against all historical evidence, the cumulative character of natural science, at least starting from the seventeenth century onwards, and the increased exactness and predictive ability of contemporary science compared to early-modern science. Historical and cultural relativism may be upheld with some arguments as far as humanities are concerned, but certainly cannot be integrally applied to scientific knowledge, for the very simple reason that the primary object of science is not the construction of the human social and cultural world, but the knowledge of the natural world in its objective reality. The history of science is evidence that this knowledge is possible and that it is constantly improved upon.

Of course, we are not holding that our current scientific knowledge is a sufficient condition to understanding past science—to assume this would be to make the naive anachronistic mistake we mentioned above—but it is surely a valuable instrument and, in any case, is an inevitable presupposition of our historiography, since we cannot remove ourselves from our culture. Our vantage point in assessing the previous stages of the history of science is a matter-of-fact which we should take advantage of. It enables us better to understand the development of natural knowledge, without compelling us anachronistically to superimpose on the past theories the categories and problems that belong to our time. Because of our historical situation we are able to detect the growth of physical science across the centuries and we can also understand the connections of scientific theories to philosophical issues which, though very closely linked to scientific problems, remained logically distinct from them and had a different historical development. While in some cases entire philosophical areas and/or specific topics have been at least partially 'absorbed' into natural science, in other cases they remained autonomous as they were not strictly relevant to the development of natural science. Moreover, philosophy itself underwent deep transformations following the rise of modern science. The long itinerary of the 'theory of knowledge' from Bacon to Kant—that Cassirer has examined with so much attention and insight in his classic *Das Erkenntnisproblem in der Philosophie und Wissenschaft der neueren Zeit*—attest to the change of philosophical thinking in connection to the rise of modern science.[2]

---

[1] Croce (1923, p. 12): "if contemporary history springs straight from life, so too does that history which is called non-contemporary, for it is evident that only an interest in the life of the present can move one to investigate past fact. Therefore this past fact does not answer to a past interest, but to a present interest in so far as it is unified with an interest of the present life".

[2] See on this topic Wagner (2002, p. 25f).

Our purpose has been to outline an interdisciplinary historical itinerary in which we have tried to highlight the connections between natural science and philosophy in the context of cosmological ideas developed during the sixteenth and seventeenth centuries. We have insisted on the significance of the relations between philosophy and science in an age when these two realms were gradually separating. Many of the results obtained by the scientists we have analysed are probably not understandable without referring to their general conception of the universe. Cosmology entails a view of the totality of nature and therefore the relation between scientific investigations and philosophical (at that time, metaphysical) assumptions is inevitable. However, it is one thing to imagine the totality of the universe only on the basis of overall metaphysical intuitions, it is quite another to establish general laws concerning a certain type of phenomena: Bruno's and Spinoza's infinitism is an instance of the first sort, while Kepler's planetary laws and Newton's law of universal gravitation are instances of the second. The discussions we have examined on the very difficult problem of inertia and the connected nature of space—two issues strictly related to cosmology—demonstrate that, progressively, problems (such as the nature of space and time) which had been regarded as philosophical were included at least tentatively within a scientific approach. This critical and conceptual distinction, that we are logically allowed to draw, does not imply that philosophical ideas and scientific research were always perceived as clearly distinct in early modern cosmological views.

To clarify further our view on this topic it may be appropriate to draw a comparison between Kepler and Leibniz: perhaps, had not Kepler believed in a harmonic universe, regulated by geometrical and musical proportions, his discoveries of the kinematical laws governing the planets and of the dynamical structure of the universe would have not been possible. It is not by chance that Kepler's third law is formulated in the *Harmonice Mundi*. Similarly, the concept of *vis viva* and its principle of conservation in Leibniz's dynamics cannot be explained prescinding from Leibniz's metaphysics, and it is a mistake to overemphasize the analogy between those concepts and our concept of energy and its conservation, which was developed only around the middle of the nineteenth century. On the other hand, the cases of Kepler and Leibniz require a careful examination to eschew the error of overestimating their similarities. Kepler had a scientific mindset: his interests concerned the material universe and he searched for a system of laws to understand it; some of these laws—such as the three Kepler's laws—are perfectly in line with our concept of scientific law; they are mathematically formulated and can be verified or falsified. As for Kepler's dynamical conceptions, they were rudimentary and imprecise, but they paved the way for Newton. Kepler's ideas of a geometrical and musical harmony of the cosmos (in his case, referred to the solar system) are wrong, but he gave a mathematical treatment of these presumed 'harmonies' in order to discover the inherent laws of the universe. In this sense, even if they appear to be a typical inheritance of Renaissance Platonism, they were elaborated in a very similar way to subsequent scientific theories. It is certain that, as a philosopher, he was a natural philosopher. Despite the fact that religious issues played a role in his view of the universe, his theory belongs to natural philosophy, and his investigative attitude is in tune with the new modern science of nature. We have recalled how he criticized the occultist Renaissance tradition: "Kepler accused Fludd of being a theosophist rather than a natural philosopher, and Kepler was right" (Copenhaver, 1998, p. 465).

When we consider Leibniz the situation is different: as is well known, he was a master of logic and mathematics, but his scientific or at least dynamic concepts do not arise completely in the context of natural philosophy. Leibniz was interested in reintroducing final causes because he conceived the universe in the light of his array of metaphysical principles (the principle of sufficient reason, of pre-established harmony, of continuity, and of the identity of indiscernibles). Leibniz interpreted the principle of *vis viva* conservation as a consequence of his key ontological concept, the monad: a spiritual, active unity encompassing the infinity in itself. Therefore, Leibniz was first of all a metaphysician with strong theological interests (think, for instance, of the voluminous *Theodicy* he published in 1710). He was also a natural philosopher, but he constantly subordinated natural philosophy to his metaphysics. Leibniz distinguished different epistemic levels (logical, mathematical, physical, metaphysical) and tried to articulate them in such a way as to make them coherent and compatible. In particular, he believed that "the ultimate reasons for mechanism and the very laws of motion" could not be found "in mathematics, that one must return to metaphysics" (Leibniz 1969, p. 655). His monadology was an attempt to ground physical science in a metaphysical worldview. As a philosopher he was much more refined than his rival Newton, but his scientific achievements are less important compared to those of Newton, and his dynamics is functional to his metaphysical views. At the same time, this connection does not imply that Leibniz's dynamics is scientifically irrelevant: the concept of *vis viva* is undoubtedly significant in the history of physics since it opened the door to a physics of principles whose approach is different from Newton's physics of forces.

In drawing these comparisons and distinctions we do maintain the point of view of the history of science and of the history of philosophy. We do not need to enter here into a complicated discussion about the well known and much equivocated difference between the "context of discovery" and the "context of justification".[3] For our purpose we need simply to observe the historical development of cosmological theories and to admit the obvious distinction between logical relations among concepts and how ideas have been developed historically. For instance, Kepler's planetary laws are interpreted by Kepler himself in the context of his own metaphysical views about the Platonic solids and the harmonic correspondences. Newton recognizes the validity of Kepler's laws without sharing Kepler's metaphysics. Newton's law of gravitation encapsulates Kepler's planetary laws and is interpreted by Newton within metaphysical assumptions that are very different from Kepler's ones. Later on, Lagrange and Laplace developed a detailed astronomical theory on the basis of Newtonian mechanics without any relation to Newton's theistic views. This historical lesson is very simple: it is always possible to link natural science to metaphysics and

[3] On the multi-layered issues involved in that distinction see Hoyningen-Huene (1987).

to give metaphysical interpretations of scientific theories; however, scientific theories are logically independent of the metaphysical views that are attached to them in various times.

When we examine the relations between philosophy and science in the early modern centuries we must be careful to neither underestimate nor overemphasize them. Furthermore, even in cases in which those relations are conspicuous in the scientists' theories, as they are in Kepler and Leibniz, their origin, development and meaning are different. It is a task of a correct interdisciplinary approach to point out homologies and differences and not to fall in an analogical vision, which presents an inaccurate and rough picture of the intricate relation between philosophy and natural science. We have tried to point out these nuances in our work.

The case of Newton is also emblematic: no one denies that Newton had strong religious interests and convictions that influenced his intellectual attitude and motivated his research psychologically. His studies on the *Book of Revelation*, and the General Scholium to the *Principia,* where he illustrated his theology and defined absolute space and time as attributes of God, testify to his interest in metaphysics and religion. But it would be a mistake to interpret his scientific concepts as deriving from his philosophical-religious convictions: according to Newton, absolute space and time fulfil a precise physical role within his theory, that is they guarantee respectively the validity of the principle of inertia (absolute space) and the absolute validity of all Newton's deductions because, in Newton's fundamental equation—i.e., considering the one-dimensional case, $F = m\frac{d^2x}{dt^2}$, or $F = m\ddot{x}$, to use Newton's notation—acceleration is the second derivative of space with respect to time. If time were not absolute and uniformly flowing this simple form of the equation would not be possible. Furthermore, in the whole of his scientific works, considering both the *Principia* and the *Optiks*, there are very few references to notions extraneous to mathematics or science. Newton's scientific results are explicable within Newton's scientific research. Therefore, the problem here becomes the link between scientific tenets and religious-philosophical convictions. What really matters is to keep firm to the distinction between Newton's science and Newton's philosophical interpretation of his own scientific theories. Whatever the genetic mutual influence between these two elements, it is undeniable that the conceptual value and consistency of Newton's scientific achievements do not require his philosophical interpretations: it is likely that Newton believed that gravity was the manifestation of God's omnipotence in the physical world,[4] but this conviction is irrelevant to the scientific significance of the law of universal gravity.

The case of Descartes is slightly different. The relations between his metaphysics and physics are highly debated, and we cannot confidently assert that his metaphysics was conceived only to provide a theoretical frame to justify his innovative physics. However, it is indisputable that his first published work was a methodical introduction to three scientific essays, all written in French,[5] and that in the only

[4] See David Gregory's record, dated 21 December 1705: cit. in Hall and Hall (1995, p. 87).

[5] The *Discours de la Méthode* published in 1637 together with *Dioptrique, Météores* and *Géométrie* as samples of the method.

systematic exposition of his philosophy—*Principia Philosophiae*—the first part is devoted to metaphysics and theory of knowledge, while the three remaining parts are entirely devoted to natural philosophy. It is also evident that Descartes' mechanistic approach depended on an epistemological project: he wanted to free science from all obscure references to the mysterious essences of objects, and from the merely verbal pseudo-explanations which characterised most aspects of astrology and Renaissance natural philosophy. From a certain perspective, even astrology and Renaissance natural philosophy can be interpreted as attempts to find laws governing the universe; but Descartes thought that such attempts were bound to fail because they misconceived the nature of matter. For him the world is a machine, it is not a sort of living being whose parts are linked by a mutual spiritual influence. The interaction of matter particles is only mechanical. His mechanistic idea of nature is both a philosophical insight and the outcome of his mathematical knowledge, because he reduced matter simply to three-dimensional space variously subdivided. His *Principia Philosophiae* are both a manifesto and the (at least embryonic) realization of these ideas. The problem is that in the *Principia* Descartes did not implement the mathematization of physical laws; despite his insistence on geometry as our only way of knowing the nature of matter, he did not describe phenomena in geometrical terms and illustrated his system of the world only in a narrative fictional way. He did not build a real physical theory, but merely outlined a conjectural sketch of a theory which was inevitably studded with errors. Huygens was the great scientist who, faithful to Descartes' mechanistic conception, developed, at least in part, the mathematical apparatus missing in the *Principia Philosophiae*. Huygens was a pure scientist, but was influenced by his convictions on the nature of physical interactions. He was surely less involved than Newton into religious, philosophical and metaphysical interests, but he remained anchored to the mechanistic world-view—a presupposition which narrowed his thinking. Newton was completely free in this respect: he acted without taking into account by which medium and in which modes an interaction (specifically gravity interaction) could be transmitted. He overcame this question and, without binding himself to any presupposition on this topic, he analysed gravity insofar as it is apparent in the universe and conceived a theory of gravity which was the milestone of physics for more than two centuries. In this case no presupposition was a good presupposition!

The genesis of scientific concepts is something different from their logical status and it is a very difficult historiographic problem to guess how much the metaphysics of a scientist influenced his cosmological and, more in general, scientific concepts and research. In every chapter of our book, we have tried to sketch an answer to this question, but this answer cannot be formulated in general terms, as it depends on the propensions and intentions of each individual scientist. All the authors we have analysed recognised that God is omniscient, almighty and perfect, but the subsidiary metaphysical principles they added to this conviction are different and influenced their scientific results in diverse ways and degrees. Moreover, there is not a one-to-one correspondence between a particular metaphysical-theological view and a scientific theory: a given metaphysics does not necessarily imply a certain scientific conception and vice versa. For example, it is true that Newton's concept of absolute space is consistent with his view of space as the *sensorium Dei*, but to believe in the existence of a *sensorium Dei* does not imply that it is 'materialized' in a physical entity with the characteristics of the absolute space as described by Newton in his physical theory. On the other hand, it is likely that the conception of absolute space influenced, in some way, his belief that space was the *sensorium Dei*. Kepler believed in the Holy Trinity, but this does not imply the identification of Sun-space of the universe-fixed stars with Father-Holy Spirit-Son. The link between the two triads is analogical. Actually, we can observe that the more metaphysical concepts become specific, and rationally well-structured, the more they influence physical conceptions. Thence, granted all the notions of Leibniz's metaphysics, his view of relative time and space follows almost naturally (though not necessarily).

From the history of cosmology we can draw a further lesson concerning the complex relations between science and metaphysics with regard to the issue of the place of man in the universe. Among the early-modern thinkers who shared a theistic world-view it was common opinion to consider the knowledge of nature as the discovery of the divine wisdom contained in creation. This theological belief conferred a religious or at least a metaphysical significance on the scientific enterprise. The 'mission' of the natural philosopher was to disclose the divine intelligence manifested in the natural works. This persuasion offered a self-justification to natural science by means of which mankind fulfilled its noblest task: to understand the manifestation of the sublime reason of the artificer and to glorify him. However, the new cosmological ideas influenced the view of mankind and of its place in nature in other ways too. In authors such as Descartes and Leibniz their metaphysical *Weltanschauungen* encompassed both the physical world and the particular position of man: man was a mixed substance in the trialist Cartesian philosophy, and a monad endowed with apperception in the Leibnitian monadology. However, if we think about Copernicus, Kepler, Huygens, Galilei, and Newton, we cannot find any overarching metaphysics that interprets the physical world and, at the same time, the ontological position of man in the universe. It is easy to see the reason for this difference: Descartes and Leibniz were proper metaphysicians, while the other above-mentioned authors focused their attention on the study of nature and on particular problems and phenomena.

This does not mean that the evolution of scientific theories did not exert an impact on the image of man. The Copernican revolution brought about, in the long run, a momentous transformation of the self-perception of mankind's position in the universe. It was a cultural process that grew slowly in time and that involved philosophers more directly than scientists. Infinity of space and/or of matter, plurality of worlds (meaning by that the existence of planetary systems around the stars in analogy with our solar system), and the existence of extraterrestrial beings were three major outcomes of Copernicanism, that did not necessarily go together.[6] Bruno was not a scientist in the modern sense of this term, but embraced these three tenets; Kepler, who was also a proper scientist, talked of the existence of the inhabitants of the Moon, but denied the infinity of the universe and the plurality of worlds for

[6] For an accurate exposition of these topics see Del Prete, (1998).

theological reasons; Galileo, who was a cautious scientist with no inclination for metaphysical speculations, did not take any definite position; Descartes' physical ontology entails the indefinite extension of the universe, which he distinguished from a positive assertion of its infinity; Descartes denied the plurality of worlds,[7] and was not interested in speculating about extraterrestrial beings. Spinoza, whose contributions to scientific investigations are very limited, was the greatest exponent of metaphysical infinitism of his age, and his view of the position of man in the universe was innovative to the point of being compatible with today's naturalism; however, he does not say anything explicit about the plurality of worlds and the existence of extraterrestrial beings, though these theses are fully compatible with his metaphysical view; Huygens talks extensively of the inhabitants of other planets, and accepts the idea of the plurality of worlds, but is more hesitant about the infinity of the universe since space is infinite, but the number of the stars or of other unknown objects could be finite; Newton's universe is immersed in the infinity of space and he admits the possibility of a plurality of worlds, but does not talk about inhabitants of other worlds. In the *Essays of Theodicy* (part I § 19) Leibniz hints at the existence of other planetary systems and of extraterrestrial beings in the context of a theodicy argument with no direct relation to cosmology: he holds that "the evil could not but seem to be almost as nothing in comparison with the good", since it is possible that the rational inhabitants of other worlds are only "blessed creatures". As for the infinity of space, he does not take a clear position ("whatever bounds are given or not given to the universe") and formulates the conjecture of an immense space beyond the region of the stars.[8] This variety of opinion is not surprising since these three topics were not logically entailed by the new physical and astronomical theories: infinitism was and still is a metaphysical assumption or conjecture, while the plurality of worlds and the existence of extraterrestrial beings were and are empirical hypotheses; the first is supported by strong analogy, and, strictly speaking, has been proved only very recently with the discovery of exoplanets, while the second is still a matter of scientific speculation.

The history of cosmology shows the common roots of science and philosophy, as well as their persisting relations even after the rise of modern science rendered scientific knowledge more and more autonomous and independent of metaphysical thought. From the eighteenth century onwards, the history of the relations between science and philosophy is an history of growing differences, which would be absurd to underestimate,[9] though ultimate issues about the ontology of physics and the fundamental features of the universe (for instance its size in space and its evolution in time) are necessarily linked to general philosophical considerations even today.

---

[7] For Descartes ' distinction between 'indefinite' and 'infinite' in relation to the universe see above, Section 5.2. As for the plurality of worlds, Descartes regarded the fixed stars as other suns, but since the matter is the same everywhere and occupies the whole space, he denied that there is a plurality of worlds separated by empty spaces and insisted on the unity of the homogeneous universe (see *Principia* II 22: AT VIII-1 52).

[8] Leibniz, (1985, p 134 f).

[9] Werner, (2002, p. 26), holds that the distinction and opposition between science and philosophy, which is so familiar to us, was definitively established only in the nineteenth century.

Our project has aimed to highlight this profound and long-standing connection, so that in our age of increasing specialized learning we remain aware of how complex and still unified is the cognitive effort of mankind to understand nature and our place in it. As we have seen, philosophy was (and is) tied to science from various perspectives, which may be grouped in two spheres: that of metaphysical interpretations and speculations, and that of epistemology, methodology and critical analysis of fundamental concepts.

As we said in our Introduction, we did not aim to outline a history of cosmology in early modern centuries either in philosophical or in scientific terms. In our project we have tried to illustrate the interrelationships of scientific issues with those wider philosophical assumptions and interpretations which accompanied them. To this end, we have focused on the main protagonists of the rise of modern science, in which these relationships emerge in all their complexity. In our twofold exposition we have tried to avoid repetitions and useless overlapping and aimed to display a rich web of concepts in their creative interplay. We have been led in this effort by the conviction that science, in all its essential articulations (mathematical, physical, biological, and philosophical), though being an historical and cultural event, is not simply that: science is the only way to dispel uncritical and delusive world-views and to gain an increasingly reliable knowledge subject to critical examination and rational validation.

Science does not have the pretension to replace the old metaphysics as a new kind of absolute knowledge, for its purposes, procedures and methods are different. Science, in particular physics and cosmology, develops on the basis of primitive assumptions (for instance, the laws and axioms of Newton or the principles of Einstein), which are presupposed to be true. A mathematical model based on these assumptions is then constructed. If the model fits the data and enables scientists to provide new correct predictions, it is accepted. Of course, there are several phenomena which escape the explanatory power of a scientific model: for example, the precession of Mercury perihelion in Newton's theory (there are similar cases in relativity theory and quantum mechanics too). This limit does not imply that the model has to be rejected, for a model does not claim to be an absolute and complete knowledge of reality itself. A model is an instrument crafted in the attempt to understand how nature works, how the universe is made, but the answers offered by the model are always restricted to a certain sphere of reality. Its aim is not to capture reality in its entirety nor to grasp the 'ultimate essence' of things, if this phrase has any meaning. Obviously, the more we build unifying theories the more we might achieve a global scientific vision of reality, but it is quite a naive idea—an idea good for a movie, at most—to think that there is a final theory, the theory of everything. Our models approximate better and better a limit—this limit we can call 'reality'—without never reaching an all-encompassing description and explanation capable of transforming the limit into an ultimate result. Perhaps we will be able to answer questions concerning the origin of the universe or the structure of black holes, but other new questions will inevitably arise. The history of science has shown that more and more comprehensive models are possible, but so far every model has left unanswered questions. This gap between our scientific models and the reality of the universe does not imply that in the universe something exists which, by definition, would be out of reach of physical knowledge. Rather, this gap is explicable in two ways: (a) it is the effect of the obvious limits of our mind, which is located in a particular and transient spatio-temporal experience; (b) as we said, given the limits of our experience, it would be idealistic naivety to assume a perfect mirroring between our mathematical-scientific constructions and external reality or, to put it better, to assume that objective reality itself is nothing other than a mathematical structure mirrored to various degrees by our intellect.[10] In its efforts to orient itself in a complicated and multifaceted reality, our mind developed scientific knowledge as its only effective instrument. An accepted scientific model is the temporary limit of our knowledge; more than one model can coexist, particularly when a significant amount of new data and the existence of unresolved problems prevent the acceptance of a unique consolidated model, as is the case in contemporary cosmology. Beyond the models, room is open for metaphysical speculations which, even when related to scientific theories, do not provide actual knowledge.

Cosmology deals with an all-encompassing reality and for this reason Kant believed that the universe as a whole was not a genuine object of cognition and could only be conceived by reason as a transcendental idea. This critical awareness had not prevented science from pushing its research further and further in a true Kantian spirit, expanding the known boundaries of the cosmos, both in time and space, and investigating the very core of matter and energy. Cosmology faces the most fundamental riddles and physics seems to be the only way of decoding at least some of them, without falling into the persistent trap of an anthropomorphic view of nature. As Heraclitus said, "*physis krypthesthai philei*", "nature loves to hide".

[10] Not all the scientists share these opinions. For example, Tegmark (2007 and 2014) thinks that the entire universe is a mathematical object and that reality is completely resolved (not to say dissolved) in a mathematical structure. Though we are convinced that any phenomenon can be scientifically understood when we are able to provide for it a mathematical model, we do not share Tegmarks' view that mathematics is inherent in the universe or that the universe is nothing other than a mathematical objectified structure. Rather, we believe that mathematics is the most suitable means we have to understand reality within a scientific approach. The relation between mathematical modelization and physical data is a primary epistemological issue and is historically a consequence of the rise of modern science, when mathematics became the fundamental theoretical method of physics while Aristotle's sharp distinction between mathematics and physics (see *Metaphysics*, VI 1) faded away to be replaced by their integration culminating in the Newtonian approach. Of course, this process does not imply a progressive identification between mathematics and physics in so far as many areas of mathematical knowledge are not connected to physics nor to other sciences.

Bibliography

Acerbi, F. (2000). Le fonti del mito platonico di Galileo. *Physis, 37*, 359–392.
Aiton, E. J. (1954). Galileo's theory of the tides. *Annals of Science, 10*(1), 44–57.
Aiton, E. J. (1960). The celestial mechanics of Leibniz. *Annals of Science, 16*(2), 65–82.
Aiton, E. J. (1962). The celestial mechanics of Leibniz in the light of Newtonian criticism. *Annals of Science, 18*(1), 1–41.
Aiton, E. J. (1964). The celestial mechanics of Leibniz: A new interpretation. *Annals of Science, 20*(2), 111–123.
Aiton, E. J. (1965). An imaginary error in the celestial mechanics of Leibniz. *Annals of Science, 21*(3), 169–173.
Aiton, E. J. (1972). *The vortex theory of planetary motions*. New York: American Elsevier Publishing Company.
Aiton, E. J. (1973). Infinitesimals and the area law. In F. Krafft, K. Meyer, & B. Stickel (Eds.), *Internationales Kepler-symposium* (pp. 285–305). Hildesheim: Gerstenberg.
Aiton, E. J. (1978). Kepler's path to the construction and rejection of his first oval orbit for Mars. *Annals of Science, 35*(2), 173–190.
Allen, D. (1984). From *vis viva* to primary force in matter. In A. Heinekamp (Ed.), *Leibniz' dynamica* (pp. 55–61). Stuttgart: Steiner.
Alquié, F. (2005). *Leçons sur Descartes. Science et métaphysique chez Descartes*. Paris: La Table Ronde.
Aquinas, Thomas (1918). *Summa contra Gentiles*. In *Opera Omnia*, editio Leonina, vol. 13. Rome: Typis Riccardi Garroni.
Armogathe, J. R. (2000). The rainbow: A privileged epistemological model. In S. Gaukroger, J. Schuster, & J. Sutton (Eds.), *Descartes' natural philosophy* (pp. 249–257). London: Routledge.
Arthur, R. (1994). Space and relativity in Newton and Leibniz. *The British Journal for the Philosophy of Science, 45*(1), 219–240.
Arthur, R. (1998). Cohesion, division and harmony: Physical aspects of Leibniz's continuum problem (1671–1686). *Perspectives on Science, 6*(1–2), 110–135.
Arthur, R. (2007). Beeckman, Descartes and the force of motion. *Journal of the History of Philosophy, XLV*, 1–28.
Assis, A. K. T. (1989). On Mach's principle. *Foundations of Physics Letters, 2*, 301–318.
Assis, A. K. T. (1995). Weber's law and Mach's principle. In J. B. Barbour & H. Pfister (Eds.), *Mach's principle from Newton's Bucket to quantum gravity* (pp. 159–171). Boston-Basel-Berlin: Birkhäuser.

P. Bussotti and B. Lotti, *Cosmology in the Early Modern Age: A Web of Ideas*, Logic, Epistemology, and the Unity of Science 56,
https://doi.org/10.1007/978-3-031-12195-1

Assis, A. K. T. (1999). Arguments in favour of action at a distance. In A. E. Chubykalo, V. Pope, & R. Smirnov-Rueda (Eds.), *Instantaneous action at a distance—Pro and Contra* (pp. 45–56). Commack: Nova Science Publishers.

Assis, A. K. T. (2003). The relationship between Mach's principle and the principle of physical proportions. In M. Sachs, A. R. Roy (Eds.), *Mach's principle and the origin of inertia* (pp. 37–44). Montreal: Apeiron.

Assis, A. K. T., & Graneau, P. (1995). The reality of Newtonian forces of inertia. *Hadronic Journal, 18*, 271–289.

Baigre, B. B. (1990). The justification of Kepler's ellipse. *Studies in History and Philosophy of Science, 21*(4), 633–664.

Barbour, J. B. (1975). Forceless Machian dynamics. *Nuovo Cimento, 26B*, 16–22.

Barbour, J. B. (1982). Relational concepts of space and time. *British Journal for the Philosophy of Science, 33*, 251–274.

Barbour, J. B. (1989). *Absolute or relative motion? A study from a Machian point of view of the discovery and the structure of dynamical theories* (Vol. 1). Cambridge: Cambridge University Press.

Barbour, J. B. (2001). *The discovery of mechanics*. Oxford: Oxford University Press.

Barbour, J. B., & Bertotti, B. (1977). Gravity and inertia in a Machian framework. *Nuovo Cimento B, 38*, 1–27.

Barbour, J. B., & Bertotti, B. (1982). Mach's principle and the structure of dynamical theories. *Proceedings of the Royal Society of London, A, 382*, 295–306.

Barbour, J. B., & Pfister, H. (Eds.). (1995). *Mach's principle from Newton's bucket to quantum gravity* (pp. 159–171). Boston-Basel-Berlin: Birkhäuser.

Barcaro, U. (1980). Il concetto di inerzia nei "Discorsi" di Galileo. *Physis, XXI, I*(3–4), 325–342.

Barker, P., & Goldstein, B. R. (2001). Theological foundations of Kepler's astronomy. *Osiris, 16*, 88–113.

Bassler, O. (2002). Motion and mind in the balance: The transformation of Leibniz's early philosophy. *Studia Leibnitiana, 34*(2), 221–231.

Beeley, P. (1999). Mathematics and nature in Leibniz's early philosophy. In S. Brown (Ed.), *The young Leibniz and his philosophy (1646–76)* (pp. 123–145). Dordrecht-Boston-London: Kluwer Academic Publishers.

Belkind, O. (2007). Newton's conceptual argument for absolute space. *International Studies in the Philosophy of Science, 21*(3), 271–293.

Belkind, O. (2012). *Physical systems. Conceptual pathways between flat space-time and matter*. Dordrecht-Heidelberg-London-New York: Springer.

Bell, A. E. (1947). *Christian Huygens and the development of science in the seventeenth century*. London: Edward Arnold.

Bentley, R. (1692–93). *A confutation of atheism from the origin and frame of the world*. London: H. Mortlock.

Berman, M. S. (2008). On the Machian origin of inertia. *Astrophysics and Space Science, 318*, 269–272.

Bernstein, H. R. (1980). *Conatus*, Hobbes and the young Leibniz. *Studies in the History and Philosophy of Science, 11*, 25–37.

Bernstein, H. R. (1981). Passivity and inertia in Leibniz's dynamics. *Studia Leibnitiana, 13*(1), 97–113.

Berry, A. (1898). *A short history of astronomy*. London: Murray.

Bertoloni Meli, D. (1993). *Equivalence and priority: Newton versus Leibniz*. Oxford: The Clarendon Press.

Bertoloni Meli, D. (2005). Who is afraid of centrifugal forces. *Early Science and Medicine, X*, 535–543.

Bertoloni Meli, D. (2006). Inherent and centrifugal forces in Newton. *Archive for History of Exact Sciences, 60*, 319–335.

Biblia Sacra iuxta vulgatam versionem. (1994). Rec. R. Weber, editionem quartam emendatam praep. R. Gryson. Stuttgart: Deutsche Bibelgesellschaft.

Biener Z. (2018). Newton's Regulae Philosophandi. In C. Smeenk & E. Schliesser (Eds.), *Oxford handbook for Isaac Newton's.* Oxford: Oxford University Press. Free at: http://philsci-archive.pitt.edu/14722/1/1-1%20Regulae%20Philosophandi%20-%20Oxford%20Handbook.pdf.

Birkenmajer, L. (1924). *Stromata Copernicana.* Studja, poszukiwania i materiały biograficzne [*Stromata Copernicana.* Studies, pieces of research and biographic material]. Kraków: Nakł. Polskiej Akademii Umiejętności.

Blumenberg, H. (2000). *The genesis of the Copernican world* (R. M. Wallace, Transl.). Cambridge (Mass.), London (England): The MIT Press.

Boccaletti, D. (2016). *Galileo and the equations of motion.* Cham-Heidelberg-New York-Dordrecht-London: Springer.

Boner, P. (2013). *Kepler's cosmological synthesis: Astrology, mechanism and the soul.* Boston: Brill.

Bordiga, G. (1925–26). Giovanni Battista Benedetti, filosofo e matematico veneziano del secolo XVI. *Atti del Reale Istituto Veneto di Scienze Lettere ed Arti, 85,* 585–754.

Bortz, F. (2013). *The sun-centered universe of Nicolaus Copernicus.* New York: Rosen Publishing.

Boudri, J. C. (2002). *What was mechanical about mechanics: The concept of force between metaphysics and mechanics from Newton to Lagrange.* Dordrecht-Boston-London: Kluwer Academic Publishers.

Bouquiaux, L. (2008). Leibniz against the unreasonable Newtonian physics. In M. Dascal (Ed.), *Leibniz: What kind of rationalist?* (pp. 99–110). New York: Springer.

Bucciantini, M. (2007). *Galileo e Keplero. Filosofia, cosmologia e teologia nell'Età della Controriforma.* Torino: Einaudi.

Burns, W. E. (2001). *The scientific revolution. An Encyclopedia.* Santa Barbara: ABC-Clio.

Burtt, E. A. (1925). *The metaphysical foundations of modern physical science: A historical and critical essay.* London: Kegan Paul, Trench, Trubner & Co.

Bussotti, P. (2011). The circulation of Kepler's cosmological ideas in Italy during Kepler's lifetime. In É. Mehl (Ed.), *Kepler: La Physique Celeste. Autour de l'*Astronomia Nova *(1609)* (pp. 209–229). Paris: Les Belles Lettres.

Bussotti, P. (2015). *The complex itinerary of Leibniz's planetary theory: Physical convictions, metaphysical principles and Keplerian inspiration.* Basel: Springer-Birkhäuser Verlag.

Bussotti, P. (2017a). Galileo e il suo concetto di inerzia. In A. Neri (Ed.), *Là dove il pensiero incontra l'esperienza,* I (pp. 51–110). Pisa: ETS.

Bussotti, P. (2017b). *Leibniz. Dal calcolo infinitesimale al linguaggio dei computer,* serie "Grandangolo Scienza", "Corriere della Sera". Milano: RCS Media Group.

Bussotti, P. (2017c). Il ruolo dello spazio assoluto nella fisica di Newton. In A. Neri (Ed.), *Là dove il pensiero incontra l'esperienza* I (pp. 111–162). Pisa: ETS.

Bussotti, P., Lotti, B. (2018). The problem of circular motion in René Descartes. *Giornale Critico della Filosofia Italiana,* settima serie vol. XIV, anno XCVII (XCIX), fasc. I, Gennaio-Aprile 2018 (pp. 76–114).

Bussotti, P., & Pisano, R. (2013). On the conceptual and civilization frames in René Descartes physical works. *Advances in Historical Studies, 2*(3), 106–125.

Bussotti, P., & Pisano, R. (2017). Historical and philosophical details on Leibniz's planetary theory as physical-structural model. In R. Pisano, M. Fichant, P. Bussotti, & A. R. E. Oliveira (eds.), *The dialogue between sciences, philosophy and engineering. New historical and epistemological insights. Homage to Gottfried W. Leibniz 1646–1716* (pp. 49–92). London: College Publications.

Butts, R. E. (1993). Whewell on Newton's rules of philosophizing. *Historical Pragmatics: Philosophical Essays., 155,* 293–312.

Camerota, M. (1992). *Gli scritti De Motu Antiquiora di Galileo Galilei. Il Ms. Gal 71: un'analisi storico-critica.* Cagliari: CUEC.

Camerota, M., Giudice, F., & Ricciardo, S. (2019). *Galileo ritrovato. La lettera a Castelli del 21 dicembre 1613.* Brescia: Morcelliana.

Camerota, M., & Helbing, M. (2000). Galileo and Pisan Aristotelianism: Galileo's *De Motu Antiquiora* and the *Questiones de Motu Elementorum* of the Pisan Professors. *Early Science and Medicine, 5*(4), 319–365.

Capecchi, D. (2012). *History of virtual works law*. Milan-Heidelberg-New York-Dordrecht-London: Springer.

Capek, M. (1966). Leibniz's thought prior to the year 1670, from atomism to a geometrical Kinetism. *Revue Internationale De Philosophie, 20*, 249–256.

Carlin, L. (2006). Leibniz on final causes. *Journal of the History of Philosophy, 44*(2), 217–233.

Cassini, A. (2005). Newton and Leibniz on Non-substantival Space. *THEORIA. An International Journal for Theory, History and Foundations of Science*, 20(1): 25–43

Cassirer, E. (1922). *Das Erkenntnisproblem in der Philosophie und Wissenschaft der neueren Zeit*. Berlin: B. Cassirer, vol. I.

Cassirer, E. (1946). Galileo's Platonism. In M. Ashley-Montagu (Ed.), *Studies and Essays in the History of Science and Learning Offered in Homage to George Sarton in the Occasion of His Sixtieth Birthday* (pp. 277–297). New York: Schuman.

Chandrasekhar, S. (1995). *Newton's Principia for the common reader*. Oxford: Clarendon Press.

Chareix, F. (2004). Huygens and mechanics. In K. Fletcher (ed.), *Proceedings of the International Conference "Titan - from discovery to encounter", 13–17 April 2004, ESTEC, Noordwijk, Netherlands*: 55–65. Noordwijk: ESA Publications division.

Chareix, F. (2006). *La philosophie naturelle de Christiaan Huygens*. Paris: Vrin.

Clarke, S. (1717). *A Collection of Papers Which passed between the late Learned Mr. Leibnitz, and Dr. Clarke, in the Years 1715 and 1716*. London.

Coffa, J. A. (1968). Galileo's concept of inertia. *Physis, X, 4*, 261–281.

Cohen, B. I. (1969). Hypotheses in Newton's Philosophy. In I. B. Cohen-Marx W. Wartofsky (eds.), *Boston Studies in the Philosophy of Science. Proceedings of the Boston Colloquium for the Philosophy of Science 1966/1968*, 5: 304–326.

Cohen, B. I. (1974). Kepler's Century: Prelude to Newton's. In A. Beer & P. Beer (Eds.), *Kepler's Four Hundred Years, Vistas in Astronomy,* 18: 3–36. Oxford, New York, Toronto, Sydney, Braunschweig: Pergamon Press.

Cohen, B. I. (1980). *The Newtonian Revolution. With Illustrations of the Transformations of Scientific Ideas*. Cambridge: Cambridge University Press.

Cohen, G. D. (1967). *The Book of Tradition by Abraham Ibn Daud*. Jewish Publication Society.

Comes, M. (2001). *Ibn Al-Hā'im's Trepidation Model. Suhayl, 2*, 291–408.

Cook, R.J. (1976). Is gravitation a results of Mach's inertial interaction? *Il Nuovo Cimento*, 35 B, 1: 25–34.

Copenhaver, B. (1998). The occultist tradition and its critics. In D. Garber & M. Ayers (eds.), *The Cambridge History of Seventeenth-Century Philosophy,* Vol. I: 454–512. Cambridge: Cambridge University Press.

Copernicus, N. (1543, 1978). *On the Revolutions*. Baltimore: The John Hopkins University Press.

Copernicus, N. (1873). *De Revolutionibus Orbium Caelestium libri VI*. Thorn: Societas Copernicana.

Copernicus, N. (1878). *De hypothesibus motuum coelestium a se constitutis Commentariolus* (ed. by M. Curtze). *Mitteilungen Des Coppernicus-Vereins Für Wissenschaft Un Kunst Zu Thorn, I*, 5–17.

Copernicus, N. (1884). *De hypothesibus motuum caelestium a se constitutis Commentariolus*. In Nicolaus Coppernicus, *Werke*, vol. II: 184–202, ed. by L. Prowe. Berlin: Weidmannsche Buchhandlung.

Copernicus, N. (1959). *Three Copernican treatises: The* Commentariolus *of Copernicus, the* Letter against Werner, *the* Narratio Prima *of Rethicus*. Translated with introduction and notes by E. Rosen. New York: Dover.

Copernicus, N. (1978). *On the Revolutions*. Edited by F. Dobrzycki. Translation and Commentary by E. Rosen. Warszawa and London: Polish Scientific Publishers and Macmillan Press.

Copernicus, N. (1979). *Opere*, a cura di F. Barone. Torino: UTET.

Copernicus, N. (1980). *On the Revolutions of the Heavenly Spheres.* In C. Ptolemy (Ed.), *Kepler* (pp. 505–838). Chicago: Encyclopaedia Britannica.
Costabel, P. (1977). Saggio critico su alcuni concetti della meccanica cartesiana. In G. Gori (a cura di), *Cartesio*: 83–102. Milano: ISEDI.
Cottingham, J. (1985), Cartesian Trialism, *Mind*, vol. 94, apr. 1985: 218–230.
Craig, E. (ed. in chief, 1998), *Routledge Encyclopedia of Philosophy*. London-New York: Routledge, 10 vols.
Croce, B. (1923). *History. Its Theory and Practice.* Author. Transl. by D. Ainslie. New York: Harcourt, Brace and Company.
Crockett, T. (2008). Space and Time in Leibniz's Early Metaphysics. *The Leibniz Review, 18*, 41–79.
D'Alembert, J. L. R. (1751). Attraction. In *Encyclopédie ou Dictionnaire Raisonné des Sciences, des Arts, et des Métiers* (Vol. I, pp. 846–855). Paris: C. Briasson, David, L. Breton, Durand.
Davidson, W. (1957). General Relativity and Mach's Principle. *Monthly Notices of the Royal Astronomical Society, 117*(2), 212–224.
Davis, A. E. L. (1992a). Kepler's "Distance Law"–Myth, not Reality. *Centaurus, 35*(2), 103–120.
Davis, A. E. L. (1992b). Kepler's Physical Framework for Planetary Motion. *Centaurus, 35*(2), 165–191.
Davis, A. E. L. (2003). The Mathematics of the Area Law: Kepler's successful proof in *Epitome Astronomiae Copernicanae* (1621). *Archive for History of Exact Sciences, 57*(5), 355–393.
Davis, A. E. L., & Field, J. V. (1995). Kepler translated. Essay review of J. Kepler, *New astronomy*. Translated, introduced and notes by W. H. Donahue, Cambridge: C.U.P. (1992). *The British Journal for the History of Science, 28*, 345–349.
De Broglie, L. (1937). *La physique nouvelle et les quanta.* Paris: Flammarion.
De Caro, M. (1992). Galileo's mathematical Platonism. In G. Czermak (Ed.), *Philosophy of mathematics* (pp. 1–9). Wien: Hoelder-Pichler-Tempsky.
De Caro, M. (2012). Galileo e il platonismo fisico-matematico. In R. Chiaradonna (ed.), *Il platonismo e le scienze* (pp. 123–142). Roma: Carocci.
De Gandt, F. (1995). *Force and geometry in Newton's Principia.* Princeton, NJ: Princeton University Press.
Delambre, J. B. J. (1821). *Histoire de l'astronomie moderne, Tome 1.* Paris: Courcier.
De Pace, A. (2009). *Niccolò Copernico e la fondazione del cosmo eliocentrico.* Milano: Mondadori.
Del Prete, A. (1998). *Universo infinito e pluralità dei mondi. Teorie cosmologiche in età moderna.* Napoli: La Città del Sole.
Derham, W. (1713). *Physico-theology, or, A demonstration of the being and attributes of God from his works of creation being the substance of XVI sermons preached in St. Mary le Bow-Church, London, at the Hon'ble Mr. Boyle's lectures in the years 1711 and 1712.* London: W. Innys.
Derham, W. (1715). *Astro-theology, or, a demonstration of the being and attributes of God: From a survey of the heavens.* London: W. Innys.
Descartes, R. (1897–1913). *Oeuvres de Descartes* (12 vols) publiées par C. Adam & P. Tannery. Paris: Vrin. Abbreviated as AT.
Descartes, R. (1637a). *La Dioptrique*, AT VI (pp. 79–228).
Descartes, R. (1637b). *Les Météores*, AT VI (pp. 230–366).
Descartes, R. (1644). *Principia philosophiae*, Latin original, AT VIII-I (pp. 1–348). French translation AT IX (pp. 1–352).
Descartes R. (1664). *Le Monde*, AT XI : 1–118.
Descartes, R. (1953). *Correspondance avec Arnauld et Morus, introduction et notes par Geneviève Lewis.* Paris: Vrin.
Descartes, R. (1970). *Philosophical letters* (A. Kenny, Transl.). Oxford: Clarendon Press.
Descartes, R. (1984–1991). *The philosophical writings of Descartes* (J. Cottingham, R. Stoothoff, & D. Murdoch, Transl.) (in 3 vols.). Cambridge-London-New York-New Rochelle-Melbourne-Sydney: Cambridge University Press.

Descartes, R. (2006). *A discourse on the method of correctly conducting one's reason and seeking truth in the sciences.* Translated with and introduced and notes by I. Mclean. Oxford: Oxford University Press.

Descartes, R. (2009). *Tutte le lettere 1619–1650,* a cura di G. Belgioioso. Testo francese, latino e olandese. Milano: Bompiani.

Dessì, P. (2012). *Causa/Effetto.* Bologna: Il Mulino.

Di Fate, V. J. (2011). Is Newton a 'Radical empiricist' about method? *Studies in History and Philosophy of Science*, Part A4 *2*(1), 28–36.

Dijksterhuis, E. J. (1961). *The mechanization of the world picture.* Oxford: Clarendon Press. (Original Dutch edition, 1950).

Di Liscia, D. A. (2007). El concepto de causalidad y el desarollo de una teoria cosmólogica en Johannes Kepler. *Anales De La Academia Nacional De Ciencias De Buenos Aires, 34*, 723–753.

Di Liscia, D. A. (2009). Kepler's A priori Copernicanism in his *Mysterium Cosmographicum.* In M. A. Granada & E. Mehl (Eds.), *New sky, new earth: Copernican revolution in Germany during the reform (1530–1630)* (pp. 283–317). Paris: Les Belles Lettres.

DiSalle, R. (1991). Conventionalism and the origins of the inertial frame concept. *Chicago Journal, 2*, 139–147.

DiSalle, R. (1994). On dynamics, indiscernibility, and spacetime ontology. *British Journal for the Philosophy of Science, 45*, 265–287.

DiSalle, R. (1995). Spacetime theory as physical geometry. *Erkenntnis, 42*(1995), 317–337.

DiSalle, R. (2002). Newton's philosophical analysis of space and time. In I. Bernard Cohen, & G. E. Smith (Eds.), *The Cambridge companion to Newton* (pp. 33–56). Cambridge: Cambridge University Press.

DiSalle, R. (2006). *Understanding space-time: The philosophical development of physics from Newton to Einstein.* Cambridge: Cambridge University Press.

DiSalle, R. (2020). Absolute space and Newton's theory of relativity. *Studies in History and Philosophy of Modern Physics, 71*, 232–244.

Dollo, C. (2003), *Galileo Galilei e la cultura della tradizione.* Soveria Mannelli: Rubbettino.

Donahue, W. H. (1981). *The dissolution of the celestial spheres.* New York: Arno Press.

Donahue, W. H. (1988). Kepler's fabricated figures: Covering up the mess in the *New astronomy. Journal for the History of Astronomy, 19*, 217–237.

Donahue, W. H. (1993). Kepler's first thoughts on oval orbits: Text, translation and commentary. *Journal for the History of Astronomy, 24*(1–2), 71–100.

Donahue, W. H. (1996). Kepler's approach to the oval of 1602, from the Mars notebook. *Journal for the History of Astronomy, 27*(4), 281–295.

Donne, J. (1996). *Selected poetry*, ed. with an Introduction and Notes by J. Carey. Oxford: Oxford University Press.

Drake, S. (Ed.) (1957). *Discoveries and opinions of Galileo.* In S. Drake (pp. 173–216). New York: Anchor-Doubleday.

Drake, S. (1964). Galileo and the Law of Inertia. *American Journal of Physics, 32*, 601–608.

Drake, S. (1970). *Galileo studies.* Ann Arbor: University of Michigan Press.

Drake, S. (1976). A further reappraisal of impetus theory: Buridan, Benedetti, and Galileo. *Studies in History and Philosophy of Science, 7*(4), 319–336.

Drake, S. (1978). *Galileo at work. His scientific biography.* Chicago and London: The University of Chicago Press.

Drake, S. (1990). *Galileo: Pioneer scientist.* Toronto: University of Toronto Press.

Dreyer, J. L. E. (1953). *History of planetary systems from Thales to Kepler.* New York: Dover.

Dubarle, D. (1964). Sur la notion Cartésienne de quantité de mouvement. In *Mélanges Alexandre Koyré* (pp. 118–128). Paris: Hermann.

Duchesneau, F. (1989). Leibniz's *Hypothesis Physica Nova*: A conjuction of models for explaining phenomena in an intimate relation. *Studies in the History and Philosophy of Science. Boston Studies in the Philosophy of Science, 116*, 153–170.

Duchesneau, F. (1994a). *La dynamique de Leibniz.* Paris: Vrin.

Duchesneau, F. (1994b). Leibniz on the principle of continuity. *Revue Internationale de Philosophie*, 48 *188*(2), 141–160.
Ducheyne, S. (2012). *The main business of natural philosophy. Isaac Newton's natural-philosophical methodology*. Dordrecht: Springer.
Dugas, R. (1954). Sur le Cartésianisme de Huygens. *Revue D'histoire Des Sciences Et De Leurs Applications, 7*(1), 22–33.
Dugas, R. (1988). *A history of mechanics*. Dover. Original edition: *Histoire de la mècanique* (1955). Neuchâtel: Éditions du Griffon.
Dumitrescu, M. (2011). L'accord entre les causes efficiente et finale dans la théorie Leibnizienne de l'harmonie préétablie. In H. Breger, J. Herbts, & S. Erdner (eds.), *Natur und Subjekt, IX internationaler Leibniz Kongress unter der Schirmerrscahft des Bundespräsidenten, Hannover 26. September bis 1. Oktober 2011. Vorträge, Teil 1* (pp. 255–261). Hannover: Hartmann.
Earman, J. (1989). *World enough and space-time. Absolute versus relational theories of space and time*. Cambridge, Massachusetts-London: The MIT Press.
Einstein, A. (1922, 2007). Ether and the theory of relativity. In M. Janssen, J. D. Norton, J. Renn, T. Sauer, & J. Stachel (eds), *The genesis of general relativity* (Vol. 250, pp. 1537–1542). Boston studies in the philosophy of science. Dordrecht: Springer.
Einstein, A. (1988, 1930). Das Raum-, Äther- und Feld-Problem der Physik. In A. Einstein, *Mein weltbild* (ed. by C. Seelig) (pp. 139–147). Frankfurt am Main-Berlin: Ullstein.
Einstein, A., & Infeld, L. (1938). *The evolution of physics. From early concepts to relativity and quanta*. Cambridge: Cambridge University Press.
Elena, A. (1983a). On the different kinds of attractive forces in Kepler. *Archive Internationales D'histoire Des Sciences, 33*, 22–29.
Elena, A. (1983b). Huygens y el Cartesianismo (A propósito de la noción de gravedad). *LULL, 5*, 5–16.
Ellis, B. D. (1962). Newton's concept of motive force. *Journal of the History of Ideas, 23*(2), 273–278.
Elzinga, A. (1972). Huygens theory of research and Descartes' Theory of knowledge. II. *Zeitschrift für allgemeine Wissenschaftstheorie/Journal for General Philosophy of Science, 3*(1), 9–27.
Emerton, N. (1989). The argument from design in early modern natural theology. *Science and Christian Belief, 1*(2), 129–147.
Enriques, F. (1906). *Problemi della scienza*. Bologna: Zanichelli.
Enriques, F. (1912). *Scienza e razionalismo*. Bologna: Zanichelli.
Enriques, F. (1936). *Il significato della storia del pensiero scientifico*. Bologna: Zanichelli.
Feyerabend, P. K. (1975). *Against method: Outline of an anarchistic theory of knowledge*. London: Verso.
Fichant, M. (1998). *Science et métaphysique dans Descartes et Leibniz*. Paris: PUF.
Field, J. V. (1982). Kepler's cosmological theories: Their agreement with observations. *The Quarterly Journal of the Royal Astronomical Society, 23*, 556–568.
Field, J. V. (1985). Giovanni Battista Benedetti on the mathematics of linear perspective. *Journal of the Warburg and Courtauld Institute, 48*, 71–99.
Field, J. V. (1988). *Kepler's geometrical cosmology*. Chicago: The University of Chicago University Press.
Field, J. V. (2009). Kepler's harmony of the world. In R. Kremer, J. Wlodarczyk (Eds). *Johannes Kepler. From Tübingen to Zagan* (pp. 11–28). Warszawa: Institute for the history of science—The Polish academy of sciences—The Copernicus centre for the interdisciplinary studies. Instytut Historii Nauki PAN.
Fludd, R. (1617–1624). *Utriusque cosmi maioris scilicet et minoris, metaphysica, physica, atque technica historia*. Oppenheim and Frankfurt: Hieronymus Gallerus for Johann Theodorus de Bry.
Fontenelle, Bernard le Bovier de (1852) *Entretiens sur la pluralité des mondes*. In *Oeuvres de Fontenelle* (pp. 33–149). Paris: Eugène Didier. (Original 1686).
Fontenelle, Bernard le Bovier de (1990). *Conversations on the plurality of worlds*. Translation by H. A. Hargreaves, Introduced by N. R. Gelbart. Berkeley: University of California Press.

Föppl, A. (1894). *Einführung in die Maxwell'sche Theorie der Elektricität*. Leipzig: Teubner.

Föppl, A. (1904a). Über einen Kreiselversuch zur Messung der Umdrehungsgeschwindigkeit der Erde. *Akademie der Wissenschaften, München, Mathematisch-Wissenschaftliche Klasse, Sitzungsberichte*, Sitzung vom 6. *Februar, 1904*, 5–28.

Föppl, A. (1904b). Über absolute und relative Bewegung. *Akademie der Wissenschaften, München, Mathematisch-Wissenschaftliche Klasse, Sitzungsberichte,* Sitzung vom 5. November 1904, 383–395.

Fouke, D. C. (1992). Metaphysics and the Eucharist in the early Leibniz. *Studia Leibnitiana*, Bd. 24, H. 2, 145–159.

Frank, P. (1909). Gibt es eine absolute Bewegung? Vortrag, gehalten am 4. Dezember, 1909. *Wissenschaftliche Beilage, 1910*, 1–10.

Frank, P., & Rothe, H. (1911). Über die Transformation der Raumzeitkoordinaten von ruhenden auf bewegte Systeme. *Annalen Der Physik, 34*, 825–855.

Freely, J. (2014). *Celestial revolutionary. Copernicus, the Man and His Universe*. London-New York: I.B. Tauris.

Freudenthal, G. (1986). *Atom and individual in the age of Newton. On the genesis of mechanistic world view*. Dordrecht: Reidel Publishing Company.

Friedlaender, B., & Friedlaender, I. (1896). *Absolute oder relative Bewegung?* Berlin: Leonhard Simion.

Friedman, M. (1983). *Foundations of space-time. Theories relativistic physics and philosophy of science*. Princeton: Princeton University Press.

Frova, A. (2006). *Anche Galileo può errare*. In *Atti di «Pianeta Galileo 2006»* (pp. 35–47). Firenze: Regione Toscana.

Gabbey, A. (1971). Force and inertia in seventeenth-century dynamics. *Studies in History and Philosophy of Science, II*, 1–67.

Gabbey, A. (1998). Force, force vive, conversation. In M. Blay & R. Halleux (Eds.), *La science classique: XVIe–XVIIIe siècle; dictionnaire critique* (pp. 524–534). Paris: Flammarion.

Gale, G. (1973). Leibniz' dynamical metaphysics and the origins of the *vis viva* controversy. *Systematics: the Journal of the Institute for the Comparative Study of History, Philosophy and the Sciences*, 11(3), 184–207.

Gale, G. (1984). Leibniz' force: Where physics and metaphysics collide. In A. Heinekamp (Ed.), *Leibniz' dynamica* (pp. 62–70). Stuttgart: Steiner.

Galilei, G. (1890–1909). *Edizione Nazionale delle Opere*. Direttore A. Favaro, (20 vols). Firenze: Barbèra. Abbreviated as EN.

Galilei, G. (1586–7/1590). *De Motu*. EN I (pp. 243–417).

Galilei, G. (1604). *De Motu accelerato*. EN II (pp. 257–266).

Galilei, G. (1610). *Sidereus Nuncius*. EN III, Parte Prima (pp. 55–96).

Galilei, G. (1613). *Istoria e dimostrazioni intorno alle macchie solari e loro accidenti*. EN V (pp. 71–249).

Galilei, G. (1616). *Discorso del flusso e reflusso del mare*. EN V (pp. 373–395).

Galilei, G. (1623). *Il Saggiatore*. EN VI (pp. 197–372).

Galilei, G. (1632). *Dialogo sopra i due massimi sistemi del mondo tolemaico e copernicano*. EN VII (pp. 21–520).

Galilei, G. (1638). *Discorsi e dimostrazioni matematiche intorno a due nuove scienze*. EN VIII (pp. 49–448).

Galilei, G. (1914). *Dialogues concerning two new sciences*. (H. Crew, & A. De Salvio, Transl.). New York: Macmillan.

Galilei, G. (1957). *The Assayer*. Selections translated by S. Drake. *Discoveries and opinions of galileo* (pp. 231–280). New York: Doubleday & Co.

Galilei, G. (1967). *Dialogue concerning the two chief world systems: Ptolemaic and Copernican* (2nd ed.). Translated and with notes by S. Drake. Foreword by A. Einstein. Berkeley and Los Angeles: University of California Press.

Galilei, G. (2009). *Capitolo Contro il portar la toga—Against the Donning of the Gown*. Italian text plus English translation by G. Bignami. Edited by L. Tongiorgi Tomasi. With essays by G. Bignami, V. Finucci, M. Rippa Bonati, F. Tognoni, L. Tongiorgi Tomasi, R. Vergara Caffarelli. Pisa: ETS. EN IX: 213–223.

Galilei, G., & Scheiner, C. (2010). *On sunspots*. Translated and introduced by E. Reeves, & A. Van Helden. Chicago and London: Chicago University Press.

Garau, R. (2014). Late-scholastic and Cartesian *conatus*. *Intellectual History Review, XXIV*, 479–494.

Garber, D. (1992). *Descartes' metaphysical physics*. Chicago and London: The University of Chicago Press.

Garber, D. (1994). Leibniz: Physics and philosophy. In L. Jolley (Ed.), *The Cambridge companion to Leibniz* (pp. 270–352). Cambridge: Cambridge University Press.

Garber, D. (2001). *Descartes embodied: Reading Cartesian philosophy through Cartesian science*. Cambridge: Cambridge University Press.

Garber, D. (2008). Dead force, infinitesimals, and the mathematicization of nature. In U. Goldenbaum & D. Jesseph (Eds.), *Infinitesimal differences: Controversies between Leibniz and his contemporaries* (pp. 281–306). Berlin: De Gruyter.

Garber, D. (2009). *Leibniz: Body, substance, monad*. Oxford: Oxford University Press.

Gaukroger, S. (Ed.). (1980). *Descartes: Philosophy, mathematics and physics*. Sussex: Harvester Press.

Gaukroger, S. (2002). *Descartes' system of natural philosophy*. Cambridge: Cambridge University Press.

Geymonat, L. (1981). *Per Galileo*. Verona: Bertani.

Ghins, M. (1990) *L'inertie et l'espace-temps absolu de Newton à Einstein: une analyse philosophique*. Bruxelles: Palais des Académies.

Gillispie, C. C. (ed. in chief, 1970–1980), *Dictionary of scientific biography* (16 vols). New York: Charles Scribner's Sons.

Giné, J. (2009). On the origin of inertia: The modified Newtonian dynamics theory. *Chaos, Solutions and Fractals, 41*(4), 1651–1660.

Gingerich, O. (1975). The origins of Kepler's third law. In A. Beer & P. Beer (Eds.), *Kepler's four hundred years, Vistas in Astronomy, 18*, 595–601.

Gingerich, O. (2004). *The book nobody read. Chasing the revolutions of Nicolaus Copernicus*. London: William Heinemann.

Gingerich, O. (2011). Kepler's trinitarian cosmology. *Science and Theology, 9*(1), 45–51.

Giorgio, E. (2011). Trägheit und *Circulatio Harmonica*: Formen der Subjektivität im "Tentamen" von 1689 und in der "Theodizee". In H. Breger, J. Herbts, & S. Erdner (Eds.), *Natur und Subjekt, IX internationaler Leibniz Kongress unter der Schirmerrscahft des Bundespräsidenten* (pp. 292–299), Hannover 26. September bis 1. Oktober 2011. Vorträge, Teil 1. Hannover: Hartmann.

Girill, T. R. (1970). Galileo and Platonistic methodology. *Journal of the History of Ideas, 31*(1970), 501–520.

Giudice, F. (2006). Isaac Newton e la tradizione dei principi attivi nella filosofia naturale inglese del XVII secolo. In C. Giuntini & B. Lotti (a cura di), *Scienza e teologia fra Seicento e Ottocento. Studi in memoria di Maurizio Mamiani* (pp. 39–55). Firenze: Leo S. Olschki.

Giudice, F. (2010). The visible world: Aspects of optics in the seventeenth century. In E. Agazzi, E. Giannetto, & F. Giudice (Eds.), *Representing light across arts and sciences: Theories and practices* (pp. 41–55). Göttingen: V&R unipress.

Giulini, G. (2002). Das Problem Der Trägheit. *Philosophia Naturalis, 39*, 343–374.

Goddu, A. (2010). *Copernicus and the Aristotelian tradition*. Leiden, Netherlands: Brill.

Goldbeck, E. (1896). *Keplers Lehre von der Gravitation*. Hildesheim-New York: Olms.

Goldstein, B. R. (1965). On the theory of trepidation according to Thābit b Qurra and Al-Zarqāllu and its implications for homocentric planetary theory. *Centaurus, 10*(4), 232–247.

Gould, J. A. (1962). The existence of absolute space. *The Ohio Journal of Science, 62*(2), 101–104.

Granada, M. A. (2009). Novelties in the heavens between 1572 and 1604 and Kepler's unified view of nature. *Journal for the History of Astronomy, 40*(4), 393–402.

Graneau, P., & Graneau, N. (2003). Machian inertia and the Isotropic universe. *General Relativity and Gravitation, 35*, 751–770.

Gregory, D. (1702). *Astronomiae physicae et geometricae elementa*. Oxford: Sheldonian Theatre.

Grünbaum, A. (1964). *Philosophical problems of space and time*. Boston: Alfred Knopf.

Gueroult, M. (1934). *Leibniz: Dynamique et métaphysique; suivi d'une note sur le principe de la moindre action chez Maupertuis*. Paris: Les Belles-Lettres.

Gueroult, M. (1953). *Descartes selon l'ordre des raisons* (2 vols). Paris: Aubier.

Guicciardini, N. (1999). *Reading the* Principia. *The debate on Newton's mathematical methods for natural philosophy from 1687 to 1736*. Cambridge: Cambridge University Press.

Guidi Itokazu, (2006). A Força que Move os Planetas: Da Noção de Species Immateriata na Astronomia de Johannes Kepler. *CHPC, 3*(16/2), 211–231.

Guidi Itokazu, (2007). Da Potência Motriz Solar Kepleriana como Emanação Imaterial. *CHPC, 3*(17/2), 303–324.

Haase, R. (1998). *Johannes Keplers Weltharmonik. Der Mensch im Geflecht von Musik, Mathematik und Astronomie*. München: Eugen Diederichs.

Hall, R. (1970). Kuhn and the Copernican Revolution. *British Journal for the Philosophy of Science, 21*(2), 196–197.

Hall, A. R. (1963). *From Galileo to Newton*. New York: Harper and Row.

Hall, A. R., & Hall, M. B. (1995 [1970]). Newton and the theory of matter. In I. B. Cohen & R. S. Westfall (Eds.), *Newton. Texts backgrounds commentaries* (pp. 72–87). New York & London: W.W. Norton and Company.

Hatfield, G. (1979). Force (God) in Descartes' physics. *Studies in History and Philosophy of Science, X*, 113–140.

Heimann, P. M. (1977, March). Leibniz and Johann Bernoulli's theory of motion. *Centaurus* (pp. 1–26).

Heimann, P. M., & McGuire, J. E. (1971). *Newtonian forces and Lockean powers concepts of matter in eighteenth-century thought*. Philadelphia: University of Philadelphia Press.

Henry, J. (2011). Gravity and *De Gravitatione*: The development of Newton's ideas on action at a distance. *Studies in History and Philosophy of Science Part A, 42*(1), 11–27.

Hill, B. (2003). Newton's *De Gravitatione et Aequipondio Fluidorum* and the Lockean four-dimensionalism. *British Journal for the History of Philosophy, 11*(2), 309–321.

Hofmann, W. (1904). *Kritische Beleuchtung der beiden Grundbegriffe der Mechanik: Bewegung und Trägheit und daraus gezogene Folgerungen betreffs der Achsendrehung der Erde und des Foucault'schen Pendelversuchs*. Wien und Leipzig: M. Kuppitsch.

Holbach, P. R. Thiry baron d' (1835: original French edition 1770). *The system of nature; or laws of the moral and physical world* (H. D. Robinson, Transl.). A new and improved edition with notes by Diderot. New York: G.W. & A.J. Matsell.

Holton, G. (1956). Johannes Kepler's universe: Its physics and metaphysics. *The American Journal of Physics, 24*(5), 340–351.

Hon, G., & Goldstein, B. R. (2008). *From Summetria to symmetry: The making of a revolutionary scientific concept*. Dordrecht: Springer.

Hoyer, U. (1979). Kepler's celestial mechanics. *Vistas in Astronomy, 23*(1), 69–74.

Hoyle, F. (1973). *Nicolaus Copernicus. An essay on his life and work*. New York: Harper and Row.

Hoyningen-Huene, P. (1987). Context of discovery and context of justification. *Studies in History and Philosophy of Science, 18*(4), 501–515.

Hugett, N. (2008). Why the parts of absolute space are immobile. *The British Journal for the Philosophy of Science, 59*(3), 391–407.

Hume, D. (1975) *Enquiries concerning human understanding and concerning the principles of morals*. Oxford: Clarendon Press. Reprint from the posthumous edition of 1777 and ed. with Introduction, comparative table of contents, and analytical index by L. A. Selby Bigge, Third Edition with Text revised and notes by P.H. Nidditch.

Huygens, C. (1888–1950). *Oeuvres complètes de Christiaan Huygens publiées par la Société Hollandaise des Sciences.* La Haye: Martinus Nijhoff; Amsterdam: Swets & Zeitlinger (abbreviation OH).
Huygens, C. (1656, posthumous 1703). *De motu corporum ex Percussione*, OH XVI (pp. 37–91). English translation by M. S. Mahoney, 1977 reviewed 1995, *On the motion of bodies resulting from impact*, freely available at www.princeton.edu/~hos/mike/texts/huygens/impact/huyimpct.html
Huygens, C. (1659a, posthumous 1703). *De Vi Centrifuga*, OH XVI (pp. 235–301). English translation by M. S. Mahoney, *On centrifugal Force*, freely available at www.princeton.edu/~hos/mike/texts/huygens/centriforce/huyforce.htm
Huygens, C. (1659b). *Systema saturnium*, OH XV (pp. 178–353).
Huygens, C. (1669) Debat de 1669 à l'Acadèmie sur les causes de la pesanteur, OH XIX (pp. 628–645).
Huygens, C. (1673). *Horologium oscillatorium.* OH XVIII (pp. 27–368). English translation: *Christiaan huygens the pendulum clock, or, Geometrical demonstrations concerning the motion of pendula as applied to clocks* by R. J. Blackwell, Introduction by H. J. M. Bos. Ames: Iowa State University Press, (1986).
Huygens, C. (1690a). *Traité de la lumière*, OH XIX (pp. 451–537). English translation by S. P. Thompson, 1912, *Treatise on light*. London: MacMillan and Co.
Huygens, C. (1690b). *Discours de la cause de la pesanteur*, OH XXI (pp. 443–488).
Huygens, C. (1698, posthumous). *Cosmotheoros*, OH XXI (pp. 633–821).
Huygens, C. (1912). *Treatise on light*. Translated by S. P. Thompson. London: Macmillan and Co.
Ihmig, K. N. (1990). Trägheit und Massebegriff bei Johannes Kepler. *Philosophia Naturalis, 27*(2), 156–205.
Iltis, C. (1971). Leibniz and the *Vis Viva* Controversy. *Isis, 62*(1), 21–35.
Iltis, C. (1973). The decline of Cartesianism in mechanics: The Leibnizian-Cartesian debates. *Isis, 64*(3), 356–373.
Jammer, M. (1957). *Concepts of force. A study in the foundation of dynamics.* Cambridge (Mass.): Harvard University Press.
Janiak, A. (2008). *Newton as philosopher*. New York: Cambridge University Press.
Jauernig, A. (2008). Leibniz on motion and the equivalence of hypotheses. *The Leibniz Review, 18*, 1–40.
Jorgensen, L. M. (2009). The principle of continuity and Leibniz's theory of consciousness. *Journal of the History of Philosophy, 47*(2), 223–248.
Kant, I. (1981). *Universal natural history and theory of the heavens.* Translation with introduction and notes by S. L. Jaky. Edinburgh: Scottish Academic Press.
Kepler, J. (1604). *Ad Vitellionem paralipomena quibus astronomiae pars optica traditur.* In KGW II.
Kepler, J. (1606). *De Stella Nova in Pede Serpentarii.* In KGW I (pp. 150–292).
Kepler, J. (1611). *Dioptrice.* In KGW IV (pp. 324–414).
Kepler, J. (1618–1621). *Epitome Astronomiae Copernicanae.* In KGW VII.
Kepler, J. (1937–2012). *Gesammelte werke.* In W. Van Dyck & M. Caspar, et al. (Eds). Revised April 2013. (10 vols). München: Deutsche Forschungsgemeinschaft und Bayerische Akademie der Wissenschaften. Beck'sche Verlagsbuchhandlung. Abbreviated as KGW.
Kepler, J. (1981). *Mysterium Cosmographicum (The Secret of the Universe)* (2nd ed., pp. 2–80, 7–128). Translation by A. M. Duncan. Introduction and commentary by E. J. Aiton. Preface by I. Bernard Cohen. New York: Abaris Book. In KGW I. KGW, VIII.
Kepler, J. (1984). *Le secret du monde.* Introduction, Traduction et notes de A. Segonds. Paris: Les Belles Lettres.
Kepler, J. (1992). *New Astronomy.* Translation by W. H. Donahue. Cambridge: The Cambridge University Press. In KGW III.
Kepler, J. (1997). *The Harmony of the World.* English translation and introduction by E. J. Aiton, A. M. Duncan, J. V. Field. Philadelphia: American Philosophical Society. In KGW VI.

Kircker, A. (1656). *Itinerarium exstaticum quo mundi opificium*. Romae: Typis Vitalis Mascardi.

Knobloch, E. (1992). Rapports historiques entre musique, mathématique et cosmologie. In *Quadrivium, Musiques et sciences* (pp. 123–167), Colloque conçu par D. Lustgarten, Cl.-H. Joubert, S. Pahaut, M. Salazar. Paris: Cité de la musique.

Knobloch, E. (1995). Harmony and cosmos: Mathematics serving a teleological understanding of the World. *Physis, 32*(1), 54–89.

Knobloch, E. (1999). Galileo and Leibniz—Different approaches to infinity. *Archive for History of Exact Sciences, 54*, 87–99.

Knobloch, E. (2002). Vielheit der Welten - extraterrestrische Existenz. In W. Voßkamp (Ed.), *Ideale Akademie - vergangene Zukunft oder konkrete Utopie* (pp. 165–186). Berlin: Akademie Berlin.

Knobloch, E. (2008). Modelle in der Geschichte der Wissenschaften. In U. Dirks & E. Knobloch (Eds.), *Modelle* (pp. 85–101). Frankfurt am Main: Peter Lang.

Knudsen, O., & Pedersen, K. M. (1969). The link between 'Determination' and conservation of motion in Descartes' dynamics. *Centaurus, 13*(2), 183–186.

Kochiras, H. (2012). Newton on matter and space. http://philsci-archive.pitt.edu/9176/1/Kochiras.NewtonOnMatter&SpaceInDeGrav(Excerpt_for_7th_Fellows_Conf).pdf

Koyré, A. (1939). *Études Galiléennes*. Paris: Hermann.

Koyré, A. (1957). *From the closed world to the infinite universe*. Baltimore: John Hopkins Press.

Koyré, A. (1965). *Newtonian studies*. Harvard (Mass.): Harvard University Press.

Koyré, A. (1973). *The astronomical revolution: Copernicus – Kepler – Borelli*. London. Methuen; Ithaca (NY): Cornell University Press. (1st ed., 1961).

Krafft, F. (1991). The new celestial physics of Johannes Kepler. In Unguru 1991 (ed.), *Physics, cosmology and astronomy, 1300–1700: Tension and accommodation* (Vol. 126, pp. 185–227). The Boston studies in the philosophy of science. Dordrecht-Boston-London: Kluwer Academic Publishers.

Kubrin, D. (1967). Newton and the cyclical cosmos: Providence and the mechanical philosophy. *Journal of the History of Ideas, 28*, 325–346.

Kuhn, T. (1957). *The Copernican Revolution: Planetary astronomy in the development of western thought*. Cambridge, MA: Harvard University Press.

Lange, L. (1885a). Über das Beharrungsgesetz. *Berichte über Verhandlungen der Königlich Sächsischen Gesellschaft der Wissenschaften* (pp. 333–351). Mathematisch-physikalische Klasse.

Lange, L. (1885b). Über die wissenschaftliche Fassung des Galileischen Beharrungsgesetzes. *Philosophische Studien, 2*, 266–297.

Lange, L. (1885c). Nochmals Über Das Beharrungsgesetz. *Philosophische Studien, 2*, 539–545.

Lange, L. (1886). *Die geschichtliche Entwicklung des Bewegungsbegriffs und ihr voraussichtliches Endergebnis. Ein Beitrag zur historischen Kritik der mechanischen Prinzipien*. Leipzig: W. Engelmann.

Lange, L. (1902). Das Inertialsystem vor dem Forum der Naturforschung. *Philosophische Studien, 20*, 1–71.

Lange, L. (2014). On the Law of inertia. Translation (by. H. Pfister) of Über das Beharrungsgesetz. *The European Physical Journal H, 39*(2), 251–262.

Lariviere, B. (1987). Leibnitian relationalism and the problem of inertia. *Canadian Journal of Philosophy, 17*(2), 437–448.

Laymon, R. (1978). Newton's Bucket experiment. *Journal of the History of Philosophy, 16*(4), 399–413.

Leibniz, G. W. (1840). *Opera philosophica Omnia*. In J. E. Erdmann (Ed.). Pars Prior. Berolini: Sumtibus G. Eichleri.

Leibniz, G. W. ([1849–1863], 1962). *Mathematische Schriften* (7 vols). In C. I. Gerhardt (Ed.) Hildesheim: Georg Olms.

Leibniz, G. W. ([1875–1890], 1978a). *Philosophische Schriften* (7 vols) In C. I. Gehrardt (Ed.). Hildesheim: Georg Olms.

Leibniz, G. W. (1896a). *New essays concerning human understanding*. Together with an Appendix consisting of some of his shorter pieces (A. G. Langley, Transl.). New York: The Macmillan Company.
Leibniz, G. W. (1903). *Opuscules et fragments inedites de Leibniz extraits des manuscrits de la Biblioteque Royale de Hannover* (ed. L. Couturat). Paris: Alcan.
Leibniz. (1923, 1938, 1950). *Sämtliche Schriften und Briefe* (abbreviated LSB). Darmstad, 1923, Leipzig, 1938, Berlin, 1950: Deutsche Akademie der Wissenschaften zu Berlin.
Leibniz, G. W. (1967). *Scritti filosofici*, a cura di D. O. Bianca. Torino: UTET.
Leibniz, G. W. (1969). *Philosophical papers and letters*. (L. E. Loemker (Ed.), Transl.). Dordrecht: Reidel.
Leibniz, G. W. (1671). *Hypothesis physica nova* (Vol. VI, pp. 17–59). In Leibniz ([1849–1863], 1962).
Leibniz, G. W. (1677). *Leibniz an Honoratus Fabri* (Vol. VI, pp. 81–106). In Leibniz ([1849–1863], 1962).
Leibniz, G. W. (1686a). *Brevis Demonstratio Erroris memorabilis Cartesii et aliorum circa Legem naturalem, secundum, quod volunt a Deo eandem semper quantitatem motus conservari, qua et in re mechanica abutuntur* plus Beilage (Vol. VI, pp. 117–123). In Leibniz ([1849–1863], 1962).
Leibniz, G. W. (1686b). *Discours de metaphysique* (Vol. IV, pp. 427–463). In Leibniz ([1875–1890], 1978).
Leibniz, G. W. (1686–1689). *De veritatibus primis* (Vol. VII, pp. 194–195). In Leibniz ([1875–1890], 1978).
Leibniz, G. W. (1689). *Tentamen de Motuum Coelestium Causis* (Erste Bearbeitung) (Vol. VI, pp. 144–161). In Leibniz ([1849–1863], 1962).
Leibniz, G. W. (1690). *De Causa gravitatis et defensio sententiae Autoris de viris Naturae Legibus contra Cartesianos* (Vol. VI, pp. 193–201). In Leibniz ([1849–1863], 1962).
Leibniz, G. W. (1690?, 1860, 1962). *Tentamen de Motuum Coelestium Causis* (Zweite Bearbeitung) (Vol. VI, pp. 161–187). In Leibniz ([1849–1863], 1962).
Leibniz, G. W. (1691?). *Essay de dynamique sur le loix du mouvement, ou il est monstré qu'il ne se conserve pas la même quantité de mouvement, mais la même force absolue, ou bien la même quantité de l'action motrice* (Vol. VI, pp. 215–230). In Leibniz ([1849–1863], 1860, 1962).
Leibniz, G. W. (1695). *Specimen dynamicum pro admirandis naturae legibus circa corporum vires et mutuas actiones detegentis et ad suas causas revocandis* (Vol. VI, pp. 234–254). Pars I et Pars II. In Leibniz ([1849–1863], 1962)
Leibniz, G. W. (1697). *De rerum originatione radicali* (Vol. VII, pp. 302–308). In Leibniz ([1875–1890], 1890).
Leibniz, G.W. (1704). *Nouveaux Essais sur l'Entendement par l'Auteur du Systeme de l'Harmonie Preestablie* (Vol. V). In Leibniz ([1875-1890], 1882, 1978).
Leibniz, G. W. (1706, 1860, 1962). *Illustratio Tentaminis de Motuum Coelestium Causis* (Vol. VI, pp. 254–280). Pars I et Pars II plus Beilage. In Leibniz ([1849–1863], 1962).
Leibniz, G.W. (1710). *Essais de Théodicée* (Vol. VI). In Leibniz ([1875–1890], 1885, 1978).
Leibniz, G. W. (1714, 2001). *La Monadologie*. Une édition électronique saisie à partir du livre de Gottfried Wilhelm Leibniz, *La Monadologie*, édition annotée par Émile Boutroux, initialement publiée chez C. Delagrave, 1881, 231 pp. http://classiques.uqac.ca/classiques/Leibniz/La_Monadologie/leibniz_monadologie.pdf
Leibniz, G. W. (1896b). *New essays concerning human understanding* (A. G. Langley, Transl.). New York: The Macmillan Company.
Leibniz, G. W. (1898). *The monadology and other philosophical writings*. Translated with introduced and notes by R. Latta. Oxford: Clarendon Press.
Leibniz, G. W. (1965). *Monadology and other philosophical essays* (P. Schrecker, & A. Martin Schrecker, Transl.). Indianapolis-New York-Kansas City: The Bobbs-Merril Company.
Leibniz, G. W. (1985). *Theodicy. Essays on the Goodness of God, the freedom of man and the origin of evil*. ed. with an introduced by A. Farrer. Translated by E. M. Huggard. Chicago and La Salle: Open Court.

Leibniz, G.W. (1989a). *Philosophical essays*. Translated by R. Ariew and D. Garber. Indianapolis & Cambridge: Hackett.

Leibniz, G. W. (1989b). *Philosophical papers and letters*. In L. E. Loemker (Ed.). Dordrecht: Kluwer.

Leibniz, G. W. (2013). *The Leibniz-De Volder Correspondence: With Selections from the Correspondence Between Leibniz and Johann Bernoulli*. Translated by P. Lodge. New Haven: Yale University Press.

Leibniz, G. W., & Clarke, S. (1957). *Correspondance Leibniz-Clarke présentée d'après les manuscripts originaux des bibliotèques de Hanovre et de Londres*. Edited by A. Robinet. Paris: Presses Universitaires de France.

Leibniz, G. W., & Clarke, S. (2000). *Correspondence*. Ed. with an Introduction by R. Ariew (Ed.). Indianapolis-Cambridge: Hackett Publishing Company.

Lindsay, R. B. (1975). The introduction of *vis viva* and related concepts. In R. B. Lindsay (Ed.), *Energy: Historical development of the concept* (pp. 108–118). Stroudsburg Pa: Dowden, Hutchinson and Ross.

Lodge, P. (1997). Force and the nature of body in "Discourse on metaphysics" 17–18. *Leibniz Society Review, 7*, 116–124.

Lodge, P. (2001). Primitive and derivative forces in Leibnizian bodies. In H. Poser, C. Asmuth, & U. Goldenbaum (Eds.), *Nihil sine ratione: Mensch, Natur und Technik im Wirken von G. W. Leibniz Schirmherrschaft; VII. Internationaler Leibniz-Kongreß* (pp. 720–727). Hannover: Gottfried-Wilhelm-Leibniz-Gesellschaft.

Look, B. C. (2011). Leibniz's theory of causation. In B. C. Look (Ed.), *The Continuum companion to Leibniz* (pp. 174–191). London: Continuum Publishing Corporation.

Lotti, B. (2006). Filosofia naturale e teologia nello *Scholium Generale* di Newton. In C. Giuntini & B. Lotti (a cura di), *Scienza e teologia fra Seicento e Ottocento. Studi in memoria di Maurizio Mamiani* (pp. 57–80). Firenze: Leo S. Olschki.

Lovejoy, A. O. (1936, 1964). *The great chain of being. A study of the history of an idea*, The William James Lectures delivered at Harvard University, 1933. Harvard: Harvard University Press.

Lucas, J. R. (1984). *Space, time, and causality*. Oxford: Oxford University Press.

Maccagni, C. (1967). *Le speculazioni giovanili "de motu" di Giovan Battista Benedetti*. Pisa: Domus Galilaeana.

Maccagni, C. (1985). *Le due edizioni della* Demonstratio proportionum motuum loculium contra Aristotelem et omnes philosophos *di Giovanni Battista Benedetti*. Istituto Veneto di Scienze, Lettere ed Arti.

Mach, E. (1919). *The science of mechanics. A critical and historical account of its development*. Chicago-London: The open court publishing Co. English translation of the fourth edition of Mach's *Die Mechanik in ihrer Entwickelung historisch-kritisch dargestellt* (1st edn., 1883).

Machamer, P. (1998). Galileo's machines, his mathematics, and his experiments. In P. Machamer (Ed.), *The Cambridge companion to Galileo* (pp. 53–79). Cambridge: Cambridge University Press.

Maitte, B. (2006). *Histoire de l'arc–en–ciel*. Paris: Seuil.

Malet, A. (1990). Gregoire, Descartes, Kepler and the law of refraction. *Archives Internationales D'histoire Des Sciences, 40*, 278–304.

Malpangotto, M. (2012). The early manuscripts of Georg von Peuerbach's *Theoricae Novae planetarum*: Presentation, description and elaboration of a book. *Revue D'histoire Des Sciences, 65*, 339–380.

Mamiani, M. (2001). To twist the meaning: Newton's regulae philosophandi revisited. In J. Z. Buchwald & I. Bernard Cohen (Eds.), *Isaac Newton's natural philosophy* (pp. 3–14). Cambridge, Mass: MIT Press.

Mancha, J. L. (1998). On Ibn al-Kammād's Table for Trepidation. *Archive for History of Exact Sciences, 52*, 1–11.

Manchak, J. (2009). On force in Cartesian physics. *Philosophy of Science, LXXVI*, 295–306.

Marcacci, F. (2018). *Cieli in contraddizione. Giovanni Battista Riccioli e il terzo sistema del mondo.* Perugia: Aguaplano.
Mathieu, V. (1986). *Introduzione a Leibniz.* Roma-Bari: Laterza.
McDonough, J. K. (2007, revised 2019). Leibniz's philosophy of physic. In E.N. Zalta (Ed.), *Stanford Encyclopedia of Philosophy*: https://plato.stanford.edu/entries/leibniz-physics/.
McGuire, J. E. (1978a). Existence, actuality and necessity: Newton on space and time. *Annals of Science, 35*, 463–508.
McGuire, J. E. (1978b). Newton on place, time and God: An unpublished source. *The British Journal for the History of Science, 11*(38), 114–129.
McGuire, J. E. (1995). *Tradition and innovation: Newton's metaphysics of nature.* Boston: Kluwer Academic Publishers.
McLaughlin, P. (1980). Force, determination and impact. *Gaukroger, 1980*, 81–112.
McLaughlin, T. (2008). Nature and inertia. *The Review of Metaphysics, 62*, 251–284.
Menéndez, V. (2004). Christiaan Huygens y sus conceptos de materia. Claves para entender las difficultades de un genio aferrado a un paradigma. In R. A.Martins, L. A. C. P. Martins, C. C. Silva, & J. M. H. Ferreira (Eds.), *Filosofia e história da ciência no Cone Sul: 3° Encontro* (pp. 479–485). Campinas: AFHIC.
Miller, R. B. (1982). *Force and substance: A study of the interrelation of dynamics and metaphysics in Leibniz.* New Brunswick, NJ: Rutgers University Press.
Minazzi, F. (1994). *Il Flauto di Popper.* Milano: Franco Angeli.
Mogens Laerke, M. (2016). Compossibility, compatibility, congruity. In G. Brown & Y. Chieck (Eds.), *Leibniz on compossibility and possible worlds* (pp. 125–144). Cham (Switzerland): Springer.
Moore, A. W. (2012). *The evolution of modern metaphysics: Making sense of things.* Cambridge: Cambridge University Press.
More, H. (1662). *A collection of several philosophical writings.* London: Print. by J. Flesher.
Mormino, G. (1993). *Penetralia Motus. La fondazione relativistica della meccanica in Christiaan Huygens, con l'edizione del Codex Hugeniorum 7A.* Firenze: La Nuova Italia.
Mormino, G. (1996). Atomismo e meccanicismo nel pensiero di Christiaan Huygens. *Rivista Di Storia Della Filosofia, 4*, 829–863.
Mormino, G. (2003). Le rôle de Dieu dans l'oeuvre scientifique et philosophique de Christiaan Huygens. *Revue D'histoire Des Sciences, 56*(1), 113–133.
Morpurgo-Tagliabue, G. (1981). *I processi di Galileo e l'epistemologia.* Roma: Armando.
Nerlich, G. (1976). *The shape of space.* Cambridge: Cambridge University Press.
Neumann, C. (1870). *Über die Principien der Galilei-Newtonschen Theorien.* Leipzig: Teubner.
Newburgh, R. (2007). Inertial forces, absolute space, and Mach's principle: The genesis of relativity. *American Journal of Physics, 75*, 427–430.
Newton, I. (1666). *De Gravitatione et Aequipondio Fluidorum.* http://www.newtonproject.ox.ac.uk/catalogue/record/THEM00093.
Newton, I. (1706). *Optice: Sive de Reflexionibus, Refractionibus, Inflectionibus et Coloribus Lucis Libri Tres.* Latine reddidit Samuel Clarke. Londini: Impensis Sam. Smith & Benj. Walford.
Newton, I. (1712?) Epistola cujusdam ad amicum. In J. Edleston (Ed.), *Correspondence of sir Isaac Newton and Professor Cotes, including letters of other eminent men* (pp. 308–314). London: Parker (1850).
Newton, I. (1721). *Opticks: Or, a treatise of the reflections, refractions, inflections and colours of light*, The third ed. corrected. London: W. and J. Innys.
Newton, I. (1726). *Philosophiae Naturalis Principia Mathematica* (3rd ed.). London: Innys.
Newton, I. (1729). *The Mathematical Principles of Natural Philosophy.* (A. Motte trans.), 2nd Latin edn. (1713). London: B. Motte.
Newton, I. (1756). *Four letters from sir Isaac Newton to doctor Bentley. Containing some arguments in proof of a Deity.* London: Print. for R. and J. Dodsley.

Newton, I. (1962). *On the gravity and equilibrium of fluids*. In A. R. Hall & M. B. Hall (Eds.), *Unpublished scientific paper of Isaac Newton* (pp. 121–156). Cambridge: Cambridge University Press

Newton, I. (1999). *The* Principia. *Mathematical principles of natural philosophy*. A new translation by I. Bernard Cohen & A. Whitman assisted by J. Budenz. Preceded by a Guide to Newton's *Principia* by I. Bernard Cohen. Berkeley-Los Angeles-London: University of California Press.

Newton, I. (2018). *Principi matematici della filosofia naturale. A cura di F.* Giudice, Torino: Einaudi.

North, J. (2008). *An illustrated history of astronomy and cosmology*. Chicago: University of Chicago Press.

Norton, J.D. (1995). Mach's principle before Einstein. In J. B. Barbour & H. Pfister (Eds.), *Mach's principle from Newton's Bucket to quantum gravity* (pp. 9–55). Boston-Basel-Berlin: Birkhäuser.

Nothaft, E. (2017). Criticism of trepidation models and advocacy of uniform precession in medieval Latin astronomy. *Archive for History of Exact Sciences, 71*, 211–244.

Okruhlik, K. (1989). The foundation of all philosophy: Newton's third rule. In J. R. Brown & J. Mittelstrass (Eds.), *An intimate relation: Studies in the history and philosophy of science* (pp. 97–114). Boston: Kluwer Academic Publishers.

Omodeo, P. D. (2014). *Copernicus in the cultural debates of the Renaissance: Reception, legacy, transformation*. Leiden: Brill.

Paley, W. (1802). *Natural theology: Or evidences of the existence and attributes of the Deity collected from the appearances of nature*. London: R. Faulder.

Palmieri, P. (1998). Re-examining Galileo's theory of the tides. *Archive for History of Exact Sciences, 53*, 223–375.

Papineau, D. (1977). The *vis viva* controversy: Do meanings matter? *Studies in History and Philosophy of Science, 8*(2), 111–142.

Parkinson, J. H. R. (1974). Science and metaphysics in Leibniz's 'Specimen inventorum.' *Studia Leibnitiana, 6*(1), 1–27.

Pascal, B. (1958). *Pascal's Pensées*. Introduction by T. S. Eliot. Translated by W. F. Trotter. New York: E. P. Dutton & Co.

Pasini, E. (2016). G. W. Leibniz's anti-death perspective: Spontaneity of death and absolute immortality. In C. Tandy (Ed.), *Death and anti-death, volume 14: Four decades after Michael Polanyi, three centuries after G. W. Leibniz* (pp. 169–186). Ann Arbour: Ria University Press.

Petroni, A. M. (1989). *I modelli l'invenzione e la scoperta. Saggio su Keplero, la rivoluzione copernicana e la «New philosophy of science»*. Milano: Franco Angeli.

Peuerbach, G. (1473). *Theoricæ novæ planetarum, id est septem errantium siderum nec non octavi seu firmamenti*. Nürnberg.

Pfister, H. (2014). Ludwig Lange and the law of inertia. *The European Physical Journal H, 39*, 245–250.

Pisano, R., & Bussotti, P. (2012). Galileo and Kepler: On *Theoremata circa Centrum Gravitatis Solidorum* and *Mysterium Cosmographicum. History Research, 2*(2), 110–145.

Pisano, R., & Bussotti, P. (2018). On the conceptualization of force in Johannes Kepler's *Corpus*: An interplay between physics/mathematics and metaphysics. In R. Pisano, J. Agassi, & D. Drodzova (Eds.), *Hypotheses and perspectives in the history and philosophy of science. Homage to Alexandre Koyré 1892–1964* (pp. 295–345). Cham: Springer.

Poincaré, H. (1902). *La science et l'hypothèse*. Paris: Flammarion.

Puryear, S. (2012). Motion in Leibniz's middle years: A compatibilist approach. In D. Garber & D. Rutherford (Eds.), *Oxford studies in early modern philosophy* (vol. VI, pp. 135–170). Oxford: Clarendon Press.

Putz, V. (2018). A theory of inertia based on Mach's principle. *Universe, 5*(8), 1–8.

Rabin, S. J. (2004, revised 2019). Nicolaus Copernicus. In E. N. Zalta (Ed.) *Stanford Encyclopedia of Philosophy*. https://plato.stanford.edu/archives/fall2019/entries/copernicus/

Rabin, S. J. (2005). Was Kepler's species immateriata substantial? *Journal for the History of Astronomy, 36*, 49–56.

Raine, D. J. (1981). Mach's principle and space-time structure. *Reports on Progress in Physics, 44*, 1151–1195.
Rajep, F. J. (1996). Al-Battānī, cosmology, and the early history of trepidation in Islam. In J. Casulleras & J. Samsó (Eds.), *From Baghdad to Barcelona: Studies in the Islamic exact sciences in honour of Prof. J. Vernet* (pp. 267–298). Barcelona: Universitat de Barcelona.
Ranea, A. G. (1986). A priori, inercia y acción motriz en la dinámica de Leibniz. *Revista De Filosofía, 26–27*, 164–169.
Ray, J. (1691). *The Wisdom of God manifested in the works of the creation being the substance of some common places delivered in the Chappel of Trinity-College, in Cambridge*. London: Samuel Smith.
Regier, J. (2013). Method and the a priori in Keplerian metaphysics. *Journal of Early Modern Studies, 2*(1), 147–162.
Reichenbach, H. (1957). *The philosophy of space and time*. New York: Dover. (Original German edition 1927).
Reichenbach, H. (1960). *The theory of relativity and a priori knowledge*. Berkeley: University of California Press. (Original German edition 1920).
Reissner, H. (1914). Über die Relativität der Beschleunigungen in der Mechanik. *Physikalishe Zeitschrift, 15*, 371–375.
Reissner, H. (1915). Über eine Möglichkeit die Gravitation als unmittelbare Folge der Relativität der Trägheit abzuleiten. *Physikalishe Zeitschrift, 16*, 179–185.
Repcheck, J. (2007). *Copernicus secret: How the scientific revolution began*. New York: Simon & Schuster.
Rheticus, G. J. (1540, 2015). *Narratio prima or first account of the books* on the Revolutions *by Nicolaus Copernicus*. With an introduction by J. Włodarczyk. Warsaw: University of Warsaw Press.
Rheticus, G. J. (1959). *Narratio Prima*. In *Three Copernican treatises*. Translated with introduction and notes by E. Rosen. Second Edition, revised with an annotated Copernicus bibliography 1939–1958. New York: Dover Books.
Roberts, J. T. (2003). Leibniz on force and absolute motion. *Philosophy of Science, 70*, 553–573.
Ronchi, V. (1939). *Storia della luce*. Bologna: Zanichelli.
Roero, C. S. (1997). Giovanni Battista Benedetti and the scientific environment of Turin in the 16th century. *Centaurus, 39*(1), 37–66.
Rosen, E. (1984). *Copernicus and the scientific revolution*. Melbourne, Florida: Robert Krieger Publishing Company.
Rosen, E. (1995). *Copernicus and his successors*. London: Hambledon Press.
Rossi, P. (1971). *Aspetti della rivoluzione scientifica*. Napoli: Morano.
Rossi, P. (1989). *La Scienza e la Filosofia dei Moderni*. Torino: Bollati Boringhieri.
Rozemond, M. (1997). Leibniz on the union of body and soul. *Archiv Für Geschichte Der Philosophie, 79*(2), 150–178.
Rozemond, M. (2009). Leibniz on final causation. In S. Newlands, L.M. Jorgensen (Eds.), *Metaphysics and good: themes from the philosophy of Robert Merryhew Adams*, pp. 272–294. Oxford: Oxford University Press.
Russell, B. (1900). *A critical exposition of the philosophy of Leibniz*. Cambridge: Cambridge University Press.
Russell, B. (1903). *The principles of mathematics*. Cambridge: Cambridge University Press.
Russell, J. L. (1964). Kepler's laws of planetary motions: 1609–1666. *British Journal for the History of Science, 2*, 1–14.
Rybka, E. (1964). *Four hundred years of the Copernican Heritage*. Kraków: Jagellonian University of Krakow.
Rynasiewicz, R. (1995). By their properties, causes and effects: Newton's Scholium on time, space, place and motion. *Studies in History and Philosophy of Science*. First part 26, *1*, 133–153. Second part 26, *2*, 295–321.

Rynasiewicz, R. (2004). Newton's views on space, time, and motion. In E. N. Zalta (Ed.), *Stanford Encyclopedia of Philosophy*. https://plato.stanford.edu/entries/newton-stm/.
Sabra, A. I. (1967). *Theories of light from Descartes to Newton*. London: Oldbourne.
Sachs, M., & Roy, A. M. (Eds.). *Mach's principle and the origin of inertia*. Montreal: Apeiron.
Samsó, J. (1994). Trepidation in al-Andalus in the 11th Century. In Idem, *Islamic astronomy and medieval Spain*. Aldershot: Ashgate-Variorum.
Schmaltz, T. M. (2008). *Descartes on causation*. Oxford: Oxford University Press.
Schofield, C. J. (1965). The geoheliocentric mathematical hypothesis in sixteenth-century planetary theory. *The British Journal for the History of Science, 2*(8), 291–296.
Schofield, C. J. (1981). *Tychonic and semi-tychonic world systems*. New York: Ayer Co Pub.
Schuster, J. (2000). Descartes opticien: The construction of the law of refraction and the manufacture of its physical rationales, 1618–1629. In S. Gaukroger, J. Schuster, & J. Sutton (Eds.), *Descartes natural philosophy* (pp. 258–312). London: Routledge.
Schuster, J. (2013). *Descartes-agonistes: Physico-mathematics, method & corpuscular-mechanism 1618–33*. Dordrecht-Heibelberg-New York-London: Springer.
Sciama, D. W. (1953). On the origin of inertia. *Monthly Notices of the Royal Astronomical Society, 113*, 34–42.
Sciama, D. W. (1957). Inertia. *Scientific American, 196*(2), 99–109.
Sciama, D. W. (1959). *The unity of the universe*. London: Faber and Faber.
Scribano, E. (2015). *Macchine con la mente. Fisiologia e metafisica tra Cartesio e Spinoza*. Roma: Carocci.
Settle, T. (1961). An experiment in the history of science. *Science, 133*, 19–23.
Shapere, D. (1974). *Galileo: A philosophical study*. Chicago: University of Chicago Press.
Shapiro, A. (1973). A study of the wave theory of light in the seventeenth century. *Archive for History of Exact Sciences, 11*, 134–266.
Shapiro, A. (1974). Light, pressure, and rectilinear propagation: Descartes' celestial optics and Newton's hydrostatic. *Studies in History and Philosophy of Science, V*, 239–296.
Shapiro, A. (1989). Huygens *Traité de la Lumière* and Newton's *Opticks*: Pursuing and eschewing hypotheses. *Notes and Records of the Royal Society of London, 43*, 223–247.
Shapiro, A. (2004). Newton's "Experimental philosophy". *Early Science and Medicine, 9*(3), 185–217.
Shea, W. R. (1991). *The magic of numbers and motion. The scientific career of René Descartes*. Science History Publications.
Simon, G. (1976). *S tructures de pensée et objets du savoir chez Kepler* (2 vols). Thèse: Philosophie: IV, 1976. Service de reproduction de thèses de l'Université de Lille III, 1979.
Simon, G. (1979). *Kepler astronome astrologue*. Paris: Gallimard.
Simon, G. (1996). *Sciences et savoirs aux XVIe et XVIIe siècles*. Villeneuve d'Ascq: Presses de Septentrion.
Simon, G. (2003). *Archéologie de la vision: l'optique, le corps, la peinture*. Paris: Seuil.
Simon, G. (2019). *Kepler, rénovateur de l'optique*. Paris: Classiques Garnier. (Original edition, 1976).
Sklar, L. (1974). *Space, time and spacetime*. Berkeley-Los Angeles-London: University of California Press.
Slowik, E. (1999). Descartes, spacetime and relational motion. *Philosophy of Science, LXVI*, 117–139.
Slowik, E. (2002). *Cartesian spacetime. Descartes physics and the relational theory of space and motion*. Berlin: Springer.
Slowik, E. (2005, 2017). Descartes physics. In E. N. Zalta (Ed.), *Stanford Encyclopedia of Philosophy*. https://plato.stanford.edu/entries/descartes-physics/2005, rev. 22 August 2017.
Smith, G.E. (2002). The methodology of the *Principia*. In I. B. Cohen & G. E. Smith (Eds.), *The Cambridge companion to Newton* (pp. 138–173). Cambridge: Cambridge University Press.
Snelders, H. A. M. (1989). Christiaan Huygens and Newton's theory of gravitation. *Notes and Records of the Royal Society of London, 43*(2), 209–222.

Snobelen, S. D. (2005). The true frame of Nature: Isaac Newton, heresy and the reformation of natural philosophy. In J. Brooke & I. Maclean (Eds.), *Heterodoxy in early modern science and religion* (pp. 223–262). Oxford: Oxford University Press.
Somervill, B. (2005). *Nicolaus Copernicus: Father of modern astronomy*. Minneapolis: Compass Point Books.
Sonego, S. (2020). *General relativity*. Lecture notes of the course of mathematical physics held at the Department of Mathematics, Informatics, Physics, University of Udine. Italy.
Spencer, Q. (2004). Do Newton's rules of reasoning guarantee truth... must they? *Studies in History and Philosophy of Science* Part A *35*(4), 759–782.
Spinoza, B. (2018). *Ethics proved in geometrical order*. Edited by J. Kisner. Translated by M. Silverthorne and M. J. Kisner. Cambridge: Cambridge University Press.
Stan, M. (2016). Huygens on inertial structure and relativity. *Philosophy of Science, 83*, 277–298.
Stein, H. (1967). Newtonian Space-Time. *Texas Quarterly, 10*(3), 174–200.
Stein, H. (1970). On the notion of field in Newton, Maxwell and beyond. In R.H. Stuewer (Ed.), *Historical and philosophical perspectives of science*, pp. 264–310. Minneapolis: University of Minneapolis Press.
Stein, H. (1977). Some philosophical prehistory of general relativity. In J. Earmann, C. Glymour, & J. Stachel, (Eds.), *Minnesota studies in the philosophy of science. Vol VIII: Foundations of space-time theories* (pp. 3–49). Minneapolis: University of Minnesota Press.
Stephenson, B. (1987). *Kepler's physical astronomy*. New York - Berlin - Heidelberg - London - Paris - Tokyo: Springer.
Stephenson, B. (1994). *The music of the heavens. Kepler's harmonic astronomy*. Princeton: Princeton University Press.
Stevenson, G. (1997). Miracles, force, and Leibnizian laws of nature. *Studia Leibnitiana, 29*, 167–188.
Stinner, A. (2001). Linking 'The Book of Nature' and the 'Book of Science': Using circular motion as an *Exemplar* beyond the textbook. *Science and Education, 10*, 323–344.
Streintz, H. (1883). *Die physikalischen Grundlagen der Mechanik*. Leipzig: Teubner.
Suisky, D. (2009). *Euler as Physicist*. Berlin: Springer.
Swerdlow, N. M. (1973). The derivation and first draft of Copernicus's planetary theory: A translation of the *Commentariolus* with commentary. *Proceedings of the American Philosophical Society, 117*(6), 423–512.
Swerdlow, N. M. (1980). Long-period Motions of the Earth in *De revolutionibus*, *Centaurus, 24*(1), 212–245.
Swerdlow, N. M., & Neugebauer, O. (1984). *Mathematical astronomy in Copernicus's De Revolutionibus: In two parts*. New York-Berlin-Heidelberg-Tokyo: Springer.
Tegmark, M. (2008). The Mathematical Universe. *Foundations of Physics, 38*, 101–150.
Tegmark, M. (2014). *Our mathematical universe. My quest for the ultimate nature of reality*. New York: Alfred A. Knopf.
Tiemersma, D. (1988). Methodological and theoretical aspects of Descartes' Treatise on the Rainbow. *Studies in History and Philosophy of Science, 19*, 347–364.
Torretti, R. (1999). *The philosophy of physics*. Cambridge: Cambridge University Press.
Torrini, M. (1993). Galileo, Platone e la filosofia. In P. Prini (a cura di), *Il neoplatonismo nel Rinascimento* (pp. 233–243). Roma: Istituto dell'Enciclopedia Italiana.
Toulmin, S. (1959). Criticism in the history of science: Newton on absolute space, time, and motion. *I. the Philosophical Review, 68*(1), 1–29.
Vailati, E. (1997). *Leibniz and Clarke. A study of their correspondence*. Oxford: Oxford University Press.
Vailati, G. (1897–1898). Le speculazione di Giovanni Battista Benedetti sul moto dei gravi.A tti dell' *Accademia delle Scienze di Torino* (Vol. 33, pp. 559–576). Reprinted in G. Vailati, Scritti di G. Vailati (pp. 161–178). Leipzig-Firenze: Barth-Seeber.
Vergara Caffarelli, R. (2007). Il principio d'inerzia negli ultimi scritti di Galilei. *Cronos, 10*, 63–88.

Vijaya, G. K. (2019). Original form of Kepler's third law and its misapplication in propositions XXXII-XXXVII in Newton's *Principia* (Book I). *Heliyon*. On line publication: https://www.cell.com/heliyon/pdf/S2405-8440(18)38696-1.pdf

Villemot, P. (1707). *Nouveau système ou nouvelle explication du movement des planets*. Lyon: Louis Declaustre.

Wagner, P. (2002). Introduction. In P. Wagner (sous la direction de), *Les philosophes et la science* (pp. 9–65). Paris: Gallimard.

Wallace, W. A. (1974). Galileo and the reasoning ex suppositione. The methodology of the Two New Sciences. In S. Cohen (Ed.), *Proceedings of the* 1974 *biennial meeting of the Philosophy of Science Association* (pp. 79–104). Dordrecht: Reidel.

Wallace, W. A. (1984). *Galileo and his sources: The Heritage of the Collegio Romano in Galileo's science*. Princeton: Princeton University Press.

Westfall, R. S. (1971a). *The construction of modern science: Mechanism and mechanic*. New York: Wiley.

Westfall, R. S. (1971b). *Force in Newton's physics. The science of dynamics in the seventeenth century*. London-New York: McDonald-Elsevier.

Westfall, R. S. (1981). *Never at rest: A biography of Isaac Newton*. Cambridge; Cambridge University Press.

Westfall, R. S. (1989). The trial of Galileo: Bellarmino, Galileo and the clash of two worlds. *Journal for the History of Astronomy, 20*(1), 1–23.

Westman, R. (2001). Kepler's early physical-astrological problematic. *Journal for the History of Astronomy, 32*, 227–236.

Westman, R. (2011). *The Copernican Question: Prognostication, skepticism, and celestial order*. Los Angeles: University of California Press.

Whiteside, D. T. (1974). Keplerian planetary eggs, laid and unlaid, 1600–1605. *Journal for the History of Astronomy, 5*, 1–21.

Will, C. M. (1993). *Theory and experiment in gravitational physics* (Revised). Cambridge: Cambridge University Press.

Williams, B. (2005 [1978]). *Descartes: The project of pure enquiry*. London: Routledge.

Wilson, C. (1968). Kepler's derivation of the elliptical path. *Isis, 59*, 5–25.

Wilson, C. (1990). Atom, Minds, and Vortices in *De Summa Rerum*: Leibniz vis-à-vis Hobbes and Spinoza. In S. Brown (Ed.), *The young Leibniz and his philosophy* (pp. 223–243). Dordrecht: Kluwer Academic Publishers.

Wisan, W. L. (1978). Galileo's scientific method: A reexamination. In R. E. Butts & J. C. Pitt (Eds.), *New perspectives on Galileo* (pp. 1–58). Dordrecht: Reidel.

Woolhouse, R. S. (2000). Leibniz's collision rules for inertialess bodies indifferent to motion. *History of Philosophy Quarterly, 17*(2), 143–157.

Yoder, J. G. (1988). *Unrolling time. Christiaan Huygens and the mathematization of nature*. Cambridge: Cambridge University Press.

Zylbersztajn, A. (1994). Newton's absolute space, Mach's principle and the possible reality of fictitious forces. *European Journal of Physics, 15*(1), 1–8.

Index of Names

P. Bussotti and B. Lotti, *Cosmology in the Early Modern Age: A Web of Ideas*, Logic, Epistemology, and the Unity of Science 56,
https://doi.org/10.1007/978-3-031-12195-1